BIRKHÄUSER

Progress in Mathematics
Volume 266

Stefano Pigola
Marco Rigoli
Alberto G. Setti

Vanishing and Finiteness Results in Geometric Analysis

A Generalization of the Bochner Technique

Birkhäuser
Basel · Boston · Berlin

Authors:

Stefano Pigola
Alberto G. Setti
Dipartimento di Fisica e Matematica
Università dell'Insubria – Como
via Valleggio 11
22100 Como
Italy
e-mail: stefano.pigola@uninsubria.it
alberto.setti@uninsubria.it

Marco Rigoli
Dipartimento di Matematica
Università di Milano
Via Saldini 50
20133 Milano
Italy
e-mail: rigoli@mat.unimi.it

2000 Mathematics Subject Classification: primary 53C21; secondary 35J60, 35R45, 53C42, 53C43, 53C55, 58J50

Library of Congress Control Number: 2007941340

Bibliographic information published by Die Deutsche Bibliothek. Die Deutsche Bibliothek lists this publication in the Deutsche Nationalbibliografie; detailed bibliographic data is available in the Internet at http://dnb.ddb.de

ISBN 978-3-7643-8641-2 Birkhäuser Verlag AG, Basel · Boston · Berlin

Basel · Boston · Berlin
P.O. Box 133, CH-4010 Basel, Switzerland
Part of Springer Science+Business Media
Printed on acid-free paper produced from chlorine-free pulp. TCF ∞
Printed in Germany

ISBN 978-3-7643-8641-2 e-ISBN 978-3-7643-8642-9

9 8 7 6 5 4 3 2 1 www.birkhauser.ch

Contents

Introduction

This book originated from a graduate course given during the Spring of 2005 at the University of Milan. Our goal was to present an extension of the original Bochner technique describing a selection of results recently obtained by the authors, in non-compact settings where in addition one didn't assume that the relevant curvature operators satisfied signum conditions. To make the course accessible to a wider audience it was decided to introduce many of the more advanced analytical and geometrical tools along the way.

The initial project has grown past the original plan, and we now aim at treating in a unified and detailed way a variety of problems whose common thread is the validity of Weitzenböck formulae.

As is well illustrated in the elegant work by H.H. Wu, [165], typically, one is given a Riemannian (Hermitian) vector bundle E with compatible fiber metric and considers a geometric Laplacian L on E which is related to the connection (Bochner) Laplacian $-tr(D^*D)$ via a fiber bundle endomorphism $\mathfrak{R}$ which is in turn related to the curvature of the base manifold M. Because of this relationship, the space of L-harmonic sections of E reflects the geometric properties of M.

To illustrate the method, let us consider the original Bochner argument to estimate the first real Betti number $b^1(M)$ of a closed oriented Riemannian manifold $(M, \langle\, ,\rangle)$.

By the Hodge–de Rham theory, $b^1(M)$ equals the dimension of the space of harmonic 1-forms $\mathcal{H}^1(M)$. A formula of Weitzenböck, independently rediscovered by Bochner, states that for every harmonic 1-form ω,

$$\frac{1}{2}\Delta |\omega|^2 = |D\omega|^2 + \mathrm{Ric}\left(\omega^\#, \omega^\#\right), \tag{0.1}$$

where Δ and Ric are the Laplace–Beltrami operator (with the sign convention $+d^2/dx^2$) and the Ricci curvature of M, respectively, D denotes the extension to 1-forms of the Levi–Civita connection, and $\omega^\#$ is the vector field dual to ω, defined by $\langle \omega^\#, X\rangle = \omega(X)$ for all vector fields X. In particular $|\omega|^2$ satisfies the differential inequality

$$\Delta |\omega|^2 - q(x) |\omega|^2 \geq 0,$$

where $q(x)/2$ is the lowest eigenvalue of the Ricci tensor at x. Thus, if $\mathrm{Ric} \geq 0$, then $|\omega|$ is subharmonic. Since M is closed, we easily conclude that $|\omega|$ =const. This can be done using two different viewpoints, (i) the L^∞ and (ii) the $L^{p<+\infty}$ one. As for (i), note that the smooth function $|\omega|$ attains its maximum at some point and, therefore, by the Hopf maximum principle we conclude that $|\omega|$ =const. In case (ii) we use the divergence theorem to deduce

$$0 = \int_M \mathrm{div}\left(|\omega|^2 \nabla |\omega|^2\right) = \int_M \left|\nabla |\omega|^2\right|^2 + \int_M |\omega|^2 \Delta |\omega|^2 \geq \int_M \left|\nabla |\omega|^2\right|^2 \geq 0.$$

This again implies $|\omega|$ =const.

Now, since Ric ≥ 0, using this information in formula (0.1) shows that ω is parallel, i.e., $D\omega = 0$. As a consequence, ω is completely determined by its value at a given point, say $p \in M$. The evaluation map $\varepsilon_p : \mathcal{H}^1(M) \to \Lambda^1\left(T_p^* M\right)$ defined by

$$\varepsilon_p(\omega) = \omega_p$$

is an injective homomorphism, proving that, in general,

$$b^1(M) = \dim \mathcal{H}^1(M) \leq m.$$

Note that (0.1) yields

$$0 = \mathrm{Ric}\left(\omega_p^{\#}, \omega_p^{\#}\right) \text{ at } p.$$

Therefore, if Ric$(p) > 0$, we get $\omega_p = 0$ which, in turn, implies $\omega = 0$. This shows that, when Ric is positive somewhere,

$$b^1(M) = \dim \mathcal{H}^1(M) = 0.$$

The example suggests that one can generalize the investigation in several directions. One can relax the assumption on the signum of the coefficient $q(x)$, consider complete non-compact manifolds, or both.

Maintaining compacteness, one can sometimes allow negative values of $q(x)$ using versions of the generalized maximum principle, according to which if $\psi \geq 0$ satisfies

$$\Delta\psi - q(x)\psi \geq 0, \tag{0.2}$$

and M supports a solution $\varphi > 0$ of

$$\Delta\varphi - q(x)\varphi \leq 0, \tag{0.3}$$

then the ratio $u = \psi/\varphi$ is constant. Combining (0.2) and (0.3) shows that ψ satisfies (0.2) with equality sign. In particular, according to (0.1), $\psi = |\omega|^2$ satisfies (0.2), and therefore, if M supports a function φ satisfying (0.3), we conclude, once again, that ω is parallel, thus extending the original Bochner vanishing result to this situation.

It is worth noting that the existence of a function φ satisfying (0.3) is related to spectral properties of the operator $-\Delta + q(x)$, and that the conclusion of the generalized maximum principle is obtained by combining (0.2) and (0.3) to show that the quotient u satisfies a differential inequality without zero-order terms; see Section 2.5 in [133].

In the non-compact setting the relevant function may fail to be bounded, and even if it is bounded, it may not attain its supremum. In the latter case, one may use a version of the maximum principle at infinity introduced by H. Omori, [124] and generalized by S.T. Yau, [167], and S.Y Cheng and Yau, [34], elaborating ideas of L.V. Ahlfors. An account and further generalizations of this technique, which however works under the assumption that $q(x)$ is non-negative, may be found in [131].

Here we consider the case where the manifold is not compact and the function encoding the geometric problem is not necessarily bounded, but is assumed to satisfy suitable L^p integrability conditions, and the coefficient $q(x)$ in the differential (in)equality which describes the geometric problem is not assumed to be non-negative.

Referring to the previous example, the space of harmonic 1-forms in L^2 describes the L^2 co-homology of a complete manifold, and under suitable assumptions it has a topological content sensitive to the structure at infinity of the manifold. It turns out to be a bi-Lipschitz invariant, and, for co-compact coverings, it is in fact a rough isometry invariant.

As in the compact case described above, one replaces the condition that the coefficient $q(x)$ is pointwise positive, with the assumption that there exists a function φ satisfying (0.3) on M or at least outside a compact set. Again, one uses a Weitzenböck-type formula to show that the geometric function $\psi = |\omega|$ satisfies a differential inequality of the form (0.2).

Combining (0.2) and (0.3) and using the integrability assumption, one concludes that either ψ vanishes and therefore the space $L^2\mathcal{H}^1(M)$ of L^2-harmonic 1-forms is trivial or that $L^2\mathcal{H}^1(M)$ is finite-dimensional.

The method extends to the case of L^p-harmonic k-forms, even with values in a fibre bundle, and in particular to harmonic maps with L^p energy density, provided we consider an appropriate multiple of $q(x)$ in (0.3), and restrict the integrability coefficient p to a suitable range. Harmonic maps in turn yield information, as in the compact case, on the topological structure of the underlying domain manifold.

This relationship becomes even more stringent in the case where the domain manifold carries a Kählerian structure. Indeed, for complex manifolds, the splitting in types allows to consider, besides harmonic maps, also pluriharmonic and holomorphic maps. If, in addition, the manifold is Kähler, the relevant Weitzenböck identity for pluriharmonic functions (which in the L^2 energy case coincides with a harmonic function with L^2 energy) takes on a form which reflects the stronger rigidity of the geometry and allows us to obtain stronger conclusions. Thus, on the one hand one can enlarge the allowed range of the integrability coefficient p, and on the other hand one may deduce structure theorems which have no analogue in the purely Riemannian case.

The extension to the non-compact case introduces several additional technical difficulties, which require specific methods and tools. The description of these is in fact a substantial part of the book, and while most, but not all, of the results are well known, in many instances our approach is somewhat original. Further, in some cases, one needs results in a form which is not easily found, if at all, in the literature.

When we feel that these ancillary parts are important enough, or the approach sufficiently different from the mainstream treatment, a fairly detailed description is given. Thus we provide, for instance, a rather comprehensive treatment of comparison methods in Riemannian geometry or of the spectral theory of Schrödinger operators on manifolds. In other situations, the relevant tools are introduced when needed. For instance this is the case of the Poincaré inequalities or of the Moser iteration procedure.

The material is organized as follows.

In Chapter 1, after a quick review of harmonic maps between Riemannian manifolds, where in particular we describe the Weitzenböck formula and derive a sharp version of Kato's inequality, we introduce the basic facts on the geometry of complex manifolds, and Hermitian bundles, concentrating on the Kähler case. Our approach is inspired by work of S.S. Chern, and is based on analyzing the Riemannian counterpart of the Kähler structure.

The same line of arguments allows us to extend a result of J.H. Sampson, [143], concerning the pluriharmonicity of a harmonic map from a compact Kähler manifold into a Riemannian target with negative Hermitian curvature to the case of a non-compact domain. This in turn yields a sharp version of a result of P. Li, [96], for pluriharmonic real-valued functions. The chapter ends with a derivation of Weitzenböck-type formulas for pluriharmonic and holomorphic maps.

Chapter 2 is devoted to a detailed description of comparison theorems in Riemannian Geometry under curvature conditions, both pointwise and integral, which will be extensively used throughout the book. We begin with general comparison results for the Laplacian and the Hessian of the distance function. The approach, which is indebted to P. Petersen's treatment, [128], is analytic in that it only uses comparison results for ODEs avoiding the use of Jacobi fields, and it is not limited to the case where the bound on the relevant curvature is a constant, but is given in terms of a suitable function G of the distance from a reference point. Some effort is also made to describe explicit bounds in a number of geometrically significant situations, namely when $G(r) = -B(1+r^2)^{\alpha/2}$, or when $G(t)$ satisfies the integrability condition $tG(t) \in L^1([0,+\infty))$ considered, among others, by U. Abresch, [1], and by S.H. Zhu, [171].

These estimates are then applied to obtain volume comparisons. Even though the method works both for upper and lower estimates, we concentrate on upper bounds, which hold under less stringent assumptions on the manifold, and in particular depend on lower bounds for the Ricci curvature alone, and do not require topological restrictions. We also describe volume estimates under integral Ricci curvature conditions which extend previous work of S. Gallot, [57], and, more recently, by Petersen and G. Wei, [129]. We then describe remarkable lower estimates for the volume of large balls on manifolds with almost non-negative Ricci curvature obtained by P. Li and R. Schoen, [95] and Li and M. Ramachandran, [98], elaborating on ideas of J. Cheeger M. Gromov and M. Taylor, [33]. These estimates in particular imply that such manifolds have infinite volume. We conclude the chapter with a version of the monotonicity formula for minimal submanifolds valid for the volume of intrinsic (as opposed to extrinsic) balls in bi-lipschitz harmonic immersions.

Chapter 3 begins with a quick review of spectral theory of self-adjoint operators on Hilbert spaces modelled after E.B. Davies' monograph, [41]. In particular, we define the essential spectrum and index of a (semibounded) operator, and apply the minimax principle to describe some of their properties and their mutual relationships. We then concentrate on the spectral theory of Schrödinger operators on manifolds, in terms of which many of the crucial assumptions of our geometrical results are formulated.

After having defined Schrödinger operators on domains and on the whole manifold, we describe variants of classical results by D. Fisher-Colbrie, [53], and Fisher-Colbrie and Schoen, [54], which relate the non-negativity of the bottom of the spectrum of a Schrödinger operator L on a domain Ω to the existence of a positive solution of the differential inequality $L\varphi \leq 0$ on Ω.

Since, as already mentioned above, the existence of such a solution is the assumption on which the analytic results depend, this relationship allows us to interpret such hypothesis as a spectral condition on the relevant Schrödinger operator. This is indeed a classical and natural feature in minimal surfaces theory where the stability, and the finiteness of the index of a minimal surface, amount to the fact that the stability operator $-\Delta - |II|^2$ has non-negative spectrum, respectively finite Morse index.

In describing these relationships we give an account of the links between essential spectrum, bottom of the spectrum, and index of a Schrödinger operator L on a manifold, and that of its restriction to (internal or external) domains. With a somewhat different approach and arguments, our presentation follows the lines of a paper by P. Berard, M.P. do Carmo and W. Santos, [13].

Chapter 4 and Chapter 5 are the analytic heart of the book. In Chapter 4 we prove a Liouville-type theorem for L^p solutions u of divergence-type differential inequalities of the form

$$u div(\varphi \nabla u) \geq 0,$$

where φ is a suitable positive function. An effort is made to state and prove the result under the minimal regularity assumptions that will be needed for geometric applications. As a consequence we deduce the main result of the chapter, namely a vanishing theorem for non-negative solutions of the Bochner-type differential inequality

$$\psi \Delta \psi + a(x)\psi^2 + A|\nabla \psi|^2 \geq 0. \tag{0.4}$$

Assuming the existence of a positive solution of the inequality

$$\Delta \varphi + Ha(x)\varphi \leq 0, \tag{0.5}$$

for a suitable constant H, one proceeds similarly to what we described above, and shows that an appropriate combination u of the function ψ and φ satisfies the hypotheses of the Liouville-type theorem.

In Chapter 6 the analytic setting is similar, one considers vector spaces of L^p-sections whose lengths satisfy the differential inequality (0.4) and proves that such spaces are finite-dimensional under the assumption that a solution φ to the differential inequality (0.5) exists in the complement of a compact set K in M. The idea of the proof is to show that there exists a constant C depending only on the geometry of the manifold in a neighborhood of K such that the dimension of every finite-dimensional subspace is bounded by C. The proof is based on a version of a lemma by Li, and uses a technique of Li and J. Wang, [104] and [105], combined with the technique of the coupling of the solutions ψ and φ which allows us to deal with L^p sections with p not necessarily equal to 2. The proof requires a number of technical results which are described in detailed, in some cases new, direct proofs.

Chapter 6 to 9 are devoted to applications in different geometric contexts. In Chapter 6 we specialize the vanishing results to the case of harmonic maps with finite L^p energy, and derive results on the constancy of convergent harmonic maps, and a Schwarz-type lemma for harmonic maps of bounded dilation. We then describe topological results by Schoen and Yau, [146], concerning the fundamental group of manifolds of non-negative Ricci curvature and of stable minimal hypersurfaces immersed in non-positively curved ambient spaces. While the main argument is the same as Schoen and Yau's, the use of our vanishing theorem allows us to relax their assumption that the Ricci curvature of the manifold is non-negative. The chapter ends by generalizing to non-compact settings the finiteness theorems of L. Lemaire, [93], for harmonic maps of bounded dilation into a negatively curved manifold, on the assumption that the domain manifold has a finitely generated fundamental group.

In Chapter 7 we use the techniques developed above to describe the topology at infinity of a Riemannian manifold M, and more specifically the number of unbounded connected components of the complement of a compact domain D in M, namely the ends of M with respect to D.

The number of ends of a manifold will in turn play a crucial role in the structure results for Kähler manifolds, and in the derivation of metric rigidity in the Riemannian setting (see Chapters 8 and 9, respectively).

The chapter begins with an account of the theory relating the topology at infinity and suitable classes of harmonic functions on the manifold as developed by Li and L.F. Tam and collaborators. At the basis of this theory is the fact that, via the maximum principle, the parabolicity/non-parabolicity of an end is intimately connected with the existence of a proper harmonic function on the end (the so-called Evans–Selberg potential of the end), or, in the non-parabolic case, of a bounded harmonic function on the end with finite Dirichlet integral. Combining these facts with the analytic results of the previous chapters in particular, we obtain that the manifold has only one, or at most finitely many non-parabolic ends, depending on spectral assumptions on the operator $L = -\Delta - a(x)$, where $-a(x)$ is the smallest eigenvalue of the Ricci tensor at x. To complete the picture, following H.-D. Cao, Y. Shen, S. Zhu, [25], and Li and Wang, [104], one shows that when the manifold supports an L^1-Sobolev inequality, then all ends are non-parabolic. This in particular applies to submanifolds of Cartan–Hadamard manifolds, provided that the second fundamental form is small in a suitable integral norm. In the chapter, using a gluing technique of T. Napier and Ramachandran, [117], we also provide the details of a construction sketched by Li and Ramachandran, [98] of harmonic functions with controlled L^2 energy growth that will be used in the structure theorems for Kähler manifolds. The last two sections of the chapter contain further applications of these techniques to problems concerning line bundles over Kähler manifolds, and to the reduction of codimension of harmonic immersions with less than quadratic p-energy growth.

In Chapter 8 we concentrate on the Kähler setting. We begin by providing a detailed description of a result of Li and Yau, [107], on the constancy of holomorphic maps with values in a Hermitian manifold with suitably negative holomorphic bisectional curvature. We then describe two variations of the result, where the conclusion is obtained under different assumptions: in the first, using Poisson equation techniques, an integral growth condition on the Ricci tensor is replaced by a volume growth condition, while in the second one assumes a pointwise lower bound on the Ricci curvature which is not necessarily integrable, together with some spectral assumptions on a variant of the operator L. We then apply this in the proof of the existence of pluri-subharmonic exhaustions due to Li and Ramachandran, [98], which is crucial in obtaining the important structure theorem of Napier and Ramachandran, [117], and Li and Ramachandran, [98].

The unifying element of Chapter 9 is the validity of a Poincaré–Sobolev inequality. In the first section, we give a detailed proof of a warped product splitting theorem of Li and Wang, [104]. There are two main ingredients in the proof. The first is to prove that the metric splitting holds provided the manifold supports a non-constant harmonic function u for which the Bochner inequality with a sharp constant in the refined Kato's inequality is in fact an equality. The second ingredient consists of energy estimates for a suitable harmonic function u on M obtained by means of an exhaustion procedure. This is the point where the Poincaré–Sobolev inequality plays a crucial role. Finally, one uses the analytic techniques of Chapter 4 to show that u is the sought-for function which realizes equality in the Bochner inequality. In the second section we begin by showing that whenever M supports an L^2 Poincaré–Sobolev-type inequality, then a non-negative L^p solution ψ of the differential inequality (0.4),

$$\psi\Delta\psi + a(x)\psi^2 + A|\nabla\psi|^2 \geq 0,$$

must vanish provided a suitable integral norm of the potential $a(x)$ is small compared to the Sobolev constant. This compares with the vanishing result of Chapter 4 which holds under the assumption that the bottom of the spectrum of $-\Delta + Ha(x)$ is non-negative. Actually, in view of the geometric applications that follow, we consider the case where M supports an inhomogeneous Sobolev inequality.

We then show how to recover the results on the topology at infinity for submanifolds of Cartan–Hadamard manifolds of Chapter 7. In fact, using directly the Sobolev inequality allows us to obtain quantitative improvements. Further applications are given to characterizations of space forms which extend in various directions a characterization of the sphere among conformally flat manifolds with constant scalar curvature of S. Goldberg, [61].

The book ends with two appendices. The first is devoted to the unique continuation property for solutions of elliptic partial differential systems on manifolds, which plays an essential role in the finite-dimensionality result of Chapter 5. Apart from some minor modifications, our presentation follows the line of J. Kazdan's paper [87].

In the second appendix we review some basic facts concerning the L^p cohomology of complete non-compact manifolds. We begin by describing the basic definitions of the L^p de Rham complex and discussing some simple, but significant examples. We then collect some classical results like the Hodge, de Rham, Kodaira decomposition, and briefly consider the role of L^p harmonic forms. Finally, we illustrate some of the relationships between L^p cohomology and the geometry and the topology of the underlying manifold both for $p = 2$ and $p \neq 2$. In particular we present (with no proofs) the Whitney-type approach developed by J. Dodziuk, [43] and V.M. Gol'dshtein, V.I. Kuz'minov, I.A. Shvedov, [63] and [64], where the topological content of the L^p de Rham cohomology is emphasized by relating it to a suitable, global simplicial theory on the underlying triangulated manifold.

The authors are grateful to G. Carron for a careful reading of the manuscript and several useful comments. It is also a pleasure to thank Dr. Thomas Hempfling of Birkhäuser for his extreme efficiency and helpfulness during the various stages of the production of this book.

Chapter 1

Harmonic, pluriharmonic, holomorphic maps and basic Hermitian and Kählerian geometry

1.1 The general setting

The aim of the chapter is to review some basic facts of Riemannian and complex geometry, in order to compute, for instance, some Bochner-type formulas that we shall need in the sequel. In doing so, we do not aim at giving a detailed treatment of the subject, but only to set down notation and relevant results, illustrating some of the computational techniques involved in the proofs.

Let $(M, \langle\,,\rangle)$ and $(N, (\,,))$ be (real) smooth manifolds of (real) dimensions m and n respectively, endowed with the Riemannian metrics $\langle\,,\rangle$ and $(\,,)$ and let $f : M \to N$ be a smooth map. The energy density $e(f) : M \to \mathbb{R}$ is the non-negative function defined on M as follows. Let $df \in \Gamma\left(T^*M \otimes f^{-1}TN\right)$ be the differential of f and set

$$e(f)(x) = \frac{1}{2}|d_x f|^2$$

where $|df|$ denotes the Hilbert-Schmidt norm of the differential map. In local coordinates $\{x^i\}$ and $\{y^\alpha\}$ respectively on M and N, $e(f)$ is expressed by

$$e(f) = \frac{1}{2}\langle\,,\rangle^{ij}\frac{\partial f^\alpha}{\partial x^i}\frac{\partial f^\beta}{\partial x^j}(\,,)_{\alpha\beta} = \frac{1}{2}\mathrm{tr}_{\langle\,,\rangle} f^*(\,,).$$

Here $f^\alpha = y^\alpha \circ f$ and $\langle\,,\rangle^{ij}$ represents the inverse of the matrix coefficient $\langle\,,\rangle_{ij} = \left\langle \partial/\partial x^i, \partial/\partial x^j \right\rangle$.

If $\Omega \subset M$ is a compact domain we use the canonical measure

$$d\mathrm{Vol}_{\langle\,,\rangle} = \sqrt{\det\langle\,,\rangle_{ij}}\; dx^1 \wedge \cdots \wedge dx^m$$

associated to $\langle\,,\rangle$ to define the energy of $f|_\Omega : (\Omega, \langle\,,\rangle) \to (N, (\,,))$ by

$$E_\Omega(f) = \int_\Omega e(f)\, d\mathrm{Vol}_{\langle\,,\rangle}.$$

Definition 1.1. A smooth map $f : (M, \langle\,,\rangle) \to (N, (\,,))$ is said to be harmonic if, for each compact domain $\Omega \subset M$, it is a stationary point of the energy functional $E_\Omega : C^\infty(M, N) \to \mathbb{R}$ with respect to variations preserving f on $\partial\Omega$.

A vector field X along f, that is, a section of the bundle $f^{-1}TN \to M$ determines a variation f_t of f by setting

$$f_t(x) = \exp_{f(x)} tX_x.$$

If X has support in a compact domain $\Omega \subset M$, then

$$\left.\frac{d}{dt}\right|_{t=0} E_\Omega(f_t) = -\int_M (\tau(f)(x), X_x) d\mathrm{Vol}_{\langle\,,\rangle}$$

where the Euler-Lagrange operator, called the *tension field* of f, is given by

$$\tau(f) = \mathrm{tr}_{\langle\,,\rangle} Ddf,$$

$Ddf \in \Gamma\left(T^*M \otimes T^*M \otimes f^{-1}TN\right)$ being the (generalized) *second fundamental tensor* of the map f. As a consequence, $\tau(f) \in \Gamma\left(f^{-1}TN\right)$ and f is harmonic if and only if

$$\tau(f) = 0 \text{ on } M.$$

In local coordinates

$$\tau(f)^\gamma = \langle\,,\rangle^{ij}\left(\frac{\partial^2 f^\gamma}{\partial x^i \partial x^j} - {}^M\Gamma_{ij}^k \frac{\partial f^\gamma}{\partial x^k} + {}^N\Gamma_{\alpha\beta}^\gamma \frac{\partial f^\alpha}{\partial x^i}\frac{\partial f^\beta}{\partial x^j}\right)$$

where ${}^M\Gamma$ and ${}^N\Gamma$ are the Christoffel symbols of the Levi–Civita connections on M and N, respectively. Thus, the harmonicity condition is represented by a system of non-linear elliptic equations.

Observe that, when $f : (M, \langle\,,\rangle) \to (N, (\,,))$ is an isometric immersion, that is, $f^*(\,,) = \langle\,,\rangle$, then $\tau(f) = m\mathrm{H}$, with H the mean curvature vector field of the immersion. It is well known that the equation $\mathrm{H} \equiv 0$ is the Euler-Lagrange equation of the volume functional

$$V_\Omega(f) = \int_\Omega d\mathrm{Vol}_{\langle\,,\rangle}$$

$\Omega \subset M$ a compact domain. Thus, an isometric immersion is *minimal* if and only if it is harmonic.

For later use, we show how to compute the tension field of $f : (M, \langle\,,\rangle) \to (N, (,))$ with the moving frame formalism. Towards this aim, let $\left\{\theta^i\right\}$ and $\{e_i\}$, $i = 1, \ldots, m$, be local ortho-normal co-frame, and dual frame, on M with corresponding Levi–Civita connection forms $\left\{\theta_j^i\right\}$. Similarly, let $\{\omega^\alpha\}, \{\varepsilon_\alpha\}, \left\{\omega_\beta^\alpha\right\}$, $1 \le \alpha, \beta, \ldots \le n$ describe, locally, the Riemannian structure of $(N, (,))$. Then

$$f^*\omega^\alpha = f_i^\alpha \theta^i$$

so that

$$df = f_i^\alpha \theta^i \otimes \varepsilon_\alpha$$

and computing the covariant derivatives

$$\text{(i) } f^{\alpha}_{ij}\theta^{j} = df^{\alpha}_{i} - f^{\alpha}_{j}\theta^{j}_{i} + f^{\beta}_{i}\omega^{\alpha}_{\beta}, \qquad \text{(ii) } f^{\alpha}_{ij} = f^{\alpha}_{ji}$$

in such a way that

$$Ddf = f^{\alpha}_{ij}\theta^{i} \otimes \theta^{j} \otimes \varepsilon_{\alpha}$$

and

$$\tau(f) = \sum_{i} f^{\alpha}_{ii}\varepsilon_{\alpha}.$$

In what follows we shall also use the next *Bochner–Weitzenböck-type formula* for harmonic maps. Since we shall prove analogous formulas in Kählerian geometry we omit here its derivation. See, e.g., [47].

Theorem 1.2. *Let* $f : (M, \langle\, ,\rangle) \to (N, (\, ,))$ *be a smooth map. Then*

$$\frac{1}{2}\Delta |df|^2 = |Ddf|^2 - \operatorname{tr}_{\langle\, ,\rangle}(D\tau(f), df) + \sum_i \left(df\left({}^{M}\mathrm{Ric}(e_i, \cdot)^{\#}\right), df(e_i)\right) - \sum_{i,j}\left({}^{N}\mathrm{Riem}(df(e_i), df(e_j))\,df(e_j), df(e_i)\right)$$

with $\{e_i\}$ *as above and* ${}^{M}\mathrm{Ric}$, ${}^{N}\mathrm{Riem}$ *respectively the Ricci tensor of* M *and the Riemannian curvature tensor of* N. *In particular, if* f *is harmonic,*

$$\frac{1}{2}\Delta |df|^2 = |Ddf|^2 + \sum_i \left(df\left({}^{M}\mathrm{Ric}(e_i, \cdot)^{\#}\right), df(e_i)\right) - \sum_{i,j}\left({}^{N}\mathrm{Riem}(df(e_i), df(e_j))\,df(e_j), df(e_i)\right).$$

Futher, assuming that f is a harmonic function the formula specializes to Bochner's formula

$$\frac{1}{2}\Delta|\nabla f|^2 = |\mathrm{Hess}\, f|^2 + \mathrm{Ric}(\nabla f, \nabla f). \tag{1.1}$$

Weitzenböck formulae will be repeatedly used in the sequel. Here we give a sharp estimate from below of the term $|Ddf|^2$. This type of estimate goes under the name of *refined Kato inequalities*. Their relevance will be clarified by their analytic consequences. For a more general and abstract treatment, we refer to work by T. Branson, [21], and by D.M.J. Calderbank, P. Gauduchon, and M. Herzlich, [24].

Proposition 1.3. *Let* $f : M \to N$ *be a harmonic map between Riemannian manifolds of dimensions* $\dim M = m$ *and* $\dim N = n$. *Then*

$$|Ddf|^2 - |\nabla |df||^2 \geq \frac{1}{(m-1)}|\nabla |df||^2$$

pointwise on the open, dense subset $\Omega = \{x \in M : |df|(x) \neq 0\}$ *and weakly on all of* M.

Remark 1.4. The dimension n of the target manifold plays no role.

Proof. It suffices to consider the pointwise inequality on Ω. Let $\{f_i^\alpha\}$ and $\{f_{ij}^\alpha\}$ be the coefficients of the (local expressions of the) differential and of the Hessian of f, respectively. Then

$$|df| = \sqrt{\sum_{\alpha,i} (f_i^\alpha)^2}$$

so that

$$\nabla |df| = \frac{\sum_i \left\{ \sum_{\alpha,j} f_{ij}^\alpha f_j^\alpha \right\} e_i}{\sqrt{\sum_{\alpha,i} (f_i^\alpha)^2}}$$

and we have

$$|Ddf|^2 - |\nabla |df||^2 = \sum_{\alpha,i,j} \left(f_{ij}^\alpha\right)^2 - \frac{\sum_i \left\{ \sum_{\alpha,j} f_{ij}^\alpha f_j^\alpha \right\}^2}{\sum_{\alpha,i} (f_i^\alpha)^2}. \tag{1.2}$$

For $\alpha = 1, \dots, n$, define

$$M^\alpha = \left(f_{ij}^\alpha\right) \in M_m(\mathbb{R}), \quad y^\alpha = (f_i^\alpha)^t \in \mathbb{R}^m.$$

Note that each matrix M^α is traceless, by harmonicity of f, and symmetric. Then (1.2) reads

$$|Ddf|^2 - |\nabla |df||^2 = \sum_\alpha \|M^\alpha\|^2 - \frac{\left|\sum_\alpha M^\alpha y^\alpha\right|^2}{\sum_\alpha |y^\alpha|^2}$$

where $\|M\|^2 = tr(MM^t)$ and $|y|$ denotes the $\mathbb{R}^m$-norm of y. We have to show that

$$\sum_\alpha \|M^\alpha\|^2 - \frac{\left|\sum_\alpha M^\alpha y^\alpha\right|^2}{\sum_\alpha |y^\alpha|^2} \geq \frac{1}{(m-1)} \frac{\left|\sum_\alpha M^\alpha y^\alpha\right|^2}{\sum_\alpha |y^\alpha|^2}.$$

This inequality is an immediate consequence of the next simple algebraic lemma. □

Lemma 1.5. *For $\alpha = 1, \dots, n$, let $M^\alpha \in M_m(\mathbb{R})$ be a symmetric matrix satisfying* $\operatorname{trace}(M^\alpha) = 0$. *Then, for every $y^1, \dots, y^n \in \mathbb{R}^m$ with $\sum_\alpha |y^\alpha|^2 \neq 0$,*

$$\sum_\alpha \|M^\alpha\|^2 - \frac{\left|\sum_\alpha M^\alpha y^\alpha\right|^2}{\sum_\alpha |y^\alpha|^2} \geq \frac{1}{(m-1)} \frac{\left|\sum_\alpha M^\alpha y^\alpha\right|^2}{\sum_\alpha |y^\alpha|^2}. \tag{1.3}$$

Moreover, suppose the equality holds. If $y^\alpha \neq 0$, then either $M^\alpha = 0$ or y^α is an eigenvector of M^α corresponding to an eigenvalue μ^α of multiplicity 1. Furthermore, the orthogonal complement $\langle y^\alpha \rangle^\perp$ is the eigenspace of M^α corresponding to the eigenvalue $-\mu^\alpha / (m-1)$ of multiplicity $(m-1)$.

Proof. First, we consider the case $\alpha = 1$. Let $\lambda_1 \leq \cdots \leq \lambda_s \leq 0 \leq \lambda_{s+1} \leq \cdots \leq \lambda_m$ be the eigenvalues of M. Without loss of generality we may assume that $\lambda_m \geq |\lambda_1|$. We are thus reduced to proving that

$$\sum_{i=1}^{m} \lambda_i^2 \geq \left(1 + \frac{1}{m-1}\right) \lambda_m^2.$$

To this end we note that, since M is traceless,

$$-\sum_{j=1}^{m-1} \lambda_j = \lambda_m \tag{1.4}$$

and therefore, from Schwarz's inequality,

$$\lambda_m^2 \leq (m-1) \sum_{j=1}^{m-1} \lambda_j^2. \tag{1.5}$$

This implies

$$\sum_{i=1}^{m} \lambda_i^2 = \lambda_m^2 + \sum_{j=1}^{m-1} \lambda_j^2 \geq \left(1 + \frac{1}{m-1}\right) \lambda_m^2,$$

as desired. Suppose now that $M \neq 0$, so that $\lambda_m > 0$, and assume that equality holds in (1.3) for some vector $y \neq 0$. Let $C \in O(m)$ be such that $CMC^t = D = diag(\lambda_1, \ldots, \lambda_m)$ and set $w = (w_1, \ldots, w_m) = Cy$. Thus

$$\left(1 + \frac{1}{m-1}\right) \lambda_m^2 \leq \sum_i \lambda_i^2 = \left(1 + \frac{1}{m-1}\right) \sum_i \left(\lambda_i \frac{w_i}{|w|}\right)^2 \leq \left(1 + \frac{1}{m-1}\right) \lambda_m^2. \tag{1.6}$$

It follows that the equality holds in (1.5) which in turn forces, according to (1.4) and (the equality case in) Schwarz's inequality,

$$\lambda_1 = \cdots = \lambda_{m-1} = \mu; \quad \lambda_m = -(m-1)\mu,$$

for some $\mu < 0$. On the other hand, (1.6) gives

$$\sum_{i=1}^{m-1} \lambda_i^2 \frac{w_i^2}{|w|^2} + \lambda_m^2 \left(\frac{w_m^2}{|w|^2} - 1\right) = 0$$

proving that $w \in \mathrm{span}\{(0, \ldots, 0, 1)^t\}$ and therefore it is an eigenvector of D belonging to the multiplicity 1 eigenvalue λ_m. It follows that $y = C^t w$ is an eigenvector of M belonging to the multiplicity 1 eigenvalue $\lambda_m = -(m-1)\mu$. Obviously, $y^\perp$ is the eigenspace corresponding to the multiplicity $(m-1)$ eigenvalue μ.

Now let α be any positive integer. We note that

$$\sum_\alpha \|M^\alpha\|^2 - \frac{\left|\sum_\alpha M^\alpha y^\alpha\right|^2}{\sum_\alpha |y^\alpha|^2} \geq \sum_\alpha \|M^\alpha\|^2 - \frac{\left(\sum_\alpha |M^\alpha y^\alpha|\right)^2}{\sum_\alpha |y^\alpha|^2}.$$

Applying the first part of the proof we get, for every $\alpha = 1, \ldots, n$,

$$|M^\alpha y^\alpha| \leq \sqrt{\frac{m-1}{m}}\, \|M^\alpha\|\, |y^\alpha| \tag{1.7}$$

which in turn, used in the above, gives

$$\begin{aligned}\sum_\alpha \|M^\alpha\|^2 - \frac{\left|\sum_\alpha M^\alpha y^\alpha\right|^2}{\sum_\alpha |y^\alpha|^2} &> \sum_\alpha \|M^\alpha\|^2 - \frac{\left(\sum_\alpha \sqrt{\frac{m-1}{m}}\, \|M^\alpha\|\, |y^\alpha|\right)^2}{\sum_\alpha |y^\alpha|^2} \\ &\geq \sum_\alpha \|M^\alpha\|^2 - \frac{m-1}{m} \frac{\sum_\alpha \|M^\alpha\|^2 \sum_\alpha |y^\alpha|^2}{\sum_\alpha |y^\alpha|^2} = \frac{1}{m} \sum_\alpha \|M^\alpha\|^2.\end{aligned}$$

Whence, rearranging and simplifying yields (1.3). To complete the proof, note that the equality in (1.3) forces equality in (1.7) and therefore the first part of the proof applies to M^α. □

1.2 The complex case

We now turn our attention to the complex case.

Definition 1.6. An almost complex manifold (M, J) is a (real) manifold together with a (smooth) tensor field $J \in \Gamma(T^*M \otimes TM)$ of endomorphisms of TM such that

$$J_p^2 = -\mathrm{id}_p \tag{1.8}$$

for every $p \in M$.

Note that (1.8) implies $\dim T_pM = 2s$.

Let $TM^{\mathbb{C}}$ denote the complexified tangent bundle of M whose fibers are $\mathbb{C} \otimes_{\mathbb{R}} T_pM$, $p \in M$. Here, $\dim_{\mathbb{C}} (\mathbb{C} \otimes_{\mathbb{R}} T_pM) = 2s$. The smooth field J can be pointwise extended $\mathbb{C}$-linearly to $T_p^{\mathbb{C}}M$ so that, again, it satisfies (1.8). It follows that J_p has eigenvalues $\imath$ and $-\imath$ and

$$T_pM^{\mathbb{C}} = T_pM^{(1,0)} \oplus T_pM^{(0,1)} \tag{1.9}$$

where $T_pM^{(1,0)}$ and $T_pM^{(0,1)}$ are the eigenspaces of the eigenvalues $\imath$ and $-\imath$, respectively, Furthermore, $v' \in T_pM^{(1,0)}$ and $v'' \in T_pM^{(0,1)}$ if and only if there exist $u, w \in T_pM$ such that

$$v' = u - iJ_pu, \quad v'' = w + iJ_pw.$$

The above decomposition induces a dual decomposition

$$T_p^*M^{\mathbb{C}} = T_p^*M^{(1,0)} \oplus T_p^*M^{(0,1)}. \tag{1.10}$$

Note that (1.9) and (1.10) hold at the bundle level. Similar decompositions are induced on tensor products and in particular on the Grassmann bundle

$$\Lambda^k T^*M^{\mathbb{C}} = \sum_{i+j=k} \Lambda^{(i,j)} T^*M^{\mathbb{C}}.$$

As we have just seen, the existence of J as in Definition 1.6 induces restrictions on M and, for instance, one can, according to the previous discussion, easily prove that an almost complex manifold (M, J) is even-dimensional and orientable. However, these conditions are not sufficient to guarantee the existence of J. Indeed, C. Ehreshmann and H. Hopf (see [154] page 217) have shown that S^4 cannot be given an almost complex structure J.

Definition 1.7. An almost Hermitian manifold $(M, \langle\,,\rangle, J)$ is an almost complex manifold (M, J) with a Riemannian metric $\langle\,,\rangle$ with respect to which J is an isometry, that is, for every $p \in M$ and every $v, w \in T_pM$,

$$\langle J_pv, J_pw\rangle - \langle v, w\rangle.$$

In what follows, we extend $\langle\,,\rangle$ complex-bilinearly to $T_pM^{\mathbb{C}}$.

Definition 1.8. The Kähler form of an almost Hermitian manifold $(M, \langle\,,\rangle, J)$ is the $(1,1)$-form defined by

$$\mathcal{K}(X, Y) = \langle X, JY\rangle$$

for each $X, Y \in TM^{\mathbb{C}}$.

Note that $d\mathcal{K} \in \Lambda^3 T^*M^{\mathbb{C}}$ can be split into types according to the decomposition in (1.10).

Definition 1.9. An almost Hermitian manifold $(M, \langle\,,\rangle, J)$ is said to be $(1,2)$-symplectic if

$$d\mathcal{K}^{(1,2)} = 0.$$

Similarly, if

$$d\mathcal{K} = 0$$

or

$$\delta\mathcal{K} = 0$$

where $\delta = -*d*$ is the co-differential acting on 2-forms (see Appendix B), the almost Hermitian manifold is said to be symplectic and co-symplectic, respectively.

Definition 1.10. Let $(M, \langle\, ,\rangle\, , J)$ be a (symplectic) almost Hermitian manifold. If the almost complex structure J is induced by a complex structure on M, that is, J is the multiplication by $\imath$ in the charts of a holomorphic atlas, then $(M, \langle\, ,\rangle\, , J)$ is called a (Kähler) Hermitian manifold.

Note that there are manifolds which cannot be given a Kählerian structure, for instance the Hopf and Calabi-Eckmann manifolds; see [35] page 69.

Given an almost complex manifold (M, J) the *Nijenhuis tensor* N is the tensor field of type $(1, 2)$ given by

$$N(X, Y) = 2\{[JX, JY] - [X, Y] - J[X, JY] - J[JX, Y]\}$$

for each vector field $X, Y \in \Gamma(TM)$, and where $[\, ,\,]$ denotes the Lie bracket.

By the Newlander-Nirenberg theorem, [118], an almost complex structure J is induced by a complex structure if and only if the Nijenhuis tensor vanishes identically.

At the cotangent bundle level, this is expressed by

$$d\omega = 0 \mod (1, 0)\text{-forms}$$

for each form ω of type $(1, 0)$. In other words the ideal generated by the $(1, 0)$-forms is a differential ideal. Note that if $\dim_{\mathbb{R}} M = 2$ this is always true (the result is due to Korn and Lichtenstein). In a way similar to that of the definition of the Kähler form, we introduce the *Ricci form* $\mathcal{R}$, that is, for every $X, Y \in TM^{\mathbb{C}}$,

$$\mathcal{R}(X, Y) = \mathrm{Ric}(JX, Y).$$

Clearly, $\mathcal{R}$ is a $(1, 1)$ form and the Kähler manifold $(M, \langle\, ,\rangle\, , J_M)$ is said to be *Kähler–Einstein* in case

$$\mathcal{R} = -\frac{\imath}{4m} S(x)\, \mathcal{K}$$

with $S(x)$ the scalar curvature.

Let $f : (M, \langle\, ,\rangle\, , J_M) \to (N, (\, ,\,)\, , J_N)$ be a smooth map between almost Hermitian manifolds. Then, df can be linearly extended to the complexified differential $df^{\mathbb{C}} : TM^{\mathbb{C}} \to TN^{\mathbb{C}}$. According to the decomposition

$$TN^{\mathbb{C}} = TN^{(1,0)} \oplus TN^{(0,1)}$$

we can write

$$df^{\mathbb{C}} = df^{(1,0)} + df^{(0,1)}.$$

Definition 1.11. A map $f : (M, \langle\, ,\rangle\, , J_M) \to (N, (\, ,\,)\, , J_N)$ between almost Hermitian manifolds is holomorphic if and only if

$$J_N \circ df = df \circ J_M.$$

This is immediately seen to be equivalent to the fact that $df^{\mathbb{C}}$ carries $(1,0)$ vectors into $(1,0)$ vectors or the pull-back of $(1,0)$ forms, under the complex linear extension $\left(f^{\mathbb{C}}\right)^*$, are $(1,0)$ forms or, finally, to the fact that $df^{(0,1)} = 0$.

On the other hand, f is said to be anti-holomorphic if

$$J_N \circ df = -df \circ J_M .$$

The basic relation between (anti-)holomorphic maps and harmonic maps is given by the following local result due to A. Lichnerowicz, [108].

Proposition 1.12. *Let $(M, \langle\,,\rangle, J_M)$ and $(N, (\,,), J_N)$ be almost Hermitian manifolds. If M is co-symplectic and N is $(1,2)$-symplectic, then any (anti-)holomorphic map $f : M \to N$ is harmonic.*

Note that, if M is symplectic, then it is also co-symplectic. We should also remark that some condition on M is necessary for a (anti-)holomorphic map to be harmonic, as an example of A. Grey shows. See [48], page 58.

We now consider the case where $(M, \langle\,,\rangle, J_M)$ is an almost Hermitian manifold and $(N, (\,,))$ is Riemannian. Given a map $f : M \to N$ we can split its generalized second fundamental tensor Ddf according to types in $T^*M^{\mathbb{C}} \otimes T^*M^{\mathbb{C}} \otimes f^{-1}TN$. We have

$$Ddf^{\mathbb{C}} = Ddf^{(2,0)} + Ddf^{(1,1)} + Ddf^{(0,2)}$$

where $Ddf^{\mathbb{C}}$ is the complex linear extension of Ddf.

Definition 1.13. The map $f : (M, \langle\,,\rangle, J_M) \to (N, (\,,))$ is said to be pluriharmonic, or $(1,1)$-geodesic, if $Ddf^{(1,1)} = 0$.

When $N = \mathbb{R}$, then $Ddf^{(1,1)}$ is a Hermitian form referred to as the Levi form of f.

Definition 1.14. We say that the function $f : (M, \langle\,,\rangle, J_M) \to \mathbb{R}$ is plurisubharmonic if all eigenvalues of its Levi form are non-negative.

Note that any pluriharmonic map is harmonic and, if the almost Hermitian manifolds $(M, \langle\,,\rangle, J_M)$ and $(N, (\,,), J_N)$ are also $(1,2)$-symplectic, then any (anti-)holomorphic map $f : M \to N$ is pluriharmonic.

Thus, the notion of pluriharmonic map lies between those of harmonic and (anti-)holomorphic maps.

In case $(M, \langle\,,\rangle, J_M)$ is almost Hermitian and $(1,2)$-symplectic, and $(N, (,))$ is Riemannian, J. Rawnsley, [136], has given the following characterization.

Theorem 1.15. *A map $f : (M, \langle\,,\rangle, J_M) \to (N, (,))$ is pluriharmonic if and only if its restriction to every complex curve in M is harmonic.*

Note that, from this it follows that if $(M, \langle\,,\rangle, J_M)$ is Kähler, then the notion of pluriharmonic map does not depend on the choice of the Kähler metric $\langle\,,\rangle$ on M.

We also note that, if $(M, \langle\,,\rangle, J_M)$ and $(N, (\,,), J_N)$ are Kähler and $f : M \to N$ is an isometry, then we can express holomorphicity of f via the system

$$\begin{cases} \mathrm{II}\,(X,Y) + \mathrm{II}\,(J_M X, J_M Y) = 0, \\ \mathrm{II}\,(X,Y) + J_N \mathrm{II}\,(X, J_M Y) = 0 \end{cases}$$

for all X, Y vector fields on M, where we have used the more familiar notation II for Ddf in the isometric case. Clearly, the first equation is nothing but the definition of a pluriharmonic map.

The notion of a pluriharmonic map has appeared in the literature in the context of the work of Y.T. Siu, [152], who used it as a bridge from harmonicity to (anti-)holomorphicity in the analysis of the strong rigidity of compact Kähler manifolds. Since then, it has been used in a variety of geometrical problems and it will be used below with the aim of providing extra geometric information.

1.3 Hermitian bundles

Later on we shall also be interested in *vector bundles* of rank q on a base manifold M. This means that we have a map

$$\pi : E \to M$$

such that the following conditions are satisfied:

(i) for each $x \in M$, $\pi^{-1}(x)$ is a real (or complex) vector space of dimension q.

(ii) E is locally a product, that is, for each $x \in M$, there exists an open neighborhood U of x and a bijection

$$\varphi_U : U \times V \to \pi^{-1}(U)$$

with V any fixed real (or complex) vector space of dimension q satisfying the condition

$$\pi \circ \varphi_U(x, v) = x,$$

for each $v \in V$.

(iii) For any two of the above neighborhoods U_1, U_2 such that $U_1 \cap U_2 \neq \emptyset$, there is a map

$$g_{U_1 U_2} : U_1 \cap U_2 \to Gl_q(\mathbb{R}) \ \ (\text{or } Gl_q(\mathbb{C}))$$

such that, for $x \in U_1 \cap U_2$, and for each $v, w \in V$,

$$\varphi_{U_1}(x, v) = \varphi_{U_2}(x, w)$$

if and only if

$$v = g_{U_1 U_2}(x)\, w.$$

Clearly, E can be given a (unique) topology and differentiable structure such that each $\left(\pi^{-1}(U), \varphi_U^{-1}\right)$ of (ii) is a local chart. The functions $g_{U_1 U_2}$ are called *transition functions* of the bundle and they satisfy

$$\begin{aligned}
g_{UU}(x) &= id \in Gl_q(\mathbb{R}), && \text{for each } x \in U,\\
g_{U_1U_2}g_{U_2U_1} &= id \in Gl_q(\mathbb{R}), && \text{for each } x \in U_1 \cap U_2,\\
g_{U_1U_2}g_{U_2U_3}g_{U_3U_1} &= id \in Gl_q(\mathbb{R}), && \text{for each } x \in U_1 \cap U_2 \cap U_3.
\end{aligned}$$

It is well known that the transition functions relative to a covering of M completely determine the bundle.

A *section* of $\pi : E \to M$ is a map $s : M \to E$ such that $\pi \circ s = id_M$. The set $\Gamma(E)$ of smooth sections of E is a vector space over $\mathbb{R}$ (or $\mathbb{C}$)

A *connection on* E is a map

$$D : \Gamma(E) \to \Gamma(T^*M \otimes E)$$

such that the following conditions are satisfied for each $s, t \in \Gamma(E)$, and for each $f \in C^\infty(M)$ (f either real- or complex-valued):

$$\begin{aligned}
(D_1) &\quad D(s+t) = Ds + Dt,\\
(D_2) &\quad D(fs) = fDs + df \otimes s.
\end{aligned}$$

Letting $X \in \Gamma(TM)$, $D_X s$ is the derivative of s in the direction of X. Note that $D_X s \in \Gamma(E)$.

It does make sense to define the *curvature transformation*

$$\tilde{K}(X,Y) : \Gamma(E) \to \Gamma(E)$$

where $X, Y \in \Gamma(TM)$ are any two vector fields of M, by setting

$$\tilde{K}(X,Y)s = D_X D_Y s - D_Y D_X s - D_{[X,Y]}s.$$

A *Riemannian vector bundle* is a smooth vector bundle with a *fibre metric* h and a *compatible connection* D, that is, if s and t are sections of $\pi : E \to M$, then, for each vector field $X \in \Gamma(TM)$,

$$Xh(s,t) = h(D_X s, t) + h(s, D_X t).$$

We will be mainly concerned with *Hermitian bundles*, that is, E is a Hermitian manifold with a connection, the Hermitian connection, which is compatible with the metric and uniquely determined by the next requirement (see, [35]).

Let $q = \dim_{\mathbb{R}} \pi^{-1}(x) = 2p$ be the real dimension of the fibres and let $\{e_a\}$, $1 \leq a, b, \ldots \leq p$, be a unitary $(1,0)$-type local frame of sections of E. Thus, indicating with $\{\mu^a\}$ the dual $(1,0)$ forms, we have

$$h = \sum_a \mu^a \otimes \bar{\mu}^a.$$

Then, the *Hermitian connection* on E is the unique connection whose connection forms μ^a_b are determined by the requirements

$$\begin{aligned} &\text{(i)}\ \ \mu^a_b + \bar{\mu}^b_a = 0, \\ &\text{(ii)}\ d\mu^a = -\mu^a_b \wedge \mu^b + \zeta^a, \end{aligned} \tag{1.11}$$

where the ζ^a are forms of $(2,0)$-type, and $^-$ denotes complex conjugation. The *curvature forms* M^a_b are then defined by the second structure equations

$$d\mu^a_b = -\mu^a_c \wedge \mu^c_b + M^a_b,$$

which are of type $(1,1)$ and satisfy

$$M^a_b + \bar{M}^b_a = 0.$$

Having set

$$M^a_b = A^a_{bc\bar{d}} \mu^c \wedge \bar{\mu}^d,$$

the metric of the bundle is said to be *Hermitian–Einstein* if

$$\sum_c A^a_{bc\bar{c}} = \lambda \delta^a_b$$

for some constant $\lambda \in \mathbb{C}$. Note that the matrix

$$\left(\sum_c A^a_{bc\bar{c}} \right)_{a,b}$$

is called the *mean curvature* and

$$\mathrm{scal}_h(x) = \sum_{a,c} A^a_{ac\bar{c}}$$

(in a unitary frame) is called the *scalar curvature* of the Hermitian bundle $\pi : E \to M$.

1.4 Complex geometry via moving frames

In what follows we shall always deal with the case where $(M, \langle\,,\rangle, J_M)$ is Kähler, while $(N, (\,,))$ or $(N, (\,,), J_N)$, the target manifolds of maps, will be Riemannian or Kählerian. Later on we shall also consider the case where $(N, (\, , \,) , J_N)$ is Hermitian. The situation, from our point of view, will be very similar to the Kähler case so, since in the Hermitian case the formalism is definitively heavier, we will not bother to provide details in derivation of the appropriate differential inequalities needed in some proofs of theorems in later chapters. We formalize the Kähler structure with a particular emphasis on its Riemannian counterpart and to do so we will use the method of moving frame. Thus, let $(M, \langle \, , \, \rangle , J_M)$ be a Kähler manifold with $s = \dim_{\mathbb{C}} M$ so that $m = 2s = \dim_{\mathbb{R}} M$. We fix the index convention $1 \leq i, j, k \ldots \leq s$. The Kähler structure of M is naturally described by a unitary coframe $\{\varphi^j\}$ of $(1,0)$-type, 1-forms giving the metric

$$\langle \, , \, \rangle = \sum_j \varphi^j \otimes \bar{\varphi}^j$$

and the corresponding Kähler connection forms $\{\varphi^i_j\}$ characterized by the property

$$\varphi^i_j + \bar{\varphi}^j_i = 0$$

and by the structure equations

$$d\varphi^j = -\varphi^j_k \wedge \varphi^k. \tag{1.12}$$

Note that, comparing with (1.11), we are now requiring that the $(2,0)$-forms ζ^a are identically zero. This can be seen to be equivalent to $d\mathcal{K} = 0$, i.e., to the condition that the complex manifold is Kähler (see Definition 1.9).

The Kähler curvature forms $\left\{\Phi^j_k\right\}$ are determined by the second structure equations

$$d\varphi^j_k = -\varphi^j_i \wedge \varphi^i_k + \Phi^j_k \tag{1.13}$$

and satisfy the symmetry relations

$$\Phi^j_k + \bar{\Phi}^k_j = 0. \tag{1.14}$$

The coefficients H^i_{jkt} of the Hermitian curvature tensor are determined by

$$\Phi^i_j = H^i_{jkt}\varphi^k \wedge \bar{\varphi}^t \tag{1.15}$$

and condition (1.14) becomes equivalent to

$$H^i_{jkt} = \overline{H^j_{itk}}.$$

Differentiating (1.12) we obtain the first complex Bianchi identities, that is,

$$\Phi^j_k \wedge \varphi^k = 0$$

while differentiating (1.13) we obtain the second complex Bianchi identities which we write in the form

$$d\Phi^j_i + \Phi^k_i \wedge \varphi^j_k - \varphi^k_i \wedge \Phi^j_k = 0.$$

We also recall that the Kähler form $\mathcal{K}$ and the Ricci form $\mathcal{R}$ are respectively given by

$$\mathcal{K} = \frac{\imath}{2} \sum_j \varphi^j \wedge \bar{\varphi}^j,$$

$$\mathcal{R} = \frac{1}{2} \sum_i H^i_{ikt} \varphi^k \wedge \bar{\varphi}^t,$$

so that $(M, \langle\, , \rangle, J_M)$ is Kähler–Einstein if and only if

$$\sum_i H^i_{ikt} = \frac{s}{8m} \delta_{kt}$$

with s the scalar curvature. In order to detect the underlying Riemannian structure we set

$$\varphi^j = \theta^j + \imath\theta^{s+j}, \tag{1.16}$$

$$\varphi^j_k = \theta^j_k + \imath\theta^{s+j}_k, \tag{1.17}$$

$$\theta^j_k = \theta^{s+j}_{s+k}, \qquad \theta^j_{s+k} = -\theta^{s+j}_k. \tag{1.18}$$

Then, the θ^j, θ^{s+j} give an orthonormal coframe for the metric $\langle\, , \rangle$ whose corresponding Levi–Civita connection forms are determined by (1.17), (1.18) and the usual skew symmetry conditions

$$\theta^a_b + \theta^b_a = 0,$$

where, from now on, we shall adhere also to the further index convention $1 \leq a, b, \ldots \leq m$. Analogously, setting

$$\Phi^k_j = \Theta^k_j + \imath\Theta^{s+k}_j, \tag{1.19}$$

$$\Theta^k_j = \Theta^{s+k}_{s+j}, \qquad \Theta^j_{s+k} = -\Theta^{s+j}_k, \tag{1.20}$$

$$0 = \Theta^a_b + \Theta^b_a, \tag{1.21}$$

the Θ^a_b's defined in (1.19), (1.20), (1.21) coincide with the corresponding curvature forms. Thus, letting R^a_{bcd} be the coefficients of the Riemannian curvature tensor (obeying the usual symmetries), for which

$$\Theta^a_b = \frac{1}{2} R^a_{bcd} \theta^c \wedge \theta^d \tag{1.22}$$

form (1.20), we obtain the Kähler symmetry relations

$$R^k_{jab} = R^{s+k}_{s+j\ ab}, \qquad R^j_{s+k\ ab} = -R^{s+j}_{kab}. \tag{1.23}$$

We use (1.15) and (1.19), (1.22) to relate Hermitian and Riemannian curvatures. We obtain

$$H^i_{jkt} = \frac{1}{2}\left(R^i_{jkt} + R^{s+i}_{j\ k+s\ t}\right) + \frac{\imath}{2}\left(R^i_{jk\ s+t} + R^{s+i}_{jkt}\right). \tag{1.24}$$

Extending $\mathbb{C}$-linearly the ((4,0)-version of the) Riemannian curvature tensor we obtain

$$\begin{aligned}R_{abcd}\theta^a\otimes\theta^b\otimes\theta^c\otimes\theta^d &= R_{i\bar{j}k\bar{l}}\varphi^i\otimes\bar{\varphi}^j\otimes\varphi^k\otimes\bar{\varphi}^l + R_{i\bar{j}\bar{k}l}\varphi^i\otimes\bar{\varphi}^j\otimes\bar{\varphi}^k\otimes\varphi^l\\ &+ R_{\bar{i}j\bar{k}l}\overline{\varphi}^i\otimes\varphi^j\otimes\overline{\varphi}^k\otimes\varphi^l + R_{\bar{i}jk\bar{l}}\overline{\varphi}^i\otimes\varphi^j\otimes\varphi^k\otimes\overline{\varphi}^l\end{aligned}$$

where

$$R_{\bar{i}j\bar{k}l} = \overline{R_{i\bar{j}k\bar{l}}},\qquad R_{\bar{i}jk\bar{l}} = \overline{R_{i\bar{j}\bar{k}l}}, \tag{1.25}$$

$$R_{i\bar{j}\bar{k}l} = -R_{i\bar{j}l\bar{k}} = -R_{\bar{j}i\bar{k}l},\qquad R_{i\bar{j}\bar{k}l} = R_{\bar{k}li\bar{j}}, \tag{1.26}$$

the remaining coefficients, for instance $R_{\bar{i}\bar{j}kl}$, being null. From (1.25), (1.26) and (1.24) we deduce

$$H^i_{jkl} = R_{i\bar{j}\bar{k}l}. \tag{1.27}$$

Recalling the first (Riemannian) Bianchi identities

$$R^a_{bcd} + R^a_{cdb} + R^a_{dbc} = 0,$$

with the aid of (1.23) we obtain

$$\sum_k R^{s+i}_{j\ s+k\ k} = \mathrm{Ric}_{s+i\ s+j} = \mathrm{Ric}_{ij}.$$

Hence, tracing (1.24) twice we obtain that the scalar curvature s is given by

$$s = 4\sum_{k,i} H^i_{ikk}.$$

Furthermore

$$\mathrm{Ric}_{s+i\ j} = -\mathrm{Ric}_{i\ s+j} = -\mathrm{Ric}_{s+j\ i} = -\mathrm{Ric}_{j\ s+i} \tag{1.28}$$

in particular, for each fixed $i = 1,\dots,s$,

$$\mathrm{Ric}_{s+i\ i} = 0.$$

Finally, the "Ricci curvature" of the Kähler manifold has components given by the Hermitian matrix

$$R_{i\bar{j}} = \sum_k H^i_{jkk} = \frac{1}{2}\mathrm{Ric}_{ij} + \frac{\imath}{2}\mathrm{Ric}_{s+i\ j}. \tag{1.29}$$

From (1.27) and (1.28) we deduce

$$\mathrm{Ric}_{s+i\ j} = \imath\sum_k\left(R_{\bar{i}jk\bar{k}} - R_{i\bar{j}\bar{k}k}\right),$$

$$\mathrm{Ric}_{ij} = \sum_k\left(R_{i\bar{j}\bar{k}k} + R_{\bar{i}jk\bar{k}}\right).$$

In particular

$$\sum_k H^i_{jkk} = R_{i\bar{j}\,\bar{k}\,k} = R_{i\bar{j}},$$

and the Ricci form can be expressed as

$$\mathcal{R} = \frac{1}{2}\sum_i R_{i\bar{j}}\varphi^k \wedge \overline{\varphi}^l.$$

Note that an orthogonal transformation U_z of $T_zM^{\mathbb{C}}$ is unitary if and only if it commutes with J_z (see, e.g., [88], p. 116), and we may therefore diagonalize the Hermitian matrix $R_{k\bar{j}}$ with a $(1,0)$-basis of the form $E_k = e_k - \imath J_z e_k = e_k - \imath e_{k+s}$, where $\{e_k,\, e_{k+s}\}$ is the orthonormal basis of T_zM dual to $\{\theta^i, \theta^{s+i}\}$. If λ_k are the corresponding eigenvalues of $R_{k\bar{j}}$, then

$$R_{k\bar{j}} = \lambda_k\delta_{kj} = \frac{1}{2}\Big(\mathrm{Ric}_{kj} + \imath \mathrm{Ric}_{k+sj}\Big)$$

which implies that

$$\mathrm{Ric}_{kj} = 2\lambda_k\delta_{kj} \ \text{ and } \ \mathrm{Ric}_{k+sj} = 0 \quad \forall k, j.$$

Further, since $\mathrm{Ric}_{k+sj+s} = \mathrm{Ric}_{kj} = 2\lambda_k\delta_{kj}$, we conclude that $2\lambda_k$ is an eigenvalue of Ric. This shows that, if

$$2R_{i\bar{j}}u^i\bar{u}^j \geq -\rho|u|^2$$

holds for every $u \in \mathbb{C}^m$, then inequality

$$\mathrm{Ric}_{i\,j}v^iv^j \geq -\rho|v|^2$$

holds for every $v \in \mathbb{R}^m$. Since the reverse implication is obviously true, we conclude that two conditions are in fact equivalent.

Let $\{e_a\}$ be the dual frame to $\{\theta^a\}$. For each $i, k = 1, \dots, s$, we consider the holomorphic 2-planes Π and $\widehat{\Pi}$ spanned by $e_i, Je_i = e_{s+i}$, and e_k, $Je_k = e_{s+k}$, respectively. Then the holomorphic bisectional curvature of Π and $\widehat{\Pi}$ is defined by

$$H^i_{ikk} = \frac{1}{4}R_{i\ s+i\ k\ s+k},$$

where, in this case, there is no summation over repeated indices. In particular, if $\Pi = \widehat{\Pi}$ we obtain the holomorphic sectional curvature of the 2-plane Π, namely,

$$H^i_{iii} = \frac{1}{4}R_{i\ s+i\ i\ s+i} = \frac{1}{4}\mathrm{Sect}\,(\Pi)$$

where, as above, there is no summation over repeated indices, and where $\mathrm{Sect}\,(\Pi)$ is the (Riemannian) sectional curvature of Π.

We say that the holomorphic bisectional curvature of M is bounded above by a function $k\,(z)$ if, for all $(1,0)$ vectors $\zeta = \xi^kE_k$, $\eta = \eta^jE_j$, at z, we have

$$\frac{1}{2}\frac{H^i_{jkl}\xi^i\overline{\xi}^j\eta^k\overline{\eta}^l}{\sum \xi^k\overline{\xi}^k \sum \eta^k\overline{\eta}^k} \leq k\,(z).$$

1.5 Weitzenböck-type formulas

We shall now derive some Weitzenböck-type formulas. First, we consider the case where $(M, \langle\, ,\rangle\, , J_M)$ is a Kähler manifold, $(N, (\, ,))$ is a Riemannian manifold, and we derive a Weitzenböck formula for pluriharmonic maps $f : M \to N$. Since it is difficult to find the formula in the literature, we will provide a rather detailed derivation. We fix the index convention $1 \leq i, j, k, \ldots \leq s = 2m = \dim_{\mathbb{C}} M$, $1 \leq \alpha, \beta, \gamma, \ldots \leq n = \dim_{\mathbb{R}} N$. For the manifold M we keep the notation previously introduced, while for N we let $\{\omega^\alpha\}, \left\{\omega^\alpha_\beta\right\}, \left\{\Omega^\alpha_\beta\right\}$ be respectively a (local) orthonormal coframe, the relative Levi–Civita connection forms and the curvature forms. For $f : M \to N$ we set

$$f^*\omega^\alpha = F^\alpha_i \varphi^i + F^\alpha_{\bar{i}} \overline{\varphi}^i \tag{1.30}$$

so that

$$df = \left(F^\alpha_i \varphi^i + F^\alpha_{\bar{i}} \overline{\varphi}^i\right) \otimes \varepsilon_\alpha$$

with $\{\varepsilon_\alpha\}$ the frame dual to $\{\omega^\alpha\}$. We observe that

$$\overline{F^\alpha_i} = F^\alpha_{\bar{i}}. \tag{1.31}$$

We take exterior differentiation of (1.30) and we use the structure equations on M and N to get

$$\left(dF^\alpha_k - F^\alpha_i \varphi^i_k + F^\beta_k \omega^\alpha_\beta\right) \wedge \varphi^k + \left(dF^\alpha_{\bar{k}} - F^\alpha_{\bar{i}} \overline{\varphi}^i_k + F^\beta_{\bar{k}} \omega^\alpha_\beta\right) \wedge \overline{\varphi}^k = 0.$$

By Cartan's lemma, there exist $F^\alpha_{ij}, F^\alpha_{\bar{i}j}, F^\alpha_{i\bar{j}}, F^\alpha_{\bar{i}\bar{j}}$ with the properties

$$F^\alpha_{\bar{k}t} = F^\alpha_{k\bar{t}},\ F^\alpha_{kt} = F^\alpha_{tk},\ F^\alpha_{\bar{k}\bar{t}} = F^\alpha_{\bar{t}\bar{k}} \tag{1.32}$$

and such that

$$dF^\alpha_k - F^\alpha_i \varphi^i_k + F^\beta_k \omega^\alpha_\beta = F^\alpha_{kt}\varphi^t + F^\alpha_{k\bar{t}}\overline{\varphi}^t, \tag{1.33}$$

$$dF^\alpha_{\bar{k}} - F^\alpha_{\bar{i}} \overline{\varphi}^i_k + F^\beta_{\bar{k}} \omega^\alpha_\beta = F^\alpha_{\bar{k}t}\varphi^t + F^\alpha_{\bar{k}\bar{t}}\overline{\varphi}^t. \tag{1.34}$$

Note that, from (1.31) we obtain

$$\overline{F^\alpha_{kt}} = F^\alpha_{\bar{k}\bar{t}}, \quad \overline{F^\alpha_{k\bar{t}}} = F^\alpha_{\bar{k}t}. \tag{1.35}$$

We observe that, comparing with the underlying Riemannian structure on M and using the notation previously introduced, we have

$$F^\alpha_k = \frac{1}{2}\left(f^\alpha_k - \imath f^\alpha_{s+k}\right), \tag{1.36}$$

$$F^\alpha_{kt} = \frac{1}{4}\left(f^\alpha_{kt} - f^\alpha_{s+k\ s+t}\right) - \frac{\imath}{4}\left(f^\alpha_{s+k\ t} + f^\alpha_{k\ s+t}\right), \tag{1.37}$$

$$F^\alpha_{k\bar{t}} = \frac{1}{4}\left(f^\alpha_{kt} + f^\alpha_{s+k\ s+t}\right) - \frac{\imath}{4}\left(f^\alpha_{s+k\ t} - f^\alpha_{k\ s+t}\right). \tag{1.38}$$

In particular,

$$Ddf^{\mathbb{C}} = F^{\alpha}_{kt}\varphi^k \otimes \varphi^t \otimes \varepsilon_\alpha + 2F^{\alpha}_{k\bar{t}}\varphi^k \otimes \overline{\varphi}^t \otimes \varepsilon_\alpha + F^{\alpha}_{\bar{k}\bar{t}}\overline{\varphi}^k \otimes \overline{\varphi}^t \otimes \varepsilon_\alpha$$

and it follows that f is pluriharmonic if and only if

$$F^{\alpha}_{k\bar{t}} = 0 = F^{\alpha}_{\bar{k}t}$$

and from (1.38) we double check, in this setting, that a pluriharmonic map is, in particular, harmonic. Indeed, the harmonicity of f is equivalent to

$$\sum_k F^{\alpha}_{k\bar{k}} = 0.$$

Note that

$$\tau(f) = 4\sum_k F^{\alpha}_{k\bar{k}}\varepsilon_\alpha.$$

Next, we take exterior differentiation of (1.33) and we use the structure equations to get

$$\begin{aligned}\left(dF^{\alpha}_{kt} - F^{\alpha}_{jt}\varphi^j_k - F^{\alpha}_{kj}\varphi^j_t + F^{\beta}_{kt}\omega^{\alpha}_{\beta}\right) \wedge \varphi^t \qquad&\\ + \left(dF^{\alpha}_{k\bar{t}} - F^{\alpha}_{j\bar{t}}\varphi^j_k - F^{\alpha}_{k\bar{j}}\overline{\varphi}^j_t + F^{\beta}_{k\bar{t}}\omega^{\alpha}_{\beta}\right) \wedge \overline{\varphi}^t&\\ = -\frac{1}{2}F^{\alpha}_i H^i_{klt}\varphi^l \wedge \overline{\varphi}^t + \frac{1}{2}F^{\beta}_k\ {}^N\mathrm{Riem}^{\alpha}_{\beta\gamma\delta}\omega^\gamma \wedge \omega^\delta.&\end{aligned}$$

We define

$$\begin{cases} F^{\alpha}_{ktl}\varphi^l + F^{\alpha}_{kt\bar{l}}\overline{\varphi}^l = dF^{\alpha}_{kt} - F^{\alpha}_{jk}\varphi^j_k - F^{\alpha}_{kj}\varphi^j_t + F^{\beta}_{kt}\omega^{\alpha}_{\beta}, \\ F^{\alpha}_{k\bar{t}l}\varphi^l + F^{\alpha}_{k\bar{t}\bar{l}}\overline{\varphi}^l = dF^{\alpha}_{k\bar{t}} - F^{\alpha}_{j\bar{t}}\varphi^j_k - F^{\alpha}_{k\bar{j}}\overline{\varphi}^j_t + F^{\beta}_{k\bar{t}}\omega^{\alpha}_{\beta}. \end{cases}$$

Using (1.30) and the above equations we deduce the commutation relations

$$\begin{aligned} F^{\alpha}_{ktl} &= F^{\alpha}_{klt} - F^{\beta}_k F^{\gamma}_t F^{\delta}_l\ {}^N\mathrm{Riem}^{\alpha}_{\beta\gamma\delta}, \\ F^{\alpha}_{k\bar{t}\bar{l}} &= F^{\alpha}_{k\bar{l}\bar{t}} - F^{\beta}_k F^{\gamma}_{\bar{t}} F^{\delta}_{\bar{l}}\ {}^N\mathrm{Riem}^{\alpha}_{\beta\gamma\delta}, \\ F^{\alpha}_{k\bar{t}l} &= F^{\alpha}_{kl\bar{t}} - \frac{1}{2}F^{\alpha}_i H^i_{klt} - \frac{1}{2}F^{\beta}_k\left(F^{\gamma}_{\bar{t}}F^{\delta}_l - F^{\delta}_{\bar{t}}F^{\gamma}_l\right)\ {}^N\mathrm{Riem}^{\alpha}_{\beta\delta\gamma}. \end{aligned}$$

Similarly, we set

$$\begin{cases} F^{\alpha}_{\bar{k}tl}\varphi^l + F^{\alpha}_{\bar{k}t\bar{l}}\overline{\varphi}^l = dF^{\alpha}_{\bar{k}t} - F^{\alpha}_{\bar{j}k}\overline{\varphi}^j_k - F^{\alpha}_{\bar{k}j}\varphi^j_t + F^{\beta}_{\bar{k}t}\omega^{\alpha}_{\beta}, \\ F^{\alpha}_{\bar{k}\bar{t}l}\varphi^l + F^{\alpha}_{\bar{k}\bar{t}\bar{l}}\overline{\varphi}^l = dF^{\alpha}_{\bar{k}\bar{t}} - F^{\alpha}_{\bar{j}\bar{t}}\overline{\varphi}^j_k - F^{\alpha}_{\bar{k}\bar{j}}\overline{\varphi}^j_t + F^{\beta}_{\bar{k}\bar{t}}\omega^{\alpha}_{\beta}. \end{cases}$$

Exterior differentiation of (1.34) and use of the structure equations yield the commutation relations

$$\begin{aligned} F^{\alpha}_{\bar{k}tl} &= F^{\alpha}_{\bar{k}lt} - F^{\beta}_{\bar{k}} F^{\gamma}_{t} F^{\delta}_{l} \; {}^{N}\mathrm{Riem}^{\alpha}_{\beta\gamma\delta}, \\ F^{\alpha}_{\bar{k}\bar{t}\bar{l}} &= F^{\alpha}_{\bar{k}\bar{l}\bar{t}} - F^{\beta}_{\bar{k}} F^{\gamma}_{\bar{t}} F^{\delta}_{\bar{l}} \; {}^{N}\mathrm{Riem}^{\alpha}_{\beta\gamma\delta}, \\ F^{\alpha}_{\bar{k}t\bar{l}} &= F^{\alpha}_{\bar{k}\bar{l}t} - \frac{1}{2} F^{\alpha}_{\bar{i}} \overline{H^{i}_{klt}} + \frac{1}{2} F^{\beta}_{\bar{k}} \left(F^{\gamma}_{\bar{l}} F^{\delta}_{t} - F^{\delta}_{\bar{l}} F^{\gamma}_{t} \right) \; {}^{N}\mathrm{Riem}^{\alpha}_{\beta\delta\gamma}. \end{aligned}$$

From (1.35) we also deduce

$$\overline{F^{\alpha}_{ktl}} = F^{\alpha}_{\bar{k}\bar{t}\bar{l}}; \; \overline{F^{\alpha}_{\bar{k}tl}} = F^{\alpha}_{k\bar{t}\bar{l}}; \; \overline{F^{\alpha}_{k\bar{t}l}} = F^{\alpha}_{\bar{k}t\bar{l}}; \; \overline{F^{\alpha}_{kt\bar{l}}} = F^{\alpha}_{\bar{k}\bar{t}l}$$

and from (1.32) we get

$$\begin{aligned} F^{\alpha}_{\bar{k}tl} &= F^{\alpha}_{t\bar{k}l}; \; F^{\alpha}_{\bar{k}t\bar{l}} = F^{\alpha}_{t\bar{k}\bar{l}}; \\ F^{\alpha}_{ktl} &= F^{\alpha}_{tkl}; \; F^{\alpha}_{kt\bar{l}} = F^{\alpha}_{tk\bar{l}}; \\ F^{\alpha}_{\bar{k}\bar{t}l} &= F^{\alpha}_{\bar{t}\bar{k}l}; \; F^{\alpha}_{\bar{k}\bar{t}\bar{l}} = F^{\alpha}_{\bar{t}\bar{k}\bar{l}}. \end{aligned}$$

We shall now use the above preliminaries to perform some computations. It is apparent that any time a harmonic map $f : (M, \langle\, , \rangle, J_M) \to (N, (\, ,))$ from a Kähler manifold to a Riemannian manifold is indeed pluriharmonic, we gain information of both geometrical and analytical flavor. Theorem 1.20 below gives a result in this direction. First we recall the following definition due to J.H. Sampson, [143].

Definition 1.16. A Riemannian manifold $(N, (\, ,))$ is said to have non-positive Hermitian curvature if

$$ {}^{N}\mathrm{Riem}_{\alpha\beta\gamma\delta} u^{\alpha} v^{\beta} \overline{u^{\gamma}} \overline{v^{\delta}} \leq 0 \tag{1.39}$$

for arbitrary complex vectors u and v.

Remark 1.17. Note that the left-hand side of (1.39) is indeed real.

Remark 1.18. If $(N, (\, ,))$ has constant, non-positive sectional curvature, then it has non-positive Hermitian curvature. Other examples are provided, for instance, by the following result of Sampson, [143].

Theorem 1.19. *Let $(N, (\, ,))$ be a Riemannian symmetric space whose irreducible local factors are all of the non-compact or Euclidean type. Then $(N, (\, ,))$ has non-positive Hermitian curvature.*

We are now in a position to state the following theorem that extends a result of Sampson, [143] to the non-compact setting.

Theorem 1.20. *Let $(M, \langle\, , \rangle, J_M)$ be a complete Kähler manifold and $(N, (\, ,))$ a Riemannian manifold of non-positive Hermitian curvature. Then any harmonic map $f : M \to N$ with energy density satisfying*

$$\left(\int_{\partial B_r} |df|^2 \right)^{-1} \notin L^1(+\infty) \tag{1.40}$$

is pluriharmonic.

Remark 1.21. As will become clear from the proof below, the harmonicity of f can be replaced by the weaker requirement that $D\tau(f) = 0$, namely, f has parallel tension field.

Remark 1.22. We remark that if f is assumed to be isometric, the conclusion of the theorem follows without additional hypotheses on the energy density from the assumption that M is Kähler and that N has non-positive (or even positive but suitably pinched) Hermitian curvature. See [52], and earlier work by [40]

Proof. For the local geometries of M and N we use the notation introduced above. Furthermore, we let

$$E_t = \frac{1}{\sqrt{2}}(e_t - ie_{s+t})$$

be the dual $(1,0)$-frame of the φ^k 's. We define the vector field

$$W = \sum_{k,t,\alpha} F^\alpha_k F^\alpha_{\bar{k}t} E_t$$

and we note that it is globally defined. With the aid of the formulas determined above and the commutation relations we obtain

$$\frac{1}{4}\mathrm{div}W = \sum_{k,t,\alpha} \left\{ F^\alpha_{\bar{k}t} F^\alpha_{k\bar{t}} + F^\alpha_k F^\alpha_{t\bar{t}\bar{k}} - F^\alpha_k F^\beta_t F^\gamma_{\bar{k}} F^\delta_{\bar{t}} \; {}^N\mathrm{Riem}^\alpha_{\beta\gamma\delta} \right\}.$$

Note that

$$D\tau(f) = 0 \Longleftrightarrow \sum_t F^\alpha_{t\bar{t}\bar{k}} = 0, \;\; \forall \alpha, k$$

and the latter is implied by harmonicity. Furthermore, (1.31) with non-positive Hermitian curvature gives

$$F^\alpha_k F^\beta_t F^\gamma_{\bar{k}} F^\delta_{\bar{t}} \; {}^N\mathrm{Riem}^\alpha_{\beta\gamma\delta} \le 0.$$

Using the divergence theorem it follows that

$$\int_{B_r} \left| Ddf^{(1,1)} \right|^2 \le \frac{1}{4} \int_{\partial B_r} \langle W, \nabla r \rangle . \tag{1.41}$$

Recalling the definition of W we have

$$\int_{\partial B_r} \langle W, \nabla r \rangle \le \left\{ \int_{\partial B_r} |df|^2 \right\}^{\frac{1}{2}} \left\{ \int_{\partial B_r} \left| Ddf^{(1,1)} \right|^2 \right\}^{\frac{1}{2}} . \tag{1.42}$$

Putting together (1.41) and (1.42) and squaring we finally get

$$\gamma(r)^2 \le \frac{1}{4} \left(\int_{\partial B_r} |df|^2 \right) \gamma'(r) \tag{1.43}$$

where we have set

$$\gamma(t) = \int_{B_r} \left| Ddf^{(1,1)} \right|^2$$

and, by the co-area formula,

$$\gamma'(r) = \int_{\partial B_r} \left| Ddf^{(1,1)} \right|^2 .$$

Next, we reason by contradiction and we suppose $Ddf^{(1,1)} \neq 0$. It follows that there exists $R > 0$ sufficiently large such that $\gamma(r) > 0$, for every $r \geq R$. Fix such an r. From (1.43) we then derive

$$\gamma(R)^{-1} - \gamma(r)^{-1} \geq 4 \int_R^r \frac{dt}{\int_{\partial B_t} |df|^2}$$

and letting $r \to +\infty$ we contradict (1.40). □

We will be interested in the following consequence that improves on a result by Li, [96], and that we state as

Corollary 1.23. *Let $(M, \langle\, ,\rangle, J_M)$ be a complete Kähler manifold. Then any harmonic function u with energy satisfying*

$$\left(\int_{\partial B_r} |\nabla u|^2 \right)^{-1} \notin L^1(+\infty)$$

is pluriharmonic.

We now derive a Weitzenböck-type formula for the energy density of a pluriharmonic map from a Kähler domain into a Riemannian target.

Theorem 1.24. *Let $(M, \langle\, ,\rangle, J_M)$ be a Kähler manifold, let $(N, (\, ,))$ be a Riemannian manifold, and let $f : M \to N$ be a pluriharmonic map. Then*

$$\begin{aligned} \Delta |df|^2 = {} & 16 \sum_{\alpha,k,t} F^\alpha_{kt} \overline{F^\alpha_{kt}} - 16 \sum_{k,t} F^\alpha_k F^\beta_t F^\gamma_{\bar{k}} F^\delta_{\bar{t}} \left({}^N\mathrm{Riem}_{\alpha\beta\gamma\delta} + {}^N\mathrm{Riem}_{\gamma\beta\alpha\delta} \right) \\ & + 16 \sum_{\alpha,k,l} R_{l\bar{k}} F^\alpha_{\bar{k}} F^\alpha_l . \end{aligned} \tag{1.44}$$

Proof. With the above notation, we let

$$u = \sum_{k,\alpha} F^\alpha_k F^\alpha_{\bar{k}} = \frac{1}{4} |df|^2 . \tag{1.45}$$

In order to compute Δu, we use the formulas

$$\begin{aligned} du &= u_t \varphi^t + u_{\bar{t}} \overline{\varphi^t}, \\ du_t - u_l \varphi^l_t &= u_{tl} \varphi^l + u_{t\bar{l}} \overline{\varphi^l}. \end{aligned} \tag{1.46}$$

Then

$$\Delta u = 4\sum_t u_{t\bar{t}}. \tag{1.47}$$

Differentiating (1.45) we obtain

$$u_t = \sum_{\alpha,k} F^\alpha_{\bar{k}} F^\alpha_{kt} + \sum_{\alpha,k} F^\alpha_k F^\alpha_{\bar{k}t}. \tag{1.48}$$

Next, we perform (1.46) and we use the structural equations to arrive at

$$u_{t\bar{l}} = \sum_{\alpha,k,t} \left\{ F^\alpha_{kt} F^\alpha_{\bar{k}\bar{l}} + F^\alpha_{\bar{k}} F^\alpha_{kt\bar{l}} + F^\alpha_{\bar{k}t} F^\alpha_{k\bar{l}} + F^\alpha_k F^\alpha_{\bar{k}t\bar{l}} \right\}. \tag{1.49}$$

Thus, according to (1.47), (1.49) and the commutation relations, we get

$$\begin{aligned}\Delta u = {} & 4\sum_{\alpha,k,t} F^\alpha_{kt} F^\alpha_{\bar{k}\bar{t}} - 4\sum_{\alpha,k,t} F^\alpha_k F^\beta_t F^\gamma_{\bar{k}} F^\delta_{\bar{t}} \left({}^N\mathrm{Riem}_{\alpha\beta\gamma\delta} + {}^N\mathrm{Riem}_{\gamma\beta\alpha\delta}\right) \\ & + 2\sum_{\alpha,k,t} \left\{ F^\alpha_{\bar{k}} F^\alpha_l H^l_{tkt} + 4F^\alpha_{\bar{k}t} F^\alpha_{k\bar{t}} + 4F^\alpha_k F^\alpha_{t\bar{t}\bar{k}} + 4F^\alpha_{\bar{k}} F^\alpha_{t\bar{t}k} \right\}.\end{aligned}$$

Note that, since f is pluriharmonic, the last three terms on the RHS of the above vanish and, furthermore,

$$F^\alpha_{\bar{k}\bar{t}} = \overline{F^\alpha_{kt}}. \tag{1.50}$$

Finally, using (1.29), we see that

$$\sum_t H^l_{tkt} = \sum_t H^l_{ktt.} = {}^M\mathrm{Ric}_{lk} + \imath\ {}^M\mathrm{Ric}_{s+l\ k} = 2R_{l\bar{k}},$$

proving the validity of (1.44). □

As an easy consequence we obtain the following corollary.

Corollary 1.25. *Let $\left(M, \langle\, ,\rangle, J_M\right)$ be a Kähler manifold with Ricci curvature satisfying*

$${}^M\mathrm{Ric} \geq -\rho(x),$$

and let $\left(N, (\, ,)\right)$ be a Riemannian manifold with non-positive Hermitian curvature. Then, for any pluriharmonic map $f: M \to N$ the following inequality holds:

$$|df|^2 \Delta |df|^2 \geq \left|\nabla |df|^2\right|^2 - 2\rho(x)|df|^4. \tag{1.51}$$

Proof. According to (1.45), $|df|^2 = 4\sum_{\alpha k} F^\alpha_{\bar{k}} F^\alpha_{\bar{k}} = 4u$. Using (1.48) and recalling that $F^\alpha_{\bar{k}} = \overline{F^\alpha_k}$ and $F^\alpha_{\bar{k}\bar{t}} = \overline{F^\alpha_{kt}}$ we obtain

$$|u_t|^2 \leq 4\sum_{\alpha k} |F^\alpha_k|^2 \sum_{\alpha k} |F^\alpha_{kt}|^2 = |df|^2 \sum_{\alpha k} |F^\alpha_{kt}|^2,$$

whence, using (1.50), we conclude that

$$\left|\nabla |df|^2\right|^2 = 4\sum |u_t|^2 \leq 16\,|df|^2 \sum_{\alpha,k,t} F^\alpha_{kt}\overline{F^\alpha_{kt}}.$$

Inserting this inequality and the curvature assumptions into (1.44) completes the proof. □

We note for future use the following

Proposition 1.26. *Let $(M, \langle\,,\rangle, J_M)$ be a Kähler manifold with Ricci curvature satisfying*

$$^M\mathrm{Ric} \geq -\rho(x).$$

Let $f : M \to \mathbb{R}$ be a pluriharmonic function which is bounded from below, and let $C > -\inf_M f$. Then $w = |\nabla \log(f+C)|^2$ satisfies

$$w\Delta w + 2\rho(x)\, w^2 \geq 2w^3 + |\nabla w|^2.$$

Proof. We may assume that f is positive on M and define

$$w = |\nabla \log f|^2 = \frac{|\nabla f|^2}{f^2}.$$

Using (1.51), we compute

$$w\Delta w \geq w\left(6w^2 - 2\rho w + \frac{1}{f^4}\frac{\left|\nabla|\nabla f|^2\right|^2}{w} - \frac{4}{f^3}\left\langle \nabla f, \nabla|\nabla f|^2\right\rangle\right).$$

Next,

$$|\nabla w|^2 = \frac{\left|\nabla|\nabla f|^2\right|^2}{f^4} - \frac{4|\nabla f|^2}{f^5}\left\langle \nabla f, \nabla|\nabla f|^2\right\rangle + \frac{4|\nabla f|^4}{f^6}\,|\nabla f|^2$$

and substituting we obtain

$$w\Delta w + 2\rho(x)\, w^2 \geq 2w^3 + |\nabla w|^2.$$ □

In a similar way, we get a Weitzenböck formula for a holomorphic map $f : M \to N$ between Kähler manifolds $(M, \langle,\rangle, J_M)$ and $(N, (,), J_N)$. We fix the index convention

$$1 \leq i, j, k, \ldots \leq s_M = \dim_{\mathbb{C}} M,$$
$$1 \leq \alpha, \beta, \gamma, \ldots \leq s_N = \dim_{\mathbb{C}} N.$$

For the manifold M we fix the local unitary coframe and related connection and curvature forms

$$\varphi^i, \varphi^i_j, \Phi^i_j,$$

while, for N, we fix the analogous coframing

$$\omega^\alpha, \omega^\alpha_\beta, \Omega^\alpha_\beta.$$

We also set $\{G_\alpha\}$ for the dual frame of $\{\omega^\alpha\}$ and let

$$\Omega^a_\beta = K^\alpha_{\beta\gamma\delta}\omega^\gamma \wedge \overline{\omega}^\delta.$$

Since $f : M \to N$ is holomorphic, we have

$$f^*\omega^\alpha = B^\alpha_i \varphi^i, \tag{1.52}$$

so that

$$df = B^\alpha_i \varphi^i \otimes G_\alpha.$$

Hence

$$u = \frac{1}{4}|df|^2 = \sum_{\alpha,i} B^\alpha_i \overline{B}^\alpha_i.$$

Differentiating (1.51) and using the structure equations on M and N, we obtain

$$\left(dB^\alpha_i - B^\alpha_j \varphi^j_i + B^\beta_i \omega^\alpha_\beta\right) \wedge \varphi^i = 0.$$

We set

$$B^\alpha_{ik}\varphi^k = dB^\alpha_i - B^\alpha_j \varphi^j_i + B^\beta_i \omega^\alpha_\beta.$$

Differentiating this latter and using once more the structure equations, we obtain

$$\left(dB^\alpha_{ik} - B^\alpha_{jk}\varphi^j_i - B^\alpha_{ij}\varphi^j_k + B^\beta_{ik}\omega^\alpha_\beta\right) \wedge \varphi^k = \left(B^\alpha_j H^j_{ikt} - B^\beta_i B^\gamma_k \overline{B}^\delta_t K^\alpha_{\beta\gamma\delta}\right)\overline{\varphi}^t \wedge \varphi^k.$$

We define

$$B^\alpha_{ikt}\varphi^t + B^\alpha_{ik\bar{t}}\overline{\varphi}^t = dB^\alpha_{ik} - B^\alpha_{jk}\varphi^j_i - B^\alpha_{ij}\varphi^j_k + B^\beta_{ik}\omega^\alpha_\beta,$$

so that, from the above we deduce

$$\begin{aligned} B^\alpha_{ikt} &= 0, \\ B^\alpha_{ik\bar{t}} &= B^\alpha_j H^j_{ikt} - B^\beta_i B^\gamma_k \overline{B}^\delta_t K^\alpha_{\beta\gamma\delta}. \end{aligned}$$

Next, we compute the Laplacian of u. We observe that

$$du = u_k\varphi^k + u_{\bar{k}}\overline{\varphi}^k$$

with, according to the previous formulas,

$$u_k = \sum_{\alpha,i} \overline{B}^\alpha_i B^\alpha_{ik}.$$

Hence, with the aid of the above calculations,

$$u_{kt}\varphi^t + u_{k\bar{t}}\overline{\varphi}^t = du_k - u_t\varphi^t_k$$

with

$$u_{k\bar{t}} = B^\alpha_{ik}\overline{B}^\alpha_{it} + \overline{B}^\alpha_i B^\alpha_{ik\bar{t}}.$$

By the definition of the Laplacian on the Kähler manifold M, we have

$$\begin{aligned}\Delta u &= 4\sum_k u_{k\bar{k}}\\ &= 4\sum_{\alpha,i,k}\left(B^\alpha_{ik}\overline{B}^\alpha_{ik} + \overline{B}^\alpha_i B^\alpha_{ik\bar{k}}\right)\\ &= 4\sum_{\alpha,i,k} B^\alpha_{ik}\overline{B}^\alpha_{ik} + 4\sum_{\alpha,i,k}\overline{B}^\alpha_i B^\alpha_j H^j_{ikk} - 4\sum_{\alpha,i,k}\overline{B}^\alpha_i B^\beta_i B^\gamma_t \overline{B}^\delta_t K^\alpha_{\beta\gamma\delta}.\end{aligned}$$

We have thus obtained the following

Theorem 1.27. *Let $(M, \langle,\rangle, J_M)$, $(N, (,), J_N)$ be Kähler manifolds and let $f : M \to N$ be a holomorphic map. Then*

$$\Delta |df|^2 = 16\sum_{\alpha,i,k} B^\alpha_{ik}\overline{B}^\alpha_{ik} + 16\sum_{\alpha,i,k} R_{i\bar{j}}\overline{B}^\alpha_i B^\alpha_j - 16\sum_{\alpha,i,k}\overline{B}^\alpha_i B^\beta_i B^\gamma_t\overline{B}^\delta_t K^\alpha_{\beta\gamma\delta}. \quad (1.53)$$

Remark 1.28. As we mentioned at the beginning of Section 1.4, the Weitzenböck formula of Theorem 1.27 holds in the more general situation where $(N, (,), J_N)$ is only Hermitian. However, the computations due to Y.C. Lu, [111], require a little more care, due to the presence of torsion terms in the first structural equations of N. Accordingly, one can deduce the following result that shall be used in Chapter 8

Corollary 1.29. *Let $(M, \langle,\rangle, J_M)$ and $(N, (,), J_N)$ be, respectively, a Kähler and a Hermitian manifold. Let $f : M \to N$ be a holomorphic map. Assume*

$$^M Ric \geq -\rho(x)$$

and suppose that N has holomorphic bisectional curvature bounded above by $k(z)$. Then

$$|df|^2\Delta|df|^2 \geq |\nabla|df|^2|^2 - 2\rho(x)|df|^4 - 2k(f(x))|df|^6. \quad (1.54)$$

Proof. With the notation introduced above, we have

$$u = \frac{1}{4}|df|^2 = \sum_{\alpha i} B^\alpha_i\overline{B}^\alpha_i \text{ and } \frac{1}{4}|\nabla u|^2 = \sum_k u_k u_{\bar{k}},$$

where $u_k = \sum_{\alpha,i} B^\alpha_{ik}\overline{B}^\alpha_i$ and $u_{\bar{k}} = \overline{u}_k$. Therefore

$$\begin{aligned}|\nabla|df|^2|^2 &= 64\sum_k u_k u_{\bar{k}}\\ &\leq 64\sum_{\alpha,i} B^\alpha_i\overline{B}^\alpha_i\sum_{\alpha,i,k} B^\alpha_{i,k}\overline{B}^\alpha_{i,k} = |df|^2\Big\{16\sum_{\alpha,i,k} B^\alpha_{i,k}\overline{B}^\alpha_{i,k}\Big\}.\end{aligned}$$

Inserting this inequality into (1.52) and using the curvature assumptions completes the proof. □

Chapter 2

Comparison Results

In this section we describe some comparison results for the Hessian and the Laplacian of the distance function and for the volume of geodesic balls under curvature conditions. In some cases, the results we are going to describe improve on classical results.

2.1 Hessian and Laplacian comparison

We begin by showing that a lower (resp. upper) bound on the radial sectional curvature of the form

$$\mathrm{Sect}_{rad} \geq -G(r(x)) \quad (\text{resp. } \mathrm{Sect}_{rad} \leq -G(r(x))) \tag{2.1}$$

implies an upper estimate for the Hessian, $\operatorname{Hess} r$, of the distance function $r(x)$ of the type

$$\operatorname{Hess}(r) \leq \frac{h'(r)}{h(r)}\big(\langle\, ,\rangle - dr \otimes dr\big) \quad \text{resp. } \operatorname{Hess}(r) \geq \frac{h'(r)}{h(r)}\big(\langle\, ,\rangle - dr \otimes dr\big) \tag{2.2}$$

for some appropriate function h. By taking traces, we will then obtain corresponding estimates for the Laplacian Δr. As we will see, an upper estimate for Δr requires only a lower bound for the radial Ricci curvature, while a lower estimate requires an upper bound for the radial sectional curvature.

To obtain these results we use an "analytic" approach inspired by P. Petersen, [128] avoiding, in this way, the "geometrical" Laplacian comparison theorem of R. Greene and H.H. Wu, [66].

We will need the following Sturm comparison result:

Lemma 2.1. *Let G be a continuous function on $[0, +\infty)$ and let ϕ, $\psi \in C^1\big([0, \infty)\big)$ with ϕ', $\psi' \in AC((0, +\infty))$ be solutions of the problems*

$$\begin{cases} \phi'' - G\phi \leq 0 & a.e.\ in\ (0, \infty), \\ \phi(0) = 0, \end{cases} \qquad \begin{cases} \psi'' - G\psi \geq 0 & a.e.\ in\ (0, \infty), \\ \psi(0) = 0,\ \psi'(0) > 0. \end{cases}$$

If $\phi(r) > 0$ for $r \in (0, T)$ and $\psi'(0) \geq \phi'(0)$, then $\psi(r) > 0$ in $(0, T)$ and

$$\frac{\phi'}{\phi} \leq \frac{\psi'}{\psi} \quad and \quad \psi \geq \phi \quad on\ (0, T). \tag{2.3}$$

Proof. Since $\psi'(0) > 0$, $\psi > 0$ in a neighborhood of 0. We observe in passing that if G is assumed to be non-negative, then integrating the differential inequality satisfied by ψ we have

$$\psi'(r) = \psi'(0) + \int_0^r G(s)\psi(s)\,ds,$$

so that ψ' is positive in the interval where $\psi \geq 0$, and we conclude that, in fact, $\psi > 0$ on $(0, +\infty)$.

In the general case where no assumption is made on the signum of G, we let $\beta = \sup\{t : \psi > 0 \text{ in } (0, t)\}$ and $\tau = \min\{\beta, T\}$, so that ϕ and ψ are both positive in $(0, \tau)$. The function $\psi'\phi - \psi\phi'$ is continuous on $[0, +\infty)$ vanishes in $r = 0$, and satisfies

$$(\psi'\phi - \psi\phi')' = \psi''\phi - \psi\phi'' \geq 0,$$

a.e. in $(0, \tau)$. Thus $\psi'\phi - \psi\phi' \geq 0$ on $[0, \tau)$, and dividing through by $\phi\psi$ we deduce that

$$\frac{\psi'}{\psi} \geq \frac{\phi'}{\phi} \quad \text{in } (0, \tau).$$

Integrating between ϵ and r $(0 < \epsilon < r < \tau)$ yields

$$\phi(r) \leq \frac{\phi(\epsilon)}{\psi(\epsilon)}\psi(r)$$

and since

$$\lim_{\epsilon \to 0+} \frac{\phi(\epsilon)}{\psi(\epsilon)} = \frac{\phi'(0)}{\psi'(0)} \leq 1,$$

we conclude that in fact

$$\phi(r) \leq \psi(r) \quad \text{in } [0, \tau).$$

Since $\phi > 0$ in $(0, T)$ by assumption, this in turn forces $\tau = T$, for otherwise, $\tau = \beta < T$, and we would have, $\phi(\beta) > 0$, while by continuity, $\psi(\beta) = 0$, which is a contradiction. □

Using the above Sturm comparison result, we deduce a comparison result for solutions of Riccati (in)equalities of the form

$$\phi' + \phi^2 = G \quad (\geq G, \ \leq G)$$

on $(0, T)$ with appropriate asymptotic behavior as $r \to 0+$. Note in this respect that the substitution $g = \phi'/\phi$ transforms the Riccati inequality into the second-order linear inequality

$$g'' = Gg \quad (\geq Gg, \ \leq Gg)$$

and conversely.

Corollary 2.2. *Let G be a continuous function on $[0,+\infty)$ and let $g_i \in AC(0,T_i)$ be solutions of the Riccati differential inequalities*

$$g_1' + \frac{g_i^2}{\alpha} - \alpha G \le 0 \qquad g_2' + \frac{g_2^2}{\alpha} - \alpha G \ge 0$$

a.e. in $(0,T_i)$ satisfying the asymptotic condition

$$g_i(t) = \frac{\alpha}{t} + O(1) \quad as\ t \to 0+,$$

for some $\alpha > 0$. Then $T_1 \le T_2$ and $g_1(t) \le g_2(t)$ in $(0,T_1)$.

Proof. Since $\tilde{g}_i = \alpha^{-1} g_i$ satisfies the conditions in the statement with $\alpha = 1$, without loss of generality we may assume that $\alpha = 1$.

Observe that the function $g_i(s) - \frac{1}{s}$ is bounded and integrable in a neighborhood of $s = 0$, and let $\phi_i \in C^1([0,T_i))$ be the positive function on $[0,T_i)$ defined by

$$\phi_i(t) = t \exp\Big\{ \int_0^t \big(g_i(s) - \frac{1}{s}\big)\, ds \Big\}.$$

Then $\phi_i(0) = 0$, $\phi_i > 0$ on $(0,T_i)$, $\phi_i' \in AC(0,T_i))$ and straightforward computations show that

$$\phi_i'(t) = g_i \phi_i(t), \quad \phi_i'(0) = 1$$

and

$$\phi_1'' \le G\phi_1 \text{ on } (0,T_1), \quad \phi_2'' \ge G\phi_2 \text{ on } (0,T_2).$$

An application of Lemma 2.1 shows that $T_1 \le T_2$ and $g_1 = \frac{\phi_1'}{\phi_1} \le \frac{\phi_2'}{\phi_2} = g_2$ on $(0,T_1)$, as required. □

After this preparation we are ready to state our comparison result for the Hessian.

Theorem 2.3. *Let $(M, \langle\, ,\rangle)$ be a complete manifold of dimension m. Having fixed a reference point $o \in M$, let $r(x) = \operatorname{dist}_M(x,o)$, and let $D_o = M \setminus cut(o)$ be the domain of the normal geodesic coordinates centered at o. Given a smooth even function G on $\mathbb{R}$, let h be the solution of the Cauchy problem*

$$\begin{cases} h'' - Gh = 0, \\ h(0) = 0,\ h'(0) = 1, \end{cases}$$

and let $I = [0,r_0) \subseteq [0,+\infty)$ be the maximal interval where h is positive. If the radial sectional curvature of M satisfies

$$\mathrm{Sect}_{rad} \ge -G(r(x)) \qquad on\ B_{r_0}(o) \tag{2.4}$$

on $B_{r_o}(o)$, then

$$\mathrm{Hess}(r) \le \frac{h'}{h}\{\langle\, ,\rangle - dr \otimes dr\} \tag{2.5}$$

on $(D_o \setminus \{o\}) \cap B_{r_0}(o)$, *in the sense of quadratic forms. On the other hand, if*

$$\mathrm{Sect}_{rad} \le -G\left(r\left(x\right)\right) \qquad on\;\; B_{r_0}(o), \tag{2.6}$$

then

$$\mathrm{Hess}\left(r\right) \ge \frac{h'}{h}\left\{\langle\,,\rangle - dr \otimes dr\right\}. \tag{2.7}$$

Proof. We essentially follow the direct approach by P. Petersen, [128], thus avoiding the classical use of Jacobi fields.

Observe first of all that $\mathrm{Hess}\,(r)(\nabla r, X) = 0$ for every $X \in T_xM$, and $x \in D_o \setminus \{o\}$. Indeed, let γ be the geodesic parametrized by arc length issuing from o with $\gamma(s_o) = x$, then γ is an integral curve of ∇r, namely, $\dot\gamma(s) = \nabla r(\gamma(s))$ so that $D_{\nabla r}\nabla r(x) = D_{\gamma(s_o)}\dot\gamma = 0$.

Next, since $\mathrm{Hess}\,(r)$ is symmetric, T_xM has an orthonormal basis consisting of eigenvectors of $\mathrm{Hess}\,(r)$. Denoting by $\lambda_{max}(x)$ and $\lambda_{min}(x)$, respectively, the greatest and smallest eigenvalues of the $\mathrm{Hess}\,(r)$ in the orthogonal complement of $\nabla r(x)$, the theorem amounts to showing that on $(D_o \setminus \{o\}) \cap B_{r_0}(o)$,

(i) if (2.4) holds, then $\lambda_{max}(x) \le \dfrac{h'}{h}(r(x))$,

(ii) if (2.6) holds, then $\lambda_{min}(x) \ge \dfrac{h'}{h}(r(x))$.

Let $x \in D_o \setminus \{o\}$, and let again γ be the minimizing geodesic joining o to x. We claim that, if (2.4) holds, then the Lipschitz function λ_{max} satisfies

$$\begin{cases} \frac{d}{ds}\left(\lambda_{\max}\circ\gamma\right) + \left(\lambda_{\max}\circ\gamma\right)^2 \le G & \text{for a.e. } s > 0, \\ \lambda_{\max}\circ\gamma = \frac{1}{s} + o\left(1\right), \text{ as } s \to 0^+. \end{cases} \tag{2.8}$$

Similarly, if (2.6) holds, then the Lipschitz function λ_{min} satisfies

$$\begin{cases} \frac{d}{ds}\left(\lambda_{\min}\circ\gamma\right) + \left(\lambda_{\min}\circ\gamma\right)^2 \ge G & \text{for a.e. } s > 0, \\ \lambda_{\min}\circ\gamma = \frac{1}{s} + o\left(1\right), \text{ as } s \to 0^+. \end{cases} \tag{2.9}$$

Since $\phi = h'/h$ satisfies

$$\phi' + \phi^2 = G \,\text{on}\; (0, r_o), \quad \phi(s) = \frac{1}{s} + 0(s) \;\text{ as } s \to 0+,$$

the required conclusion follows at once from Corollary 2.2. It remains to prove that λ_{max} and λ_{min} satisfy the required differential inequalities. To this end, given a smooth real function u, denote by $\mathrm{hess}\,(u)$ the $(1,1)$ symmetric tensor field defined by

$$\mathrm{hess}\left(u\right)\left(X\right) = D_X\nabla u,$$

so that

$$\mathrm{Hess}\left(u\right)\left(X,Y\right) = \left\langle \mathrm{hess}\left(u\right)\left(X\right), Y\right\rangle.$$

By definition of covariant derivative in $TM^* \otimes TM$,

$$D_X(\text{hess}(u))(Y) = D_X\big[\text{hess}(u)(Y)\big] - \text{hess}(u)(D_X Y),$$

so that, recalling the definition of the curvature tensor, we deduce the Ricci commutation rule

$$D_X(\text{hess}(u))(Y) - D_Y(\text{hess}(u))(X) = R(X,Y)\nabla u.$$

Now, choose $u = r(x)$, $X = \nabla r$, and let γ be the minimizing geodesic joining o to $x \in D_o \setminus \{o\}$. For every unit vector $Y \in T_x M$ such that $Y \perp \dot{\gamma}(s_o)$, define a vector field $Y \perp \dot{\gamma}$, by parallel translation along γ. Since, as noted above, $\text{hess}\,(r)(\nabla r) \equiv 0$, we compute

$$\begin{aligned}
D_{\dot{\gamma}(t_0)}\big[\text{hess}\,(r)(Y)\big] &= D_{\dot{\gamma}(t_0)}(\text{hess}\,(r))(Y) + \text{hess}\,(r)(D_{\dot{\gamma}(t_0)}Y) \\
&= D_{\nabla(r)}(\text{hess}\,(r))(Y) \\
&= D_Y(\text{hess}\,(r))(\nabla r) + R(\nabla r, Y)\nabla r \\
&= D_Y[\text{hess}\,(r)(\nabla r)] - \text{hess}\,(r)(D_Y \nabla r) - R(Y,\nabla r)\nabla r \\
&= -\text{hess}\,(r)(\text{hess}\,(r)(Y)) - R(Y,\nabla r)\nabla r,
\end{aligned}$$

that is,

$$D_{\dot{\gamma}(t_o)}[\text{hess}\,(r)(Y)] + \text{hess}\,(r)(\text{hess}\,(r)(Y)) - R(Y,\nabla r)\nabla r.$$

Since Y is parallel,

$$\frac{d}{dt}\langle \text{hess}\,(r)(Y), Y\rangle = \langle D_{\dot{\gamma}}[\text{hess}\,(r)(Y)], Y\rangle,$$

and we conclude that

$$\frac{d}{ds}\left(\text{Hess}\,(r)\,(\gamma)\,(Y,Y)\right) + \langle \text{hess}\,(r)(\gamma)(Y), \text{hess}\,(r)(\gamma)(Y)\rangle = -\text{Sect}_\gamma\,(Y \wedge \dot{\gamma})\,. \tag{2.10}$$

Now assume that $\text{Sect}_{rad} \geq -G\,(r\,(x))$. Note that, for any unit vector field $X \perp \nabla r$,

$$\text{Hess}\,(r)\,(X,X) \leq \lambda_{\max}.$$

Thus, if Y is chosen so that, at s_0,

$$\text{Hess}\,(r)\,(\gamma)\,(Y,Y) = \lambda_{\max}\,(\gamma\,(s_0))\,,$$

then the function

$$\text{Hess}\,(r)\,(\gamma)\,(Y,Y) - \lambda_{\max} \circ \gamma$$

attains its maximum at $s = s_0$ and, if at this point λ_{max} is differentiable, then its derivative vanishes:

$$\left.\frac{d}{ds}\right|_{s_0} \text{Hess}\,(r)\,(\gamma)\,(Y,Y) - \left.\frac{d}{ds}\right|_{s_0} \lambda_{\max} \circ \gamma = 0.$$

Whence, using (2.10), we obtain, at s_0,

$$\frac{d}{ds}\left(\lambda_{\max}\circ\gamma\right)+\left(\lambda_{\max}\circ\gamma\right)^2\leq G,$$

which is the desired inequality stated in (2.8). The asymptotic behavior of $\lambda_{\max}\circ\gamma$ near $s=0^+$ follows from the fact that

$$\mathrm{Hess}\,(r)=\frac{1}{r}\left(\langle\,,\rangle-dr\otimes dr\right)+o\left(1\right),\ r\to 0^+,$$

as one can verify by a standard computation in normal coordinates at $o\in M$. The argument in the case where $\mathrm{Sect}_{rad}\leq -G$ is completely similar. □

As mentioned above, by taking traces in Theorem 2.3 we immediately obtain corresponding estimates for Δr. In particular, if $\mathrm{Sect}_{rad}\leq -G(r(x))$ it follows that

$$\Delta r(x)\geq (m-1)\frac{h'(r(x))}{h(r(x))}$$

on $(D_o\backslash\{o\})\cap B_{r_0}(o)$. Clearly the corresponding upper estimate holds if we assume instead that the radial sectional curvature is bounded below by $-G$. In this case however, the conclusion holds under the weaker assumption that the radial Ricci curvature is bounded from below by $-(m-1)G(r(x))$. Indeed we have the following Laplacian comparison theorem,

Theorem 2.4. *Maintaining the notation of the previous theorem, assume that the radial Ricci curvature of M satisfy*

$$Ric_{(M,\langle\,,\rangle)}(\nabla r,\nabla r)\geq -(m-1)G(r) \tag{2.11}$$

for some function $G\in C^0([0,+\infty))$, and let $h\in C^2([0,+\infty))$ be a solution of the problem

$$\begin{cases} h''-Gh\geq 0,\\ h(0)=0,\ h'(0)=1.\end{cases} \tag{2.12}$$

Then the inequality

$$\Delta r(x)\leq (m-1)\frac{h'(r(x))}{h(r(x))} \tag{2.13}$$

holds pointwise on $M\setminus(cut(o)\cup\{o\})$, and weakly on all of M.

Proof. Let $[0,r_0)\subseteq[0,+\infty)$ be the maximal interval where h is positive. Note that comparing h with the solution of the differential equation associated to (2.12) and using the remark at the beginning of the proof of Lemma 2.1 shows that if G is non-negative, then $r_o=+\infty$.

As in the proof of Theorem 2.3, let $D_o=M\setminus cut(o)$ be the maximal star-shaped domain of the normal coordinates at o. Fix any $x\in D_o\cap(B_{r_o}(o)\setminus\{o\})$

and let $\gamma : [0, l] \to M$ be the minimizing geodesic from o to x parametrized by arc-length. Set

$$\varphi(s) = (\Delta r) \circ \gamma(s), \quad s \in (0, l].$$

We claim that φ satisfies

$$\begin{cases} \text{i)} & \varphi(s) = \frac{m-1}{s} + o(1), \quad \text{as } s \to 0^+, \\ \text{ii)} & \varphi' + \frac{1}{m-1}\varphi^2 \leq (m-1)G, \quad \text{on } (0, l]. \end{cases} \tag{2.14}$$

Indeed (2.14) i) follows from the well-known fact that

$$\Delta r = \frac{m-1}{r} + o(1), \quad \text{as } r \to 0^+.$$

As for (2.14) ii), note that by tracing in (2.10) we deduce that

$$\frac{d}{dt}(\Delta r \circ \gamma) + |\operatorname{Hess} r|^2(\gamma) = -\operatorname{Ric}(\nabla r, \nabla r)(\gamma).$$

Using the elementary inequality

$$\frac{(\Delta r)^2}{m-1} \leq |\operatorname{Hess}(r)|^2,$$

which in turn follows easily from the Cauchy–Schwarz inequality, we deduce that

$$\frac{d}{dt}(\Delta r \circ \gamma) + \frac{(\Delta r \circ \gamma)^2}{m-1} \leq -\operatorname{Ric}(\nabla r, \nabla r)(\gamma). \tag{2.15}$$

Inequality (2.14) ii) follows from the assumption on Ric. Arguing as in the proof of Theorem 2.3 shows that (2.13) holds pointwise on $D_o \cap (B_{r_o}(o) \setminus \{o\})$.

Note now that a computation in polar geodesic coordinates shows that

$$\Delta r \circ \gamma(t) = \frac{1}{\sqrt{g(t,\theta)}} \frac{\partial \sqrt{g(t,\theta)}}{\partial t}$$

where $\theta = \gamma'(0)$ and $g(r, \theta)$ is the determinant of the metric in geodesic polar coordinates. Thus (2.13) can be rewritten in the form

$$\frac{1}{\sqrt{g(t,\theta)}} \frac{\partial \sqrt{g(t,\theta)}}{\partial t} \leq (m-1)\frac{h'(t)}{h(t)}$$

whence, integrating and using the asymptotic behavior of h and $\sqrt{g}$ as $t \to 0+$, show that for every unit length $\theta \in T_oM$,

$$\sqrt{g(t,\theta)} \leq h(t) \qquad \forall t < \min\{r_o, c(\theta)\}$$

where $c(\theta)$ denotes the distance of o from $cut(o)$ along the geodesic γ_θ. Since $\sqrt{g(t,\theta)} > 0$ if (t, θ) belongs to the domain of the geodesic polar coordinates, while, if $r_o < +\infty$, then $h(r_o) = 0$, we deduce that for all θ $c(\theta) \le r_o$, and therefore $D_o \subset B_{r_o}(o)$.

Thus (2.13) holds pointwise on $M \setminus (\{o\} \cup cut(o))$, and it remains to prove that the inequality holds weakly on all of M. This is guaranteed by the following lemma. □

Lemma 2.5. *Set $D_o = M \backslash cut(o)$ and suppose that*

$$\Delta r \le \alpha(r) \quad \textit{pointwise on } \Omega \backslash \{o\} \tag{2.16}$$

for some $\alpha \in C^0\big((0, +\infty)\big)$. Let $v \in C^2(\mathbb{R})$ be non-negative and set $u(x) = v(r(x))$ on M. Suppose either

$$\text{i) } v' \le 0 \quad \textit{or} \quad \text{ii) } v' \ge 0. \tag{2.17}$$

Then we respectively have

$$\text{i) } \Delta u \ge v''(r) + \alpha(r)v'(r); \quad \text{ii) } \Delta u \le v''(r) + \alpha(r)v'(r) \tag{2.18}$$

weakly on M.

Proof. Let E_o be the maximal star-shaped domain on which $\exp_o$ is a diffeomorphism onto its image, so that $D_o = \exp(E_o)$ and we have $cut(o) = \partial\left(\exp_o(E_o)\right)$. Since E_o is a star-shaped domain, we can exhaust E_o by a family $\{E_o^n\}$ of relatively compact, star-shaped domains with smooth boundary. We set $\Omega^n = \exp_o(E_o^n)$ so that

$$\overline{\Omega}^n \subset \Omega^{n+1} \text{ and } \cup_n \Omega^n = D_o.$$

The fact that each E_o^n is star-shaped implies

$$\frac{\partial r}{\partial \nu_n} > 0, \quad \text{on } \partial\Omega^n \tag{2.19}$$

where ν_n denotes the outward unit normal to $\partial\Omega^n$. Now, we assume the validity of (2.17) i). Since $r \in C^\infty(\Omega^n \backslash \{o\})$, computing we get

$$\Delta u \ge v'' + \alpha(r)v' \quad \text{pointwise on } \Omega^n \backslash \{o\}. \tag{2.20}$$

Let $0 \le \varphi \in C_0^\infty(M)$. We claim that, $\forall n$,

$$\int_{\Omega^n} u\Delta\varphi \ge \int_{\Omega^n} \left(v'' + \alpha(r)v'\right)\varphi + \varepsilon_n$$

with $\varepsilon_n \to 0$ as $n \to +\infty$. Since $M = \Omega \cup cut(o)$ and $cut(o)$ has measure 0, inequality (2.18) i) will follow by letting $n \to +\infty$. To prove the claim we fix $\delta > 0$ small and we apply the second Green formula on $\overline{\Omega^n} \backslash B_\delta(o)$ to obtain

$$\int_{\Omega^n \backslash B_\delta(o)} u\Delta\varphi = \int_{\Omega^n \backslash B_\delta(o)} \varphi\Delta u - \int_{\partial\Omega^n \cup \partial B_\delta(o)} \left(\varphi \frac{\partial u}{\partial \nu_n} - u\frac{\partial \varphi}{\partial \nu_n}\right) \tag{2.21}$$

where ν_n is the outward unit normal to $\partial\Omega^n \cup \partial B_\delta(o)$. We note that, according to (2.17) i) and (2.19),

$$\frac{\partial u}{\partial \nu_n} = v'(r)\frac{\partial r}{\partial \nu_n} \leq 0 \quad \text{ on } \partial\Omega_n.$$

Using this, (2.20) and (2.21), we obtain

$$\int_{\Omega^n} u\Delta\varphi \geq \int_{\Omega^n} \left(v'' + \alpha(r)v'\right)\varphi + \varepsilon_n + I_\delta$$

with

$$\varepsilon_n = \int_{\partial\Omega^n} u\frac{\partial \varphi}{\partial \nu_n},$$

$$I_\delta = \int_{B_\delta(o)} \left[u\Delta\varphi - \left(v'' + \alpha(r)v'\right)\varphi\right] - \int_{\partial B_\delta(o)} \left[u\frac{\partial \varphi}{\partial r} - \varphi\frac{\partial u}{\partial r}\right].$$

Clearly, $I_\delta \to 0$ as $\delta \to 0^+$. On the other hand, since $\varphi \in C_0^\infty(M)$ and $cut(o)$ has measure 0, using the divergence and Lebesgue theorems we see that, as $n \to +\infty$,

$$\varepsilon_n = \int_{\Omega^n} \operatorname{div}(u\nabla\varphi) \to \int_\Omega \operatorname{div}(u\nabla\varphi) = \int_M \operatorname{div}(u\nabla\varphi) = 0.$$

This proves the claim and the validity of (2.18) i).

The case of (2.17) ii) and (2.18) ii) can be dealt with in a similar way. □

Remark 2.6. We note that, for the above proofs to work, it is not necessary that (2.11) holds on the entire M. Indeed, for instance, if (2.11) is valid on $B_R(o)$, then (2.13) holds on $B_R(o)\setminus(\{o\} \cup cut(o))$ and weakly on $B_R(o)$.

We also remark that in the course of the proof we have shown that if the solution h of (2.12) vanishes at r_o, then $D_o \subset B_{r_o}(o)$ and therefore $M \subset \overline{B}_{r_o}(o)$. This easily yields the classical Bonnet-Myers theorem, stating that if $\operatorname{Ric} \geq (m-1)B^2$, then M is compact with diameter at most $\sqrt{\pi}/B$.

Remark 2.7. We note for future use that a modification of the above argument shows that on $M \setminus \{o\}$ the singular part of the distribution Δr is negative, and therefore it is the opposite of a positive measure concentrated on the cut locus. Indeed, let ϕ be a smooth, non-negative test function with support contained in $M \setminus \{o\}$. Arguing as above we may write

$$(\phi, \Delta r) = \int_M r\Delta\phi = \lim_n \Big(\int_{\Omega^n} \phi\Delta r + \int_{\partial\Omega^n} r\langle\nabla\phi, \nu\rangle - \int_{\partial\Omega^n} \phi\langle\nabla r, \nu\rangle\Big).$$

As $n \to +\infty$, the first term on the right-hand side tends to $\int_{E_o} \phi\Delta r$, and, as noted in the above proof, the second term tends to zero. Thus the limit of the third term exists, and we have

$$(r, \Delta\phi) - \int_{E_o} \phi\Delta r = (\phi, (\Delta r)_{sing}) = -\lim_n \int_{\partial\Omega^n} \phi\langle\nabla r \nu\rangle,$$

and since $\langle\nabla r, \nu\rangle \geq 0$, the limit is non-negative, as claimed.

In order to apply Theorem 2.4 one needs to find solutions of (2.12). We begin with the following fairly general result.

Lemma 2.8. *Suppose that G is a positive C^1 function on $[0,+\infty)$ such that*

$$\inf_{[0,+\infty)} \frac{G'}{G^{3/2}} > -\infty. \tag{2.22}$$

Then, there exists $D > 0$ sufficiently large that the function h defined by

$$h(r) = \frac{1}{D\sqrt{G(0)}} \left\{ e^{D\int_0^r \sqrt{G(s)}ds} - 1 \right\} \tag{2.23}$$

is a solution of (2.12).

Proof. Indeed, it is a simple matter to check that if h is as in the statement, then $h(0) = 0$, $h'(0) = 1$ and furthermore

$$h'' - G\,h \geq \frac{G}{\sqrt{G(0)}} \left[\inf_{[0,+\infty)} \frac{G'}{2G^{3/2}} + D - \frac{1}{D} \right]$$

so that, by (2.22), (2.12) holds provided $D > 0$ is sufficiently large. □

Remark 2.9. Assumption (2.22) implies $G(r)^{1/2} \notin L^1(+\infty)$. It is then a simple matter to check that

$$\frac{h'(r)}{h(r)} \sim \begin{cases} \frac{1}{r} & \text{as } r \to 0, \\ D\,G(r)^{1/2} & \text{as } r \to +\infty. \end{cases} \tag{2.24}$$

Remark 2.10. When we can find an explicit solution of the problem

$$\begin{cases} h'' - h\,G = 0, & \text{on } [0,+\infty), \\ h(0) = 0, \; h'(0) = 1, \end{cases} \tag{2.25}$$

then (2.13) yields a better estimate. For instance if $G(r) \equiv B^2$, $B > 0$,

$$h(r) = B^{-1} \sinh(Br)$$

satisfies (2.25) and we obtain

$$\Delta r \leq (m-1)B \coth(Br) \quad \text{weakly on } M \tag{2.26}$$

and pointwise on $M \backslash (\{o\} \cup cut(o))$. This estimate will be repeatedly used in the sequel.

We next describe some upper and lower estimates for h'/h and for h obtained in [20] for the case where G has the form $G(r) = B^2(1+r^2)^{\alpha/2}$ for some constants $B > 0$ and $\alpha \geq -2$. We begin with upper bounds.

Proposition 2.11. *Assume h is a solution of*

$$\begin{cases} h'' - B^2(1+r^2)^{\alpha/2}h = 0, \\ h(0) = 0 \quad h'(0) = 1, \end{cases} \tag{2.27}$$

where $B > 0$ and $\alpha \geq -2$. Set

$$B' = \begin{cases} B & \text{if } \alpha > -2, \\ \frac{1+\sqrt{1+4B^2}}{2} & \text{if } \alpha = -2. \end{cases} \tag{2.28}$$

Then

$$\frac{h'}{h}(r) \leq B' r^{\alpha/2}(1+o(1)) \qquad \text{as } r \to +\infty. \tag{2.29}$$

Moreover there exists a constant C such that, when $r > 1$,

$$h(r) \leq C \begin{cases} \exp\left(\dfrac{2B'}{2+\alpha}(1+r)^{1+\alpha/2}\right) & \text{if } \alpha \geq 0, \\ r^{-\alpha/4}\exp\left(\dfrac{2B'}{2+\alpha} r^{1+\alpha/2}\right) & \text{if } -2 < \alpha < 0, \\ r^{B'} & \text{if } \alpha = -2. \end{cases} \tag{2.30}$$

Proof. The case where $\alpha \geq 0$ was already treated in [134]. It is a simple matter to check that the function ψ defined by

$$\psi(r) = B^{-1}\sinh\left(\frac{2B}{2+\alpha}\left[(1+r)^{1+\alpha/2} - 1\right]\right)$$

satisfies $\psi''/\psi \geq B^2(1+r^2)^{\alpha/2}$, $\psi(0) = 0$ and $\psi'(0) = 1$. The estimates (2.29) and (2.30) follow at once.

Next assume that $-2 < \alpha < 0$. Denoting by I_ν the modified Bessel function of order ν, it may be verified that the function defined by

$$\psi(r) = r^{1/2} I_{\frac{1}{2+\alpha}}\left(\frac{2B}{2+\alpha} r^{1+\alpha/2}\right),$$

is a positive C^1 solution on $[0, +\infty)$ of the (singular) differential equation

$$\psi'' = B^2 r^\alpha \psi$$

satisfying $\psi(0) = 0$ ([91], page 106). Since $r^\alpha \geq (1+r^2)^{\alpha/2}$, the argument above shows that $h'/h \leq \psi'/\psi$. Using the recurrence relation

$$I_\nu = \frac{1}{2}\left(I_{\nu+1} + I_{\nu-1}\right)$$

and the asymptotic representation

$$I_\nu(r) = \frac{e^r}{\sqrt{2\pi r}}(1+o(1)) \qquad \text{as } r \to +\infty$$

([91], page 110 and page 123, respectively), it is not hard to show that

$$\frac{\psi'}{\psi}(r) = Br^{\alpha/2} - \frac{\alpha}{4}r^{-1} + O\left(r^{-2-\alpha/2}\right) \qquad \text{as } r \to +\infty$$

and (2.29) follows. Since we also have $\psi'(0) = C$ where $C = C_{\alpha,B}$ is a positive constant depending on α and B, we have $h \le C^{-1}\psi$ and (2.30) is again a consequence of the asymptotic representation of I_ν.

Finally, if $\alpha = -2$, we consider the function defined by

$$\psi(r) = r^\beta.$$

A simple computation shows that

$$\psi''/\psi \ge B^2(1+r^2)^{-1}$$

holds provided $\beta = \left[1+\sqrt{1+4B^2}\right]/2$, showing that (2.29) holds. The validity of (2.30) is then established by integrating h'/h over $[1, r]$. □

Next we consider lower bounds. To illustrate the method, assume that $h'' = G(r)h$ and that $G(r)$ is non-increasing. Then

$$h''(r) = G(r)h(r) \ge G(R)h(r) \qquad \forall\, r \in [0, R].$$

Thus, the comparison argument used in the proof of Lemma 2.1 applied to the function $\phi(r) = G(R)^{-1/2}\sinh\left(G(R)^{1/2}r\right)$ implies that

$$\frac{h'}{h}(r) \ge G(R)^{1/2}\tanh\left(G(R)^{1/2}r\right).$$

Suppose now that $G(r)$ has the form $G(r) = B^2(1+r^2)^{\alpha/2}$. The condition that $G(r)$ be decreasing requires that we restrict ourselves to the case where $-2 \le \alpha \le 0$.

The case where $\alpha = 0$ is trivial since then $G(r)$ is constant, and we have the equality $h(r) = B^{-1}\sinh(Br)$.

If $-2 < \alpha < 0$, the argument above shows that

$$\frac{h'}{h} \ge B(1+r^2)^{\alpha/4} \ge Br^{\alpha/4}\left[1+\frac{\alpha}{4}r^{-2}\right] \qquad r \gg 1,$$

where the last inequality follows by expanding the function $(1+r^{-2})^{\alpha/4}$. Integrating over $[R, r]$, $R \gg 1$, we obtain

$$\begin{aligned} h(r) &\ge h(R)\exp\left(B\int_R^r s^{\alpha/4}[1+\alpha/(4s^2)]\right) ds \\ &\ge C\exp\left(\frac{2B}{2+\alpha}r^{1+\alpha/2}\right). \end{aligned}$$

Finally, if $\alpha = -2$, we consider the function $\phi(r) = (1+r)^{B'}$, with B' as in (2.28). Then

$$\frac{\phi''}{\phi}(r) = \frac{B'(B'-1)}{(1+r)^2} \geq \frac{B^2}{1+r^2}.$$

Thus $(h'\phi - h\phi')'(r) > 0$, $r > 0$, and since $(h'\phi - h\phi')(0) = 1$, we conclude that

$$\frac{h'}{h}(r) \geq B'(1+r)^{-1},$$

and then integrating

$$h(r) \geq C(1+r)^{B'}.$$

The above considerations prove the following

Proposition 2.12. *Let h be the solution of (2.27) with $-2 \leq \alpha \leq 0$, and let B' be defined as in (2.28). Then*

$$\frac{h'}{h}(r) \geq B' r^{\alpha/2}(1+o(1)) \qquad \text{as } r \to +\infty, \tag{2.31}$$

and there exists a constant C such that when $r > 1$

$$h(r) \geq C \begin{cases} \exp\left(\dfrac{2B'}{2+\alpha} r^{1+\alpha/2}\right) & \text{if } -2 < \alpha \leq 0, \\ r^{B'} & \text{if } \alpha = -2. \end{cases} \tag{2.32}$$

We conclude this section by considering the case where G is a non-negative non-increasing function satisfying the condition

$$\int_0^{+\infty} tG(t)\,dt < +\infty. \tag{2.33}$$

According to a terminology introduced by U. Abresch [1] for the sectional curvature, if a manifold $(M, \langle\, ,\, \rangle)$ satisfies

$$\operatorname{Sect}(x) \geq -G(r(x)) \quad \text{resp.} \quad \operatorname{Ric}(\nabla r, \nabla r) \geq -(m-1)G(r(x))$$

with G as above, then one says that M has asymptotically non-negative sectional, resp. Ricci, curvature. Note that under these assumption, G decays faster than quadratically at infinity, and in particular, condition (2.22) does not hold. This case is dealt with in the following lemma (see also [125], Lemma 1, and [171], Lemma 2.1).

Lemma 2.13. *Assume that $h \in C^2([0, +\infty)$ is the solution of the problem*

$$\begin{cases} h'' - G(r)h = 0, \\ h(0) = 1, \ h'(0) = 1, \end{cases} \tag{2.34}$$

where G is a continuous non-negative function on $[0, +\infty)$. Let ψ be defined by the formula

$$\psi(r) = \int_0^r e^{\int_0^s uG(u)\,du} ds.$$

Then we have the estimates

$$\text{(i)}\ \ 1 \le h'(r) \le \psi'(r) \quad \text{(ii)}\ \ \frac{1}{r} \le \frac{h'}{h} \le \frac{\psi'}{\psi} \ \ \textit{and}\ \ \text{(iii)}\ \ r \le h(r) \le \psi(r). \tag{2.35}$$

In particular, if

$$\int_0^{+\infty} tG(t)\,dt = b_0 < +\infty, \tag{2.36}$$

then, for every $r \ge 0$,

$$1 \le h'(r) \le e^{b_0} \quad \textit{and} \quad r \le h(r) \le e^{b_0} r. \tag{2.37}$$

Proof. Since the function $\phi(t) = t$ satisfies

$$\begin{cases} \phi'' - G(r)\phi \le 0, \\ \phi(0) = 1,\ \phi'(0) = 1, \end{cases}$$

applying Lemma 2.1 we easily obtain the left-hand side inequalities in (2.35) (ii) and (iii), which in turn imply (i).

The upper bounds are proved in a similar way. Indeed, since

$$\psi(r) \le r e^{\int_0^r sG(s)\,ds},$$

a straightforward computation shows that the function ψ satisfies

$$\begin{cases} \psi'' - G(r)\psi \ge 0, \\ \psi(0) = 1,\ \psi'(0) = 1, \end{cases}$$

and therefore we obtain the right-hand side inequalities in (2.35). It is clear that if (2.36) holds, then (2.35) yields (2.37) □

2.2 Volume comparison and volume growth

The following result is a somewhat generalized version of what is known in the literature as the Bishop-Gromov volume comparison theorem. In fact, on manifolds with Ricci curvature bounded from below, R. Bishop proved an upper estimate for the volume of balls not intersecting the cut locus of their centers, in terms of the volumes of corresponding balls in space-forms, [10]. Later, M. Gromov extended the estimate including the cut locus and improved the result by showing a monotonicity behavior.

Theorem 2.14. *Let $(M, \langle\,,\rangle)$ be a complete, m-dimensional Riemannian manifold satisfying*

$$\operatorname{Ric}(x) \geq -(m-1)\,G(r(x)) \text{ on } M \tag{2.38}$$

for some non-negative function $G \in C^0([0,+\infty))$, where $r(x) = \operatorname{dist}(x,o)$ is the distance from a fixed reference origin $o \in M$. Let $h(t) \in C^2([0,+\infty))$ be the non-negative solution of the problem

$$\begin{cases} h''(t) - G(t)\,h(t) = 0, \\ h(0) = 0 \; h'(0) = 1. \end{cases} \tag{2.39}$$

Then, for almost every $R > 1$, the function

$$R \mapsto \frac{\operatorname{vol}\partial B_R(o)}{h(R)^{m-1}} \quad \text{is non-increasing} \tag{2.40}$$

and

$$\operatorname{vol}\partial B_R(o) \leq c_m h(R)^{m-1} \tag{2.41}$$

where c_m is the volume of the unit sphere in $\mathbb{R}^m$. Moreover,

$$R \mapsto \frac{\operatorname{vol}B_R(o)}{\int_0^R h(t)^{m-1}\,dt} \tag{2.42}$$

is a non-increasing function on $(0,+\infty)$.

Proof. In case o is a pole of M one simply integrates the radial vector field

$$X = h(r(x))^{-m+1}\nabla r$$

on concentric balls $B_R(o)$, and uses the divergence and the Laplacian comparison theorems. However, in general, objects are non-smooth and inequalities are intended in the sense of distributions. Therefore, we have to take some extra care.

The Laplacian comparison theorem asserts that

$$\Delta r(x) \leq (m-1)\frac{h'(r(x))}{h(r(x))} \tag{2.43}$$

pointwise on the open, star-shaped, full-measured set $M \setminus \operatorname{cut}(o)$ and weakly on all of M. Thus, for every $0 \leq \varphi \in Lip_c(M)$,

$$-\int \langle\nabla r, \nabla\varphi\rangle \leq (m-1)\int \frac{h'(r(x))}{h(r(x))}\varphi. \tag{2.44}$$

For any $\varepsilon > 0$, consider the radial cut-off function

$$\varphi_\varepsilon(x) = \rho_\varepsilon(r(x))\,h(r(x))^{-m+1} \tag{2.45}$$

where ρ_ε is the piecewise linear function

$$\rho_\varepsilon(t) = \begin{cases} 0 & \text{if } t \in [0, r), \\ \frac{t-r}{\varepsilon} & \text{if } t \in [r, r+\varepsilon), \\ 1 & \text{if } t \in [r+\varepsilon, R-\varepsilon), \\ \frac{R-t}{\varepsilon} & \text{if } t \in [R-\varepsilon, R), \\ 0 & \text{if } t \in [R, \infty). \end{cases} \tag{2.46}$$

Note that

$$\nabla\varphi_\varepsilon = \left\{-\frac{\chi_{R-\varepsilon,R}}{\varepsilon} + \frac{\chi_{r,r+\varepsilon}}{\varepsilon} - (m-1)\frac{h'(r(x))}{h(r(x))}\rho_\varepsilon\right\} h(r(x))^{-m+1}\,\nabla r,$$

for a.e. $x \in M$, where $\chi_{s,t}$ is the characteristic function of the annulus $B_t(o) \setminus B_s(o)$. Therefore, using φ_ε into (2.44) and simplifying, we get

$$\frac{1}{\varepsilon}\int_{B_R(o)\setminus B_{R-\varepsilon}(o)} h(r(x))^{-m+1} \leq \frac{1}{\varepsilon}\int_{B_{r+\varepsilon}(o)\setminus B_r(o)} h(r(x))^{-m+1}.$$

Using the co-area formula we deduce that

$$\frac{1}{\varepsilon}\int_{R-\varepsilon}^{R} \mathrm{vol}\partial B_t(o)\, h(t)^{-m+1} \leq \frac{1}{\varepsilon}\int_{r}^{r+\varepsilon} \mathrm{vol}\partial B_t(o)\, h(t)^{-m+1}$$

and, letting $\varepsilon \searrow 0$,

$$\frac{\mathrm{vol}\partial B_R(o)}{h(R)^{m-1}} \leq \frac{\mathrm{vol}\partial B_r(o)}{h(r)^{m-1}} \tag{2.47}$$

for a.e. $0 < r < R$. Letting $r \to 0$, and recalling that $h(r) \sim r$ and $\mathrm{vol}\,\partial B_r \sim c_m r^{m-1}$ as $r \to 0$, we conclude that, for a.e. $R > 0$,

$$\mathrm{vol}\partial B_R(o) \leq c_m h(R)^{m-1}, \text{ a.e. } R > 0.$$

To prove the second statement, we note that it was observed by M. Gromov, see [33], that for general real-valued functions $f(t) \geq 0$, $g(t) > 0$,

$$\text{if } t \to \frac{f(t)}{g(t)} \text{ is decreasing, then } t \to \frac{\int_0^t f}{\int_0^t g} \text{ is decreasing.}$$

Indeed, since f/g is decreasing, if $0 < r < R$,

$$\int_0^r f \int_r^R g = \int_0^r g\frac{f}{g}\int_r^R g \geq \frac{f(r)}{g(r)}\int_0^r g\int_r^R g \geq \int_0^r g\int_r^R g\frac{f}{g} = \int_0^r g\int_r^R f,$$

whence

$$\int_0^r f\int_0^R g = \int_0^r f\int_0^r g + \int_0^r f\int_r^R g \geq \int_0^r f\int_0^r g + \int_0^r g\int_r^R f = \int_0^r g\int_0^R f.$$

In particular, applying this observation to (2.47) and using the co-area formula we deduce that

$$r \to \frac{\mathrm{vol} B_r(o)}{\int_0^r h(t)^{m-1}\,dt} \quad \text{is decreasing,}$$

as required to conclude the proof. □

The following straightforward consequence of Theorem 2.14 is useful in applications.

Corollary 2.15. *Let $(M, \langle\,,\rangle)$ be a complete, m-dimensional manifold satisfying (2.38) for some non-negative function $G(t) \in C^0([0,\infty))$.*

(i) *Having fixed $x_0 \in M$ and $0 < \bar{R} \le r(x_0)$, define*

$$sn_H(t) = \begin{cases} H^{-1}\sinh(Ht) & \text{if } H > 0, \\ t & \text{if } H = 0, \end{cases} \tag{2.48}$$

with

$$H^2 = \max_{B_{\bar{R}}(x_0)} G(r(x)) = \max_{[r(x_0)-\bar{R},\, r(x_0)+\bar{R}]} G(t). \tag{2.49}$$

Then, the function

$$R \mapsto \frac{\mathrm{vol} B_R(x_0)}{\int_0^R sn_H(t)^{m-1}\,dt}$$

is non-increasing on $[0, \bar{R}]$.

(ii) *If $G(t)$ is non-decreasing, then, for every $x_0 \in M$, the function*

$$R \mapsto \frac{\mathrm{vol} B_R(x_0)}{\int_0^R h_{x_0}(t)^{m-1}\,dt}$$

is non-increasing on $(0, +\infty)$. Here, h_{x_0} is the solution of the problem

$$\begin{cases} h''(t) - G_{x_0}(t)\,h(t) = 0, \\ h(0) = 0 \;\; h'(0) = 1, \end{cases}$$

where $G_{x_0}(t) = G(t + r(x_0))$.

Remark 2.16. Using the first part of the statement, J. Cheeger, M. Gromov and M. Taylor, [33], were able to deduce sharp volume growth estimates under point-inhomogeneous curvature conditions. See Theorem 2.26 and Corollary 2.27 below. See also Remark 2.23.

Proof. Part (i) is just a local version of Theorem 2.14. As for part (ii) , let $G(t)$ be non-decreasing. Fix $x_0 \in M$, set $r_{x_0} = \mathrm{dist}(x_0, x)$, and observe that, by the triangle inequality,

$$\mathrm{Ric}(x) \ge -(m-1)\,G(r(x)) \ge -(m-1)\,G_{x_0}(r_{x_0}(x)).$$

Therefore, the assumptions of Theorem 2.14 are met with respect to the origin x_0. □

In the previous section, we made several choices of the lower Ricci bound $G(t)$ in such a way that the solutions $g(t)$ of the corresponding problem (2.39) can be explicitly estimated. In view of Theorem 2.14, these estimates can now be applied to get upper volume bounds. We limit ourselves to "polynomial situations" (see Proposition 2.11 and Lemma 2.13 above).

Corollary 2.17. *Let $(M, \langle\,,\rangle)$ be a complete, m-dimensional manifold satisfying*

$$\mathrm{Ric}\,(x) \geq -\,(m-1)\,G\,(r\,(x)) \quad on\ M$$

for some non-negative function $G \in C^0([0,+\infty))$, where $r(x) = \mathrm{dist}(x,o)$ is the distance from a fixed reference origin $o \in M$.

(i) *Assume $G(t) = B^2\left(1+t^2\right)^{-1}$, with $B > 0$. Then, for every $R > r >> 1$,*

$$\frac{\mathrm{vol}B_R(o)}{\mathrm{vol}B_r(o)} \leq C\left(\frac{R}{r}\right)^{(m-1)B'+1}, \qquad \mathrm{vol}\partial B_r(o) \leq C r^{(m-1)B'},$$

where

$$B' = \frac{1+\sqrt{1+4B^2}}{2}$$

and $C > 0$ is a suitable constant.

(ii) *Assume $\int_0^{+\infty} tG(t)\,dt = b_0 < +\infty$. Then, for every $R > r > 1$,*

$$\frac{\mathrm{vol}B_R(o)}{\mathrm{vol}B_r(o)} \leq e^{(m-1)b_0}\left(\frac{R}{r}\right)^m, \qquad \mathrm{vol}\partial B_r(o) \leq c_m e^{(m-1)b_0} r^{m-1},$$

where, as in (2.41), c_m is the $(m-1)$-volume of the unit sphere in $\mathbb{R}^m$.

The volume estimates described above depend on uniform bounds on the Ricci curvature. We next consider a situation in which the Ricci curvature satisfies some L^p-integrability conditions, and describe upper bounds for the volume growth of balls obtained by P. Petersen and G. Wei in [129], who consider the slightly less general case where the function G below is a non-negative constant and M is compact. Previous related results have been obtained by S. Gallot, [57], Li and Yau, [107] and D. Yang, [166].

As above, we assume that G is non-negative and continuous on $[0,+\infty)$ and that $h(t) \in C^2([0,+\infty))$ is the non-negative solution of the problem

$$\begin{cases} h''(t) - G(t)\,h(t) = 0, \\ h(0) = 0\ h'(0) = 1. \end{cases}$$

Letting $r(x)$ be the distance function from the reference point o, we also define

$$\psi(x) = \max\{0, \Delta r(x) - (m-1)\frac{h'(r(x))}{h(r(x))}\} \tag{2.50}$$

in the domain D_o of the normal coordinates at o, and 0 on the cut locus, and let

$$\rho(x) = \max\{0, -\mathrm{Ric}(\nabla r, \nabla r) - (m-1)G(r(x))\}, \tag{2.51}$$

so that if $\rho \equiv 0$, that is $\mathrm{Ric}(\nabla r, \nabla r) \geq -(m-1)G(r(x))$, then, by the Laplacian comparison theorem, $\psi(x) \equiv 0$. For ease of notation, in the course of the arguments that follow we will denote by

$$A_G(r) = c_m h(r)^{m-1} \quad \text{and} \quad V_G(r) = c_m \int_0^r h(t)^{m-1} dt \tag{2.52}$$

the measures of the sphere and of the ball of radius r centered at the pole in the m-dimensional model manifold M_G with radial Ricci curvature equal to $-(m-1)G$, that is the manifold that is diffeomorphic to $\mathbb{R}^m$ and whose metric is given, in geodesic polar coordinates by

$$\langle\, , \rangle_G = dr^2 + h(r)^2 (\, ,)_{\mathbb{S}^{m-1}}.$$

We begin with the following lemma, which is a minor modification of Lemma 2.1 in [129].

Lemma 2.18. *Maintaining the notation introduced above, for every fixed* $p > 1/2$ *we have*

$$\frac{d}{dR} \frac{\mathrm{vol}\, B_R(o)}{V_G(R)} \leq \frac{c_m R A_G(R)}{V_G(R)^{1+1/2p}} \Big(\frac{\mathrm{vol}\, B_R}{V_G(R)}\Big)^{1-1/2p} \Big(\int_{B_R} \psi^{2p}\Big)^{1/2p}. \tag{2.53}$$

Proof. According to Remark 2.7, for every non-negative Lipschitz function φ compactly supported in $M \setminus \{o\}$ we have

$$-\int_M \langle \nabla r, \nabla \varphi \rangle = (\varphi, \Delta r) = \int_{E_o} \varphi \Delta r + (\varphi, (\Delta r)_{sing}) \leq \int_{E_o} \varphi \Delta r,$$

where D_o is the domain of the normal geodesic coordinates at o. Applying this inequality to the function $\varphi_\varepsilon(x) = \rho_\varepsilon(r(x))\, h(r(x))^{-m+1}$ where ρ_ε is the Lipschitz cut-off function defined in (2.46), arguing as in the proof of Theorem 2.14, and using the fact that h is non-decreasing, and the definition of ψ, we deduce that for a.e. $0 < r < R$,

$$\begin{aligned}\frac{\mathrm{vol}\, \partial B_R}{A_G(R)} - \frac{\mathrm{vol}\, \partial B_r}{A_G(r)} &\leq \int_{B_R \setminus B_r} h(r(x))^{-m+1} \Big[\Delta r - (m-1)\frac{h'(r(x))}{h(r(x))}\Big] \\ &\leq h(r)^{-m+1} \int_{B_R \setminus B_r} \psi.\end{aligned}$$

Using Hölder inequality and the definition of A_G we obtain

$$\begin{aligned}A_G(r)\mathrm{vol}\, \partial B_R - A_G(R)\mathrm{vol}\, \partial B_r &\leq c_m A_G(R) \int_{B_R \setminus B_r} \psi(x) \\ &\leq c_m A_G(R) \big(\mathrm{vol}\, B_R\big)^{1-1/2p} \Big(\int_{B_R} \psi^{2p}\Big)^{1/2p}.\end{aligned}$$

Therefore, applying the co-area formula, and using the above inequality we obtain

$$\begin{aligned}\frac{d}{dR}\Big(\frac{\operatorname{vol}B_R}{V_G(R)}\Big) &= \frac{V_G(R)\operatorname{vol}\partial B_R - A_G(R)\operatorname{vol}B_R}{V_G(R)^2}\\ &= V_G(R)^{-2}\int_0^R \big(A_G(r)\operatorname{vol}\partial B_R - A_G(R)\operatorname{vol}\partial B_r\big)\,dr\\ &\le V_G(R)^{-2}\int_0^R c_m A_G(R)\big(\operatorname{vol}B_R\big)^{1-1/2p}\Big(\int_{B_R}\psi^{2p}\Big)^{1/2p}dr\\ &= \frac{c_m R A_G(R)}{V_G(R)^{1+1/2p}}\Big(\frac{\operatorname{vol}B_R}{V_G(R)}\Big)^{1-1/2p}\Big(\int_{B_R}\psi^{2p}\Big)^{1/2p}\end{aligned}$$

as required. □

As noted before the statement of the lemma, if the Ricci tensor satisfies the inequality $\operatorname{Ric}(\nabla r, \nabla r) \ge -(m-1)G(r(x))$, then $\psi \equiv 0$, and we recover the fact that the function

$$r \mapsto \frac{\operatorname{vol}\partial B_r(o)}{A_G(r)}$$

is a decreasing function of r.

The following lemma (see Lemma 2.2 in [129]) allows to estimate the $2p$-norm of ψ over B_R in terms of the p-norm of ρ.

Lemma 2.19. *For every $p > m/2$ there exists a constant $C = C(m,p)$ such that for every R,*

$$\int_{B_R}\psi^{2p} \le C\int_{B_R}\rho^p$$

with $\rho(x)$ as defined in (2.51).

Proof. Integrating in polar geodesic coordinates we have

$$\int_{B_R} f = \int_{S^{m-1}} d\theta \int_0^{\min\{R,c(\theta)\}} f(t\theta)\omega(t\theta)dt$$

where ω is the volume density with respect to Lebesgue measure $dtd\theta$, and $c(\theta)$ is the distance from o to the cut locus along the ray $t \to t\theta$. It follows that it suffices to prove that for every $\theta \in S^{m-1}$,

$$\int_0^{\min\{R,c(\theta)\}}\psi^{2p}(t\theta)\omega(t\theta)dt \le C\int_0^{\min\{R,c(\theta)\}}\rho^p(t\theta)\omega(t\theta)dt. \tag{2.54}$$

An easy computation which uses (2.15) yields

$$\frac{\partial}{\partial t}\{\Delta r - (m-1)\frac{h'}{h}\} \le -\frac{(\Delta r)^2}{m-1} - Ric(\nabla r, \nabla r) - (m-1)\Big\{\frac{h''}{h} - \Big(\frac{h'}{h}\Big)^2\Big\}.$$

Thus, recalling the definitions of ψ and ρ, we deduce that the locally Lipschitz function ψ satisfies the differential inequality

$$\psi' + \frac{\psi^2}{m-1} + 2\frac{h'}{h}\psi \leq \rho$$

on the set where $\rho > 0$ and a.e. on $(0, +\infty)$. Multiplying through by $\psi^{2p-2}\omega$ and integrating, we obtain

$$\int_0^r \psi'\psi^{2p-2}\omega + \frac{1}{m-1}\int_0^r \psi^{2p}\omega + 2\int_0^r \frac{h'}{h}\psi^{2p-1}\omega \leq \int_0^r \rho\psi^{2p-1}\omega. \tag{2.55}$$

On the other hand, integrating by parts, and recalling that $\omega^{-1}\partial\omega/\partial t = \Delta r \leq \psi + (m-1)\frac{h'}{h}$ and that $\psi(t\theta) = 0$ if $t \geq c(\theta)$, yield

$$\begin{aligned}\int_0^r \psi'\psi^{2p-2}\omega &= \frac{1}{2p-1}\psi(r)^{2p-1}\omega(r\theta) - \frac{1}{2p-1}\int_0^r \psi^{2p-1}\Delta r\omega \\ &\geq -\frac{1}{2p-1}\int_0^r \psi^{2p-1}\big(\psi + (m-1)\frac{h'}{h}\big)\omega.\end{aligned}$$

Substituting this into (2.55), and using Hölder inequality we obtain

$$\begin{aligned}\Big(\frac{1}{m-1} - \frac{1}{2p-1}\Big)\int_0^r \psi^{2p}\omega + \Big(2 - \frac{m-1}{2p-1}\Big)\int \psi^{2p-1}\omega \\ \leq \int_0^r \rho\psi^{2p-1}\omega \\ \leq \Big(\int_0^r \rho^p\omega\Big)^{1/p}\Big(\int_0^r \psi^{2p}\omega\Big)^{(p-1)/p},\end{aligned}$$

and, since the coefficient of the first integral on the left-hand side is positive, by the assumption on p, while the second summand is non-negative, rearranging and simplifying we conclude that (2.54) holds with

$$C(m,p) = \Big(\frac{1}{m-1} - \frac{1}{2p-1}\Big)^{-p}. \qquad \square$$

We are now ready to state the announced volume comparison theorem under L^p Ricci curvature assumptions.

Theorem 2.20. *Keeping the notation introduced above, if $p > m/2$, then for every $0 < r < R$,*

$$\frac{\operatorname{vol} B_R(o)}{V_G(R)} - \frac{\operatorname{vol} B_r(o)}{V_G(r)} \leq \Big(\frac{1}{2p}\int_r^R f(t)dt\Big)^{2p}, \tag{2.56}$$

where

$$f(t) = \frac{c_m t A_G(t)}{V_G(t)^{1+1/2p}}\Big(\int_{B_t} \rho^p\Big)^{1/2p}. \tag{2.57}$$

Moreover f is integrable in $r=0$ and

$$\frac{\operatorname{vol} B_R(o)}{V_G(R)} \le \Big(1+\frac{1}{2p}\int_0^R f(t)dt\Big)^{2p}, \tag{2.58}$$

and

$$\begin{aligned}\frac{\operatorname{vol} \partial B_R(o)}{A_G(R)} &\le \Big(1+\frac{1}{2p}\int_0^R f(t)dt\Big)^{2p} \\ &\quad + \frac{c_m R}{V_G(R)^{1/2p}}\Big(\int_{B_R}\rho^p\Big)^{1/2p}\Big(1+\frac{1}{2p}\int_0^R f(t)dt\Big)^{2p-1}.\end{aligned} \tag{2.59}$$

Proof. Set

$$y(r) = \frac{\operatorname{vol} B_r(o)}{V_G(r)}.$$

According to (2.53) in Lemma 2.18, Lemma 2.19 and (2.57) we have

$$\begin{cases} y'(t) \le f(t)y(t)^{1-1/2p}, \\ y(t)\sim 1 \text{ as } t\to 0+, \quad y(t)>0 \text{ if } t>0,\end{cases}$$

whence, integrating between r and R we obtain

$$y(R)^{1/2p} - y(r)^{1/2p} \le \frac{1}{2p}\int_r^R f(t)\,dt,$$

that is, (2.56). Since

$$A_G(t)\sim c_m t^{m-1} \quad \text{and} \quad \int_{B_t}\rho^p = O(t^m) \text{ as } t\to 0,$$

f is integrable in $t=0$, and letting $r\to 0$ we obtain (2.58).

On the other hand, according to (2.53) and Lemma 2.19,

$$\frac{\operatorname{vol}\partial B_R}{A_G(R)} \le \frac{\operatorname{vol} B_R}{V_G(R)} + \frac{c_m R}{V_G(R)^{1/2p}}\Big(\int_{B_t}\rho^p\Big)^{1/2p}\Big(\frac{\operatorname{vol} B_R}{V_G(R)}\Big)^{1-1/2p}$$

and the conclusion follows inserting (2.58). □

Corollary 2.21. *Keeping the notation introduced above, we have:*

(i) *Let $G=0$, so that $h(t)=t$, $A_G(t)=c_m t^{m-1}$ and*

$$\rho = \max\{0, -\operatorname{Ric}(\nabla r,\nabla r)\} = \operatorname{Ric}_-.$$

(1) *If* $\mathrm{Ric}_- \in L^p(M)$ *for some* $p > m/2$, *then there exist constants* C_1 *and* C_2, *depending only on* m *and* p, *such that*

$$\operatorname{vol} B_R \le \frac{c_m}{m}\Big[1 + C_1 ||\mathrm{Ric}_-||^{1/2}_{L^p(M)} R^{1-m/2p}\Big]^{2p} R^m,$$

and

$$\operatorname{vol} \partial B_R \le c_m \Big[1 + C_2 ||\mathrm{Ric}_-||^{1/2}_{L^p(M)} R^{1-m/2p}\Big]^{2p} R^{m-1}.$$

(2) *(cf.* [107], *Corollary* 1.2*) If*

$$\int_{B_R} \mathrm{Ric}^p_- = o\big(R^\gamma\big) \quad as \quad t \to +\infty, \tag{2.60}$$

for some $p > m/2$ *and* $\gamma > 0$, *then*

$$\operatorname{vol} B_R = o(R^{2p+\gamma}) \quad as \quad R \to +\infty,$$

and

$$\operatorname{vol} \partial B_R = o(R^{2p+\gamma-1}) \quad as \quad R \to +\infty.$$

(ii) *Let* $G = H^2 > 0$ *be constant, so that* $h(t) = H^{-1} \sinh Ht$, *and*

$$V_G(t) \sim \frac{c_m}{(m-1)2^{m-1}H^m} e^{(m-1)Ht} \quad as \quad t \to +\infty.$$

If $\rho \in L^p(M)$ *for some* $p > m/2$, *then there exist constants* C_3 *and* C_4 *depending only on* p, m *and* H *such that*

$$\operatorname{vol} B_R \le \big(1 + C_3 ||\rho||^{1/2}_{L^p(M)}\big)^{2p} V_G(R),$$

and

$$\operatorname{vol} \partial B_R \le \big(1 + C_4 ||\rho||^{1/2}_{L^p(M)}\big)^{2p} A_G(R),$$

Proof. Assume that $G = 0$ and that $\mathrm{Ric}_- \in L^p$. Then we may estimate

$$f(t) = \frac{c_m t A_G(t)}{V_G(t)^{1+1/2p}} \Big(\int_{B_t} \mathrm{Ric}^p_-\Big)^{1/2p} \le C ||\mathrm{Ric}_-||^{1/2}_{L^p(M)} t^{-m/2p}$$

and (i) (1) follows at once from (2.58) and (2.59). Similarly, if (2.60) holds, then

$$f(t) = o\big(t^{(\gamma-m)/2p}\big) \quad \text{as} \quad t \to +\infty,$$

and (i) (2) follows as before from (2.58) and (2.59). Finally, if $G = H^2 > 0$ and $\rho \in L^p(M)$, we may estimate

$$f(t) \le C ||\rho||^{1/2}_{L^p(M)} \max\{1, t\} V_G(t)^{-1/2p}$$

and since the right-hand side is integrable on $[0, +\infty)$, the required conclusion follows once again from (2.58) and (2.59). □

The Bishop-Gromov comparison theorem enables one to get an upper estimate of the volumes of geodesic balls, assuming a lower control on the Ricci tensor. However, there are special situations where lower volume estimates can be deduced from the same curvature conditions. To see this, let us first consider the next important consequences of Theorem 2.14, as was pointed out in [33].

Corollary 2.22. *Let* $(M, \langle\, ,\rangle)$ *and* h *be as in Theorem* 2.14.

(i) *For every* $r_1 \leq r_2 \leq r_3$,

$$\frac{\mathrm{vol}B_{r_3}(o) - \mathrm{vol}B_{r_2}(o)}{\int_{r_2}^{r_3} h(t)^{m-1}\, dt} \leq \frac{\mathrm{vol}B_{r_1}(o)}{\int_0^{r_1} h(t)^{m-1}\, dt}. \tag{2.61}$$

(ii) *Assume that the lower Ricci curvature bound is a constant function* $G(t) \equiv H^2 \geq 0$. *Let* $x, y \in M$, *set* $d = \mathrm{dist}(x,y)$ *and take* $r + R < d$. *Then*

$$\mathrm{vol}B_R(x) \frac{\int_0^r sn_H(t)^{m-1}\, dt}{\int_{d-R}^{d+R} sn_H(t)^{m-1}\, dt} \leq \mathrm{vol}B_r(y), \tag{2.62}$$

where sn_H *is defined in* (2.48).

Remark 2.23. As will be clear from the proof, property (2.61) can be localized according to the first part of Corollary 2.22. In this situation, condition (2.61) becomes: for every $r_1 \leq r_2 \leq r_3 \leq \bar{R}$,

$$\frac{\mathrm{vol}B_{r_3}(x_0) - \mathrm{vol}B_{r_2}(x_0)}{\int_{r_2}^{r_3} sn_H(t)^{m-1}\, dt} \leq \frac{\mathrm{vol}B_{r_1}(x_0)}{\int_0^{r_1} sn_H(t)^{m-1}\, dt}$$

with H defined in (2.49).

Proof. We first consider case (i). A repeated use of Theorem 2.14 gives

$$\begin{aligned}
\mathrm{vol}B_{r_3}(o) - \mathrm{vol}B_{r_2}(o) &= \mathrm{vol}B_{r_3}(o) - \frac{\mathrm{vol}B_{r_2}(o)}{\int_0^{r_2} h(t)^{m-1}\, dt} \int_0^{r_2} h(t)^{m-1}\, dt \\
&\leq \mathrm{vol}B_{r_3}(o) - \frac{\mathrm{vol}B_{r_3}(o)}{\int_0^{r_3} h(t)^{m-1}\, dt} \int_0^{r_2} h(t)^{m-1}\, dt \\
&= \frac{\mathrm{vol}B_{r_3}(o)}{\int_0^{r_3} h(t)^{m-1}\, dt} \int_{r_2}^{r_3} h(t)^{m-1}\, dt \\
&\leq \frac{\mathrm{vol}B_{r_1}(o)}{\int_0^{r_1} h(t)^{m-1}\, dt} \int_{r_2}^{r_3} h(t)^{m-1}\, dt,
\end{aligned}$$

proving (2.61).

We now consider case (ii). The desired inequality easily follows from (2.61). Indeed, observe that

$$B_R(x) \subset B_{R+d}(y) \setminus B_{d-R}(y).$$

Therefore, recalling that $d - R \geq r$ and using (2.61), we conclude that

$$\begin{aligned}\mathrm{vol}B_R(x) &\leq \mathrm{vol}B_{d+R}(y) - \mathrm{vol}B_{d-R}(y)\\ &\leq \int_{d-R}^{d+R} sn_H(t)^{m-1}\,dt \frac{\mathrm{vol}B_r(y)}{\int_0^r sn_H(t)^{m-1}\,dt}.\end{aligned}$$

□

Now, by way of example, assume that $(M, \langle\,,\rangle)$ is a complete manifold of non-negative Ricci curvature. Using (2.62) of Corollary 2.22 with $H = 0$, so that $\mathrm{sn}_H(t) = t$, and radii $d = s - 1$, $r = s - 2$, and $R = 1$, we obtain

$$\mathrm{vol}B_s(y) \geq \frac{(s-2)^m}{s^m - (s-2)^m} \mathrm{vol}B_1(x) \geq Cs,$$

for every $s >> 1$ and for some constant $C > 0$. Namely, a complete manifold of non-negative Ricci curvature has at least a linear volume growth. This result was originally due to E. Calabi and S.T. Yau. Similar conclusions can be reached in a slightly more general situation. The starting point is that, in the above Ricci curvature assumption,

$$\frac{\mathrm{vol}B_R(x)}{\mathrm{vol}B_r(x)} \leq \left(\frac{R}{r}\right)^m$$

for every $x \in M$, $R \geq r > 0$. Thus, in particular, $(M, \langle\,,\rangle)$ enjoys the *doubling property*. This means that, for some (hence any) $\alpha > 1$, there is a constant $D_\alpha > 1$, such that

$$\mathrm{vol}B_{\alpha R}(x) \leq D_\alpha \mathrm{vol}B_R(x)$$

regardless of $x \in M$ and $R > 0$. We recall the following characterization of manifolds enjoying the doubling property.

Lemma 2.24. *A complete manifold $(M, \langle\,,\rangle)$ has the doubling property if and only if for some (hence any) $\alpha > 1$ there exists a constant $D_\alpha > 1$ such that the following holds. Let $A \subset M$ be a bounded set. Then, for every ball $B_r(x)$ centered at $x \in A$ and of radius r satisfying*

$$r < \delta_x(A) = \sup_{y \in A} \mathrm{dist}(x, y),$$

we have

$$\frac{\mathrm{vol}(A)}{\mathrm{vol}B_r(x)} \leq D_\alpha \left(\frac{\delta_x(A)}{r}\right)^{\log_\alpha D_\alpha}. \tag{2.63}$$

In particular, for every $0 < r \leq R$ we have

$$\frac{\mathrm{vol}B_R(x)}{\mathrm{vol}B_r(x)} \leq D_\alpha \left(\frac{R}{r}\right)^{\log_\alpha D_\alpha}.$$

Proof. We only need to show that if M has the doubling property, then the property holds. Assume therefore that $(M, \langle\,,\rangle)$ satisfies the doubling property with constants $\alpha, D_\alpha > 1$. Fix $B_r(x)$ as in the statement and consider $j \in \mathbb{N}$ satisfying

$$\alpha^{j-1} \leq \frac{\delta_x(A)}{r} \leq \alpha^j.$$

Note that

$$A \subset B_{\delta_x(A)}(x) \subset B_{r\alpha^j}(x).$$

Therefore, using the doubling property,

$$\mathrm{vol}(A) \leq \mathrm{vol}B_{r\alpha^j}(x) \leq D_\alpha^j \mathrm{vol}B_r(x).$$

To conclude, note that

$$D_\alpha^{j-1} = \left(\alpha^{j-1}\right)^{\log_\alpha D_\alpha} \leq \left(\frac{\delta_x(A)}{r}\right)^{\log_\alpha D_\alpha}.$$

□

Direct use of this Lemma gives a (quantitative) volume growth result which is true regardless of any monotonicity property.

Proposition 2.25. *Let $(M, \langle\,,\rangle)$ be a complete manifold with the doubling property and let $x \in M$. Then, there exist explicit constants $C = C(x, \alpha) > 0$ and $k = k(D_\alpha, \alpha) > 0$ such that, for every $R > 1$,*

$$\mathrm{vol}B_R(x) \geq CR^k. \tag{2.64}$$

Proof. Fix $x \in M$ and, for every j, choose $y_j \in \partial B_{\frac{\alpha^{j-1}+\alpha^j}{2}}(x)$. Then, we have the inclusions

$$B_{\alpha^j}(x) \setminus B_{\alpha^{j-1}}(x) \supset B_{\frac{\alpha^j-\alpha^{j-1}}{2}}(y_j); \qquad B_{\frac{\alpha^j+3\alpha^{j-1}}{2}}(y_j) \supset B_{\alpha^{j-1}}(x).$$

Using Lemma 2.24, we deduce

$$\begin{aligned}
\mathrm{vol}B_{\alpha^j}(x) - \mathrm{vol}B_{\alpha^{j-1}}(x) &\geq \mathrm{vol}B_{\frac{\alpha^j-\alpha^{j-1}}{2}}(y_j) \\
&\geq D_\alpha^{-1}\left(\frac{\alpha^j-\alpha^{j-1}}{\alpha^j+3\alpha^{j-1}}\right)^{\log_\alpha D_\alpha} \mathrm{vol}B_{\frac{\alpha^j+3\alpha^{j-1}}{2}}(y_j) \\
&\geq D_\alpha^{-1}\left(\frac{\alpha-1}{\alpha+3}\right)^{\log_\alpha D_\alpha} \mathrm{vol}B_{\alpha^{j-1}}(x).
\end{aligned}$$

Therefore

$$\mathrm{vol}B_{\alpha^j}(x) \geq E\mathrm{vol}B_{\alpha^{j-1}}(x),$$

where we have set

$$E = 1 + \left(\frac{\alpha-1}{\alpha^2+3\alpha}\right)^{\log_\alpha D_\alpha} > 1.$$

Iterating this inequality j-times gives

$$\mathrm{vol}B_{\alpha^j}(x) \geq E^j \mathrm{vol}B_1(x),$$

and choosing $k = \log_\alpha E > 0$ we finally obtain

$$\mathrm{vol}B_{\alpha^j}(x) \geq \alpha^{jk}\mathrm{vol}B_1(x) = \left(\alpha^{j+1}\right)^k \frac{\mathrm{vol}B_1(x)}{\alpha^k}. \tag{2.65}$$

Now, for any given $R > 1$, let j satisfy $\alpha^j < R < \alpha^{j+1}$. From (2.65) we conclude the validity of (2.64) with $C = \mathrm{vol}B_1(x)/\alpha^k$. □

The last result of the section is the following impressive volume growth property of complete manifolds with almost non-negative Ricci curvature (in the sense specified in (2.66), (2.67) see also Corollary 2.27 below). It is a contribution of P. Li and R. Schoen, [95], and P. Li and M. Ramachandran, [98]. Its proof relies heavily on the work by Cheeger-Gromov-Taylor [33].

Theorem 2.26. *For every $m \geq 2$ there exists a constant $\varepsilon = \varepsilon(m) > 0$ such that the following holds. Let $(M, \langle\,,\rangle)$ be a complete manifold satisfying*

$$\mathrm{Ric} \geq -(m-1)\,G(r(x)) \quad \text{on } M \tag{2.66}$$

where $G(t) \in C^0([0,+\infty))$ is a positive function such that

$$G(t) = \frac{\varepsilon^2}{t^2} \text{ for } t >> 1, \tag{2.67}$$

and $r(x) = \mathrm{dist}(x, o)$, for some reference origin $o \in M$. Then, for every $d > 1$,

$$\mathrm{vol}\left(B_{\frac{r(x)}{d}}(x)\right) \to +\infty, \text{ as } r(x) \to +\infty. \tag{2.68}$$

Proof. We divide the quite involved proof in several steps.

First Step. Fix

$$0 < \xi < 1, \qquad \beta > \frac{2}{\left(2^{1/m} - 1\right)} > 1.$$

If $\varepsilon = \varepsilon(m, \xi, \beta) > 0$ is sufficiently small, then the following holds: For every $x \in M$ choose a geodesic γ parametrized by arc length which realizes the distance between $o = \gamma(0)$ and $x = \gamma(r(x))$. Define a set of values $t_0 < t_1 < \cdots < t_k \leq r(x)$ by

$$t_0 = 0; \qquad t_i = (\beta + 1)\sum_{j=0}^{i-1} \beta^j, \tag{2.69}$$

t_k being the largest value so that $t_k < r(x)$. Define $x_j = \gamma(t_j)$. Then, if k is large enough,

$$\mathrm{vol}B_{\beta^k}(x_k) \geq C_1 \left(\frac{\beta^m}{(\beta+2)^m - \beta^m}\right)^{k(1-\xi)} \mathrm{vol}B_1(o), \tag{2.70}$$

for some constant $C = C(\beta) > 0$.

Indeed, observe that $\{x_i\}$ form a set of points along γ with the property that

$$\begin{cases} \text{(i)} & r(x_i) = t_i = (\beta+1)(\beta-1)^{-1}(\beta^i - 1), \\ \text{(ii)} & \operatorname{dist}(x_i, x_{i+1}) = t_{i+1} - t_i = \beta^i + \beta^{i+1}, \\ \text{(iii)} & \operatorname{dist}(x_k, x) = r(x) - t_k < t_{k+1} - t_k = \beta^k + \beta^{k+1}. \end{cases} \tag{2.71}$$

Thus, the geodesic balls $\{B_{\beta^i}(x_i)\}$ are disjoint and their closures cover the set $\gamma([0, t_k + \beta^k))$. Furthermore

$$B_{\beta^{i-1}}(x_{i-1}) \subset B_{\beta^i + 2\beta^{i-1}}(x_i) \setminus B_{\beta^i}(x_i). \tag{2.72}$$

Setting

$$H_i^2 = \max_{B_{\beta^i+2\beta^{i-1}}(x_i)} G(r(y)),$$

from Corollary 2.22 (i) and (2.72), we deduce that

$$\begin{aligned} \operatorname{vol} B_{\beta^i}(x_i) &\geq T_i \left\{ \operatorname{vol} B_{\beta^i+2\beta^{i-1}}(x_i) - \operatorname{vol} B_{\beta^i}(x_i) \right\} \\ &\geq T_i \operatorname{vol} B_{\beta^{i-1}}(x_{i-1}) \end{aligned}$$

where

$$\begin{aligned} T_i &= \frac{\displaystyle\int_0^{\beta^i} H_i^{-1} \sinh^{m-1}(H_i t)\, dt}{\displaystyle\int_{\beta^i}^{(\beta^i+2\beta^{i-1})} H_i^{-1} \sinh^{m-1}(H_i t)\, dt} \\ &= \frac{\displaystyle\int_0^{\beta^i H_i} \sinh^{m-1} t\, dt}{\displaystyle\int_{\beta^i H_i}^{(\beta^i+2\beta^{i-1})H_i} \sinh^{m-1} t\, dt}. \end{aligned} \tag{2.73}$$

Iterating this inequality we conclude that

$$\operatorname{vol} B_{\beta^k}(x_k) \geq \prod_{i=1}^{k} T_i \operatorname{vol} B_1(o). \tag{2.74}$$

The validity of (2.70) will follow once we show that, for every i, hence k, large enough,

$$T_i \geq \left(\frac{\beta^m}{(\beta+2)^m - \beta^m} \right)^{1-\xi}. \tag{2.75}$$

Note that, according to (2.66) and (2.67), for sufficiently large $i \geq i_0 = i_0(\beta, m)$, we have

$$\operatorname{Ric} \geq -(m-1)H_i^2 \quad \text{on } B_{\beta^i+2\beta^{i-1}}(x_i)$$

with

$$H_i = \frac{\varepsilon}{\operatorname{dist}\left(o, \partial B_{\beta^i + 2\beta^{i-1}}(x_i)\right)} = \varepsilon \frac{(\beta - 1)}{2\beta^{i-1} - \beta - 1}.$$

In particular, for every $i \geq i_0$,

$$\left(\beta^i + 2\beta^{i-1}\right) H_i = \varepsilon \frac{(\beta+2)(\beta-1)}{2 - \frac{\beta+1}{\beta^{i+1}}} \leq \varepsilon A,$$

where

$$A = A(\beta, i_0) = \frac{(\beta-1)(\beta+2)}{2 - \frac{\beta+1}{\beta^{i_0-1}}}. \tag{2.76}$$

Let $0 < \delta = \delta(m, \beta, \xi) < 1$ be defined by the equation

$$\left(\frac{1}{1+\delta}\right)^{m-1} = \left\{\frac{(\beta+2)^m - \beta^m}{\beta^m}\right\}^{\xi}.$$

The definition is consistent because our assumption that $\beta > 2/\left(2^{1/m} - 1\right)$ guarantees that the right-hand side is < 1. If $\varepsilon = \varepsilon(m, \delta) > 0$ is chosen small enough, we can approximate

$$t \leq \sinh t \leq t(1+\delta) \text{ on } [0, \varepsilon A]$$

which in turn, used in (2.73), gives

$$\begin{aligned} T_i &\geq \left(\frac{1}{1+\delta}\right)^{m-1} \frac{(\beta^i H_i)^m}{[(\beta^i + 2\beta^{i-1})H_i]^m - (\beta^i H_i)^m} \\ &= \left(\frac{1}{1+\delta}\right)^{m-1} \frac{\beta^m}{(\beta+2)^m - \beta^m}, \end{aligned}$$

proving (2.75).

Second Step. There exists a constant $0 < \alpha = \alpha(\beta) < 1$ such that, for $r(x) >> 1$, hence $k >> 1$,

$$\operatorname{vol}\left(B_{\alpha r(x)}(x)\right) \geq \operatorname{vol} B_{\beta^k}(x_k). \tag{2.77}$$

To see this, first observe that, by the triangle inequality,

$$B_{\operatorname{dist}(x_k, x) + \beta^k}(x) \supset B_{\beta^k}(x_k). \tag{2.78}$$

On the other hand, x and x_k lie on the minimizing geodesic γ from o to x, so that

$$\frac{\operatorname{dist}(x_k, x) + \beta^k}{r(x)} = \frac{\operatorname{dist}(x_k, x) + \beta^k}{\operatorname{dist}(x_k, x) + r(x_k)}.$$

Since, by (2.71) (i), $r(x_k) \geq \beta^k$, the function

$$t \mapsto \frac{t + \beta^k}{t + r(x_k)}$$

is monotone increasing. Whence, using (2.71) (iii),

$$\begin{aligned}\frac{\operatorname{dist}(x_k,x)+\beta^k}{r(x)} &= \frac{\operatorname{dist}(x_k,x)+\beta^k}{\operatorname{dist}(x_k,x)+r(x_k)}\\ &\le \frac{2\beta^k+\beta^{k+1}}{\beta^k+\beta^{k+1}+\frac{(\beta+1)(\beta^k-1)}{\beta-1}}\\ &= \frac{2+\beta}{1+\beta+\frac{\beta+1}{\beta-1}\left(1-\beta^{-k}\right)}.\end{aligned}$$

Note that

$$\lim_{k\to+\infty}\frac{2+\beta}{1+\beta+\frac{\beta+1}{\beta-1}\left(1-\beta^{-k}\right)} = \frac{2+\beta}{1+\beta+\frac{\beta+1}{\beta-1}} < 1.$$

Therefore, up to choosing k (hence $r(x)$) large enough, say $k \ge k_0 = k_0(\beta) > 0$, we have

$$\frac{\operatorname{dist}(x_k,x)+\beta^k}{r(x)} \le \alpha,$$

with

$$\alpha = \frac{2+\beta}{1+\beta+\frac{\beta+1}{\beta-1}\left(1-\beta^{-k_0}\right)} < 1.$$

It follows that

$$B_{\alpha r(x)}(x) \supset B_{\operatorname{dist}(x_k,x)+\beta^k}(x)$$

which in turn, combined with (2.78), proves (2.77).

Third Step. Let $d > 1$ be fixed. There exists a constant $C_2 = C_2(\alpha, d, \varepsilon) > 0$ such that, for $r(x) >> 1$,

$$\operatorname{vol} B_{\frac{r(x)}{d}}(x) \ge C_2 \operatorname{vol}\left(B_{\alpha r(x)}(x)\right). \tag{2.79}$$

Indeed, suppose $d > 1/\alpha$, for otherwise there would be nothing to prove. Note that, for every $y \in B_{\alpha r(x)}(x)$,

$$\operatorname{Ric}(y) \ge -(m-1)\frac{\varepsilon^2}{r(x)^2(1-\alpha)^2},$$

provided $r(x) >> 1$. Therefore, by Corollary 2.15 with $H = \varepsilon/r(x)(1-\alpha)$ we deduce

$$\frac{\operatorname{vol} B_{\frac{r(x)}{d}}(x)}{\operatorname{vol}\left(B_{\alpha r(x)}(x)\right)} \ge C_2 := \frac{\int_0^{\frac{\varepsilon}{d(1-\alpha)}} \sinh^{m-1}(t)\,dt}{\int_0^{\frac{\alpha\varepsilon}{(1-\alpha)}} \sinh^{m-1}(t)\,dt}.$$

Fourth Step. Putting together (2.70), (2.77) and (2.79) gives

$$\operatorname{vol} B_{\frac{r(x)}{d}}(x) \ge C_3\left(\frac{\beta^m}{(\beta+2)^m-\beta^m}\right)^{k(1-\xi)} \operatorname{vol} B_1(o)$$

with $C_3 = C_1 C_2 > 0$. Since, by our choice of $\beta > 2/\left(2^{1/m} - 1\right)$,

$$\left(\frac{\beta^m}{(\beta+2)^m - \beta^m} \right) > 1,$$

the validity of (2.68) follows by letting $r(x)$, hence k, tend to $+\infty$. □

We point out the following consequence of the above proof.

Corollary 2.27. *Suppose we are given*

$$m \geq 2, \qquad 0 < \xi < 1, \qquad \beta > \frac{2}{\left(2^{1/m} - 1\right)} > 1, \qquad d > 1.$$

Then, there exists $\varepsilon = \varepsilon(m, \xi, \beta) > 0$ sufficiently small such that the following holds. Let $(M, \langle\, ,\rangle)$ be a complete manifold satisfying the curvature conditions (2.66) and (2.67) above. Then

$$\mathrm{vol} B_{\frac{r(x)}{d}}(x) \geq C r(x)^{\log_\beta E}, \quad as\ r(x) \to +\infty,$$

where

$$E = \left(\frac{\beta^m}{(\beta+2)^m - \beta^m} \right)^{(1-\xi)} > 1$$

and $C = C(m, \beta, \xi, d) > 0$ is a suitable constant.

Proof. Let $x \in M$ be such that $r(x) >> 1$. Recall that, by definitions of k and $\{t_j\}$ in the First Step, we have

$$t_{k+1} = \frac{(\beta+1)\left(\beta^{k+1} - 1\right)}{\beta - 1} \geq r(x).$$

Therefore

$$k \geq \log_\beta r(x) + \log_\beta \left(\frac{\beta - 1}{\beta + 1} \right) - 1.$$

On the other hand, in the Fourth Step we obtained, for $r(x) >> 1$,

$$\mathrm{vol} B_{\frac{r(x)}{d}}(x) \geq C_3 \left(\frac{\beta^m}{(\beta+2)^m - \beta^m} \right)^{k(1-\xi)} \mathrm{vol} B_1(o).$$

It follows that

$$\mathrm{vol} B_{\frac{r(x)}{d}}(x) \geq C E^{\log_\beta r(x)},$$

where we have set

$$C = C_3 \mathrm{vol} B_1(o) \left(\frac{\beta^m}{(\beta+2)^m - \beta^m} \right)^{(1-\xi)\left\{\log\left(\frac{\beta-1}{\beta+1}\right) - 1\right\}}$$

and

$$E = \left(\frac{\beta^m}{(\beta+2)^m - \beta^m} \right)^{(1-\xi)}.$$

□

2.3 A monotonicity formula for volumes

In the previous sections we obtained volume growth estimates under curvature conditions. In the present section we replace the curvature conditions with the assumption that the manifold M is isometrically immersed in an appropriate ambient space N.

In this context, we recall that if M is an m-dimensional manifold minimally immersed into $\mathbb{R}^n$, denoting by $B_R^{\mathbb{R}^n}(p)$ the Euclidean ball of radius R centered at $p \in \mathbb{R}^n$, then the monotonicity formula states that

$$\frac{\operatorname{vol}\left(M \cap B_R^{\mathbb{R}^n}(x_o)\right)}{R^m}$$

is a non-decreasing function of R (see, e.g., [38]), and therefore

$$\operatorname{vol}\left(M \cap B_R^{\mathbb{R}^n}(x_o)\right) \geq c_m R^m,$$

where c_m is the volume of the unit sphere in $\mathbb{R}^m$. Such estimates have proven to be important in a number of applications, including the regularity theory of harmonic maps. A variation of the argument shows that similar monotonicity formulas hold replacing the "exterior ball" $B_R^{\mathbb{R}^n}(x_o)$ with the intrinsic ball $B_R(p)$, $p \in M$. As a matter of fact, we have the following slightly more general result concerning the volume growth of manifolds which admit a bi-Lipschitz, harmonic immersion into a Cartan–Hadamard space. Before stating the theorem, we recall that a geodesic ball $B_\rho^N(q)$ in N is said to be regular if it does not intersect the cut locus of q, and, having denoted by k an upper bound for the sectional curvature of N on $B_\rho^N(q)$, one has $\sqrt{k}\rho < \pi/2$.

Theorem 2.28. *Let $(M, \langle\,,\rangle)$ be a complete, non-compact Riemannian manifold of dimension $\dim M = m$, immersed into a complete manifold $(N, (\,,))$ via $f : M \to N$. Assume that either N is Cartan–Hadamard, or that $f(M)$ is contained in a regular geodesic ball. Furthermore, suppose that the immersion f is harmonic and bi-Lipschitz so that there exist positive constants A and B such that*

$$A\langle X, X\rangle \leq \big(df(X), df(X)\big) \leq B\langle X, X\rangle \tag{2.80}$$

for every $X \in TM$. Fix an origin $o \in M$ and denote $\rho(y) = \operatorname{dist}_N(y, f(o))$. Next, set

$$k = \sup_{f(M)} {}^N\mathrm{Sect}$$

and define sn_k to be the unique solution of the Cauchy problem

$$\begin{cases} \ddot{\mathrm{sn}}_k + k\,\mathrm{sn}_k = 0, \\ \mathrm{sn}_k(0) = 0;\ \dot{\mathrm{sn}}_k(0) = 1. \end{cases}$$

Set also

$$\mathrm{cn}_k(t) = \dot{\mathrm{sn}}_k(t).$$

Then, there exists a constant B_1,

$$B_1 \begin{cases} = 1 & \text{if } f \text{ is isometric,} \\ \leq \sqrt{mB} & \text{otherwise,} \end{cases} \tag{2.81}$$

such that

$$R \longmapsto \frac{\int_{B_R(o)} \mathrm{cn}_k (\rho \circ f)}{\left\{ \mathrm{sn}_k \left(\sqrt{B} R \right) \right\}^{\frac{mA}{B_1 \sqrt{B}}}} \tag{2.82}$$

is a non-decreasing function. In particular, in case N *is Cartan–Hadamard, for every* $R \geq R_0 > 0$ *it holds that*

$$vol B_R(o) \geq C \begin{cases} R^{\frac{mA}{B_1 \sqrt{B}}} & \text{if } k = 0, \\ \sinh^{\frac{mA}{B_1 \sqrt{B}} - 1} \left(\sqrt{-kB} R \right) \tanh \left(\sqrt{-kB} R \right) & \text{if } k < 0, \end{cases} \tag{2.83}$$

for a suitable constant $C = C(R_0, k, A, B, m) > 0$.

Proof. Consider the function

$$F(x) = \mathrm{in}_k \circ \rho \circ f(x)$$

where we have set

$$\mathrm{in}_k(t) = \int_0^t \mathrm{sn}_k(s)\, ds.$$

The chain rule for the Hessian yields

$${}^M\Delta F(x) = \sum_{i=1}^m {}^N\mathrm{Hess}\,(\mathrm{in}_k \circ \rho)\,(df(e_i), df(e_i)) + d\,(\mathrm{in}_k \circ \rho)\,(\tau(f))$$

where $\{e_i\}$ is any local orthonormal frame filed in $(M, \langle\, , \rangle)$. Since f is harmonic, we have

$$\tau(f) = 0,$$

and therefore

$${}^M\Delta F(x) = \sum_{i=1}^m {}^N\mathrm{Hess}\,(\mathrm{in}_k \circ \rho)\,(df(e_i), df(e_i)). \tag{2.84}$$

Note that, by Hessian comparison,

$$\begin{aligned} {}^N\mathrm{Hess}\,(\mathrm{in}_k \circ \rho) &= \ddot{\mathrm{in}}_k(\rho)\, d\rho \otimes d\rho + \dot{\mathrm{in}}_k(\rho)\ {}^N\mathrm{Hess}\,(\rho) \\ &= \mathrm{cn}_k(\rho)\, d\rho \otimes d\rho + \mathrm{sn}_k(\rho)\ {}^N\mathrm{Hess}\,(\rho) \\ &\geq \mathrm{cn}_k(\rho)\, d\rho \otimes d\rho + \mathrm{sn}_k(\rho)\ \frac{\mathrm{cn}_k(\rho)}{\mathrm{sn}_k(\rho)} \{(\, ,) - d\rho \otimes d\rho\} \\ &= \mathrm{cn}_k(\rho)\,(\, ,). \end{aligned}$$

Therefore, using the fact that f is bi-Lipschitz and $\mathrm{cn}_k(\rho) > 0$,

$$\begin{aligned} RHS(2.84) &\geq \sum_{i=1}^{m} \mathrm{cn}_k(\rho(f))\left(df(e_i), df(e_i)\right) \\ &\geq mA\,\mathrm{cn}_k(\rho(f)). \end{aligned}$$

It follows that

$$^M\Delta F \geq mA\,\mathrm{cn}_k(\rho(f)). \tag{2.85}$$

Starting from (2.85), we now apply the divergence theorem on the (intrinsic) geodesic ball $B_R(o)$ of M and use Schwarz inequality and Gauss' lemma to obtain

$$\begin{aligned} mA\int_{B_R(o)} \mathrm{cn}_k(\rho(f)) &\leq \int_{B_R(o)} {}^M\Delta F \\ &= \int_{\partial B_R(o)} \left\langle {}^M\nabla F, \frac{\partial}{\partial r}\right\rangle \\ &\leq \int_{\partial B_R(o)} \left|{}^M\nabla F\right|_M. \end{aligned} \tag{2.86}$$

It is readily seen that

$$\left|{}^M\nabla F\right|_M \leq B_1\,\mathrm{sn}_k(\rho(f)) \tag{2.87}$$

with B_1 satisfying (2.81). Indeed,

$$\begin{aligned} \left|{}^M\nabla F\right|_M &= \dot{\mathrm{in}}_k(\rho(f))\left|{}^M\nabla(\rho\circ f)\right|_M \\ &= \mathrm{sn}_k(\rho(f))\left|{}^M\nabla(\rho\circ f)\right|_M. \end{aligned}$$

Moreover, with respect to local orthonormal frames $\{e_i\}$ on M, $\{E_A\}$ on N with dual frames $\{\theta^i\}$,$\{\Theta^A\}$, we have

$$d\rho = \rho_A\Theta^A,\ df = f_i^A\theta^i \otimes E_A$$

so that

$$^M\nabla(\rho\circ f) = \rho_A f_i^A e_i.$$

This latter implies

$$\left|{}^M\nabla(\rho\circ f)\right|_M = |D\mathbf{y}|_{\mathbb{R}^m}$$

where

$$D = \left(f_i^A\right) \in M_{n,m}(\mathbb{R}), \qquad \mathbf{y} = (\rho_A)^t \in \mathbb{R}^n.$$

Clearly, in case f is isometric, $D = \left(\delta_i^A\right)$, while, in the general bi-Lipschitz case,

$$|D| = \sqrt{\sum \langle df(e_i), df(e_i)\rangle_N} = |df| \leq \sqrt{m}B.$$

Now recall that, by Gauss' lemma,

$$|\mathbf{y}|_{\mathbb{R}^n} = \left|{}^N\nabla\rho\right|_N = 1,$$

and therefore

$$|D\mathbf{y}|_{\mathbb{R}^m} \leq \sup_{|\mathbf{x}|=1} |D\mathbf{x}|_{\mathbb{R}^m} \begin{cases} = 1 & f \text{ isometric,} \\ \leq \sqrt{mB} & \text{otherwise.} \end{cases}$$

This proves the claimed inequality (2.87) with

$$B_1 = \sup_M \sup_{|\mathbf{x}|=1} |D\mathbf{x}|_{\mathbb{R}^m}.$$

Inserting (2.87) into (2.86) gives

$$mA \int_{B_R(o)} \mathrm{cn}_k\left(\rho\left(f\right)\right) \leq B_1 \int_{\partial B_R(o)} \mathrm{sn}_k\left(\rho\left(f\right)\right). \tag{2.88}$$

Let us elaborate the RHS of this latter. Observe that the positive function

$$t \longmapsto \frac{\mathrm{sn}_k\left(t\right)}{\mathrm{cn}_k\left(t\right)} \text{ is increasing.} \tag{2.89}$$

Furthermore, from the very definitions of dist_N and dist_M, we see that

$$\rho\left(f\left(x\right)\right) \leq \sqrt{B}\,\mathrm{dist}_M\left(x,o\right). \tag{2.90}$$

Indeed,

$$\begin{aligned} \rho\left(f\left(x\right)\right) &= \mathrm{dist}_N\left(f\left(x\right), f\left(o\right)\right) \\ &= \inf_{\substack{\Gamma\subset N \\ \Gamma(0)=f(o),\ \Gamma(1)=f(x)}} \int_0^1 \left|\dot{\Gamma}\left(t\right)\right|_N \\ &\leq \inf_{\substack{\gamma\subset M \\ \gamma(0)=o,\ \gamma(1)=x}} \int_0^1 \left|\frac{d}{dt} f\circ\gamma\left(t\right)\right|_N \\ &\leq \sqrt{B} \inf_{\substack{\gamma\subset M \\ \gamma(0)=o,\ \gamma(1)=x}} \int_0^1 \left|\dot{\gamma}\left(t\right)\right|_M \\ &= \sqrt{B}\,\mathrm{dist}_M\left(x,o\right). \end{aligned}$$

Combining (2.90) with (2.89) we deduce

$$\begin{aligned} RHS\left(2.88\right) &= \int_{\partial B_R(o)} \frac{\mathrm{sn}_k\left(\rho\left(f\right)\right)}{\mathrm{cn}_k\left(\rho\left(f\right)\right)} \mathrm{cn}_k\left(\rho\left(f\right)\right) \\ &\leq B_1 \frac{\mathrm{sn}_k\left(\sqrt{B}R\right)}{\mathrm{cn}_k\left(\sqrt{B}R\right)} \int_{\partial B_R(o)} \mathrm{cn}_k\left(\rho\left(f\right)\right) \end{aligned}$$

so that

$$mA \int_{B_R(o)} \mathrm{cn}_k(\rho(f)) \leq B_1 \frac{\mathrm{sn}_k\left(\sqrt{B}R\right)}{\mathrm{cn}_k\left(\sqrt{B}R\right)} \int_{\partial B_R(o)} \mathrm{cn}_k(\rho(f)) . \tag{2.91}$$

Using the co-area formula, (2.91) can be written in the form

$$\frac{d}{dR} \log \left(\frac{\int_{B_R(o)} \mathrm{cn}_k(\rho(f))}{\mathrm{sn}_k^{\frac{mA}{B_1\sqrt{B}}}\left(\sqrt{B}R\right)} \right) \geq 0 \tag{2.92}$$

which implies the validity of (2.82). □

Chapter 3

Review of spectral theory

In this chapter we collect the results from spectral theory that will be used in the sequel, concentrating our attention on the spectral theory of Schrödinger operators on Riemannian manifolds. The vanishing and finite-dimensionality results which we will present in the next chapters are based on assumptions on the spectrum of suitable Schrödinger operators. And some of the geometric conditions which we will encounter in later chapters can be interpreted in spectral sense.

Our presentation is based on the delightful book by E.B. Davies, [41], and on the paper by P. Bérard, M. do Carmo and W. Santos, [13] (see also Reed and Simon, [137], [138]).

3.1 The spectrum of a self-adjoint operator

Let $T : \mathcal{D}(T) \subseteq \mathcal{X} \to \mathcal{X}$ be a possibly unbounded densely defined operator on a complex Banach space $\mathcal{X}$. We recall the resolvent set of T is the set of point λ in $\mathbb{C}$ such that $(\lambda - T)$ has dense range $\mathcal{R}(\lambda - T)$, and $(\lambda - T)^{-1}$ extends to a bounded operator on $\mathcal{X}$ (note that if in addition T is closed, then necessarily $\mathcal{R}(\lambda - T) = \mathcal{X}$). The spectrum $\sigma(T)$ of T is the complement of the resolvent set.

In particular, if $ker(\lambda - T) \neq \{0\}$ we say that λ is an eigenvalue of T, $ker(\lambda - T)$ is the eigenspace belonging to λ, and its dimension is the multiplicity of the eigenvalue.

The point spectrum $\sigma_p(T)$ of T is the set of all eigenvalues, the discrete spectrum $\sigma_d(T)$ is the set of all eigenvalues of finite multiplicity which are isolated points of the spectrum, and the essential spectrum $\sigma_{ess}(T)$ is the complement of the discrete spectrum, and consists of eigenvalues of infinite multiplicity together with cluster points of the spectrum. Both the spectrum, and the essential spectrum are closed.

Assume now that T is a densely defined operator defined on a Hilbert space $\mathcal{H}$, endowed with inner product $(,)$. It's adjoint operator is defined by $\mathcal{D}(T^*) = \{y \in \mathcal{H} : x \mapsto (Tx, y)$ is a bounded linear functional on $\mathcal{D}(T)\}$ and T^*y is defined via the Riesz representation theorem by the identity $(x, T^*y) = (Tx, y)$ for all $x \in \mathcal{D}(T)$. If T is closed and densely defined, then so is T^*.

Then T is symmetric if $T \subseteq T^*$, that is $(Tx, y) = (x, Ty)$ for all $x, y \in \mathcal{D}(T)$ and it is self-adjoint if $T = T^*$. Finally T is essentially self-adjoint on $\mathcal{D}$ if its closure $\bar{T}$ is self-adjoint, and since $(\bar{T})^* = T^*$ this amounts to $\bar{T} = T^*$. In this case $\bar{T}$ is the only self-adjoint extension of T.

We have the following criteria of self-adjointness, respectively essential self-adjointness.

Theorem 3.1. *Let $T : \mathcal{D} \to \mathcal{H}$ be a symmetric operator. Then the following are equivalent:*

(i) *T is self-adjoint (resp. T is essentially self-adjoint);*

(ii) *T is closed and $ker(T^* \pm i) = \{0\}$ (resp. $ker(T^* \pm i) = \{0\}$);*

(iii) *$\mathcal{R}(T \pm i) = \mathcal{H}$ (resp. $\overline{\mathcal{R}(T \pm i)} = \mathcal{H}$).*

In general, for a symmetric operator T on $\mathcal{D} \subseteq \mathcal{H}$, we define its deficiency indices $d_\pm = \dim ker(T^* \pm i)$. Then T is essentially self-adjoint if and only if $d_+ = d_- = 0$, while it has at least one self-adjoint extension if and only if $d_+ = d_-$.

In the case of differential operators on bounded domains (of $\mathbb{R}^n$ or of a manifold), the following sufficient condition is often useful.

Theorem 3.2. *Let T be a symmetric operator with domain $\mathcal{D}(T)$, and let $\{f_n\}$ be a complete orthonormal set in $\mathcal{H}$ such that, for every n, $f_n \in \mathcal{D}(T)$ and $Tf_n = \lambda_n f_n$ for some λ_n, which is necessarily real since T is symmetric. Then T is essentially self-adjoint on $\mathcal{D}(T)$ and the spectrum of $\bar{T} = T^*$ is the closure in $\mathbb{R}$ of the set $\{\lambda_n\}$.*

The condition of the theorem is satisfied if for some $\lambda \in \mathbb{R}$ the operator $(\lambda - T)$ is invertible with compact inverse. In the case of second-order differential operators on bounded domains on $\mathbb{R}^n$, the compactness of the inverse may often be deduced from the Rellich embedding theorem.

We next consider multiplication operators on L^2 spaces, which provide the basic example of self-adjoint (and in fact normal) operators, and are the building blocks of one form of the spectral theorem.

Let (X, μ) be a measure space. For every measurable function $f : X \to \mathbb{C}$ define a, possibly unbounded, operator M_f on $L^2(X, \mu)$ as follows:

$$\mathcal{D}(M_f) = \{\psi \in L^2 \,:\, f\psi \in L^2\},$$
$$M_f\psi = f\psi \text{ for every } \psi \in \mathcal{D}(M_f).$$

The operator M_f is then said to be the multiplication operator associated to f. Under the assumption that μ is σ-finite we have

(i) for every measurable almost everywhere finite f, M_f is closed and densely defined;

(ii) M_f is bounded if and only if $f \in L^\infty(X, \mu)$ and in this case $|||M_f||| = ||f||_{L^\infty}$;

(iii) $(M_f)^* = M_{\bar{f}}$ so that M_f is self-adjoint if and only if f is real-valued;

(iv) $\sigma(M_f) = \text{essrange}(f) = \{\lambda \in \mathbb{C} : \forall\epsilon > 0,\ \mu(f^{-1}(B(\lambda, \epsilon))) > 0\}$, where $B(\lambda, \epsilon)$ is the ball in $\mathbb{C}$ of radius ϵ centered in λ;

(v) λ is an eigenvalue if and only if $\mu(f^{-1}(\{\lambda\}) > 0$;

(vi) By (v) and (iv) , isolated points in $\sigma(M_f)$ are necessarily eigenvalues;

(vii) for every Borel subset E of $\mathbb{C}$, let $P(E)\phi = \chi_{f^{-1}(E)}\phi$. Then $P(E)$ is the orthogonal projection on the subspace $\{\phi \in L^2 : (M_f\phi, \phi) \in E\}$;

(viii) $\lambda \in \sigma_{ess}(M_f)$ if and only if for every $\epsilon > 0$ the $\mathcal{R}(P(B(\lambda, \epsilon)))$ is infinite-dimensional.

The spectral theorem allows us to deduce properties of arbitrary self-adjoint operators from those of multiplication operators.

Theorem 3.3. *Let $T : \mathcal{H} \to \mathcal{H}$ be a a self-adjoint operator on a separable Hilbert space $\mathcal{H}$. Then there exist a measure space (X, μ), a unitary operator $U : \mathcal{H} \to L^2(X, \mu)$ (that is U is a bijection and preserves the inner product) and a measurable almost everywhere finite function f on X such that*

$$UTU^{-1} = M_f\,.$$

Using the spectral theorem, one can easily establish the following properties:

(i) $\lambda \in \sigma(T)$ if and only if there exists a sequence $\{\phi_n\} \in \mathcal{D}(T)$ such that $||\phi_n|| = 1$ and $\lim_n ||(\lambda - T)\phi_n|| = 0$.

(ii) $\lambda \in \sigma_{ess}(T)$ if and only if there exists a sequence $\{\phi_n\} \in \mathcal{D}(T)$ such that

  (a) $\lim_n ||(\lambda - T)\phi_n|| = 0$;

  (b) $\{\phi_n\}$ is an orthonormal system (or $\{\phi_n\}$ does not have any converging subsequence).

  In this case $\{\phi_n\}$ is said to be a characteristic sequence for λ.

(iii) The essential spectrum $\sigma_{ess}(T)$ is empty if and only if there exists a complete orthonormal system $\{\phi_n\}$ in $\mathcal{H}$ such that

  (a) each ϕ_n is an eigenfunction belonging to the eigenvalue λ_n;

  (b) $|\lambda_n| \to +\infty$ as $n \to +\infty$.

(iv) T is bounded if and only if $\sigma(T)$ is a bounded subset of $\mathbb{R}$, and T is compact if and only if $\sigma_{ess}(T) = \{0\}$ and there exists a complete orthonormal set $\{\phi_n\}$ of eigenfunctions belonging to the eigenvalue λ_n and $|\lambda_n| \to 0$.

Corollary 3.4. *Let T be a self-adjoint operator, and assume that for some λ the resolvent $(\lambda - T)^{-1}$ is compact. Then T has discrete spectrum consisting of a sequence of eigenvalues repeated according to multiplicity $\{\lambda_n\}$ such that $|\lambda_n| \to +\infty$ as $n \to +\infty$.*

We concentrate our attention on operators bounded from below. A symmetric operator T defined on $\mathcal{D}$ is bounded from below, or semibounded, if there exists $c \in \mathbb{R}$ such that

$$(T\phi, \phi) \geq c||\phi|| \quad \text{for all } \phi \in \mathcal{D}.$$

This condition implies that the dimension $\dim ker(\lambda - T^*)$ is constant in $\mathbb{C}\setminus[c, +\infty)$ and we have the following proposition (see, e.g., [138], pp. 136–137).

Proposition 3.5. *Let T be a semibounded symmetric operator on $\mathcal{D}$. Then T has equal deficiency indices, and therefore admits at least one self-adjoint extension. Furthermore, if $ker(\lambda - T^*) = \{0\}$ for some real λ, then T is essentially self-adjoint on $\mathcal{D}$.*

In the investigations of several questions related to the spectrum of a self-adjoint operator, as well as problems like the existence of self-adjoint extensions of differential operators, quadratic forms associated to the operator are a very useful tool.

A quadratic form defined on a dense domain $\mathcal{D}$ in $\mathcal{H}$ is a map $Q : \mathcal{D} \times \mathcal{D} \to \mathbb{C}$, which is linear in the first entry and conjugate linear in the second entry. We say that Q is

- Hermitian if $Q(\phi, \psi) = \overline{Q(\psi, \phi)}$;
- bounded if $|Q(\phi, \phi)| \leq C||\phi||^2$ for some constant C;
- bounded from below, or semibounded, if it is Hermitian, and $Q(\phi, \phi) \geq c||\phi||^2$ for some c.

If Q is semibounded, the infimum of Q is defined by

$$\inf_{0 \neq \phi \in \mathcal{D}} \frac{Q(\phi, \phi)}{||\phi||^2},$$

and Q is said to be positive if its infimum is ≥ 0.

If Q is semibounded with bound c, then the Hermitian form

$$(\phi, \psi)_Q = Q(\phi, \psi) + (1 - c)(\phi, \psi)_{\mathcal{H}}$$

is an inner product on $\mathcal{D}(Q)$, and Q is said to be closed if $\mathcal{D}$ is a Hilbert space with respect to $(\phi, \psi)_Q$. From the definition of $(\ ,\ )_Q$ it is not difficult to see that a form Q is closed if and only if

$$\begin{aligned} &\text{for every } \{\phi_n\} \in \mathcal{D}(Q) \text{ such that} \\ &\phi_n \to \phi \text{ in } \mathcal{H} \text{ and } Q(\phi_n - \phi_m, \phi_n - \phi_m) \to 0 \text{ as } n, m \to +\infty, \\ &\text{then } \phi \in \mathcal{D}(Q) \text{ and } Q(\phi_n - \phi, \phi_n - \phi) \to 0. \end{aligned} \tag{3.1}$$

A form Q which is bounded from below on a domain $\mathcal{D}$ is said to be closable if there exists a closed form Q' defined on a subspace $\mathcal{D}' \supseteq \mathcal{D}$ in $\mathcal{H}$ which extends Q. Note that the abstract completion of $(\mathcal{D}, (\ ,\ )_Q)$ can always be defined. The issue here is the possibility of identifying it with a subspace of $\mathcal{H}$.

If Q is a semibounded quadratic form, a subspace $\mathcal{D}' \subseteq \mathcal{D}(Q)$ is a core for Q if $\mathcal{D}'$ is dense in $\mathcal{D}(Q)$ with respect to the inner product to $(\ ,\ )_Q$. It is clear from the definition that if Q is a closable form, then $\mathcal{D}(Q)$ is a core for its closure.

Now let T be a self-adjoint operator on $\mathcal{H}$. We may associate to it the Hermitian quadratic form Q defined via the spectral theorem as follows: identify T with the multiplication operator M_f defined on $L^2(X, \mu)$, then

$$\mathcal{D}(Q) = \{\phi \in L^2(X, \mu) \, : \, |f|^{1/2}\phi \in L^2(X, \mu)\}$$
$$Q(\phi, \psi) = \int_X f\phi\bar{\psi}\, d\mu.$$

Since T is self-adjoint, f is real-valued. Then Q is a Hermitian form on the dense domain $\mathcal{D}(Q) \supset \mathcal{D}(T)$ and $Q(\phi, \psi) = (T\phi, \psi)$ if $\phi \in \mathcal{D}(T)$. Further, T is bounded, if and only if Q is bounded, i.e., $|Q(\phi, \phi)| \leq C||\phi||^2$ for some constant C, and T is bounded from below if and only if Q is bounded from below, and we have

$$\inf_{0\neq\phi\in\mathcal{D}(Q)} \frac{Q(\phi, \phi)}{||\phi||^2} = \inf \sigma(T).$$

Note that if T is bounded from below by c, then $T + c$ is non-negative, and it has a unique self-adjoint non-negative square root $(T + c)^{1/2}$, and in this case

$$\mathcal{D}(Q) = \mathcal{D}((T + c)^{1/2}) \quad Q(\phi, \psi) = ((T + c)^{1/2}\phi, (T + c)^{1/2}\psi) - c(\phi, \psi).$$

Furthermore, T can be characterized in terms of Q as follows: $\phi \in \mathcal{D}(T)$ if and only if $\phi \in \mathcal{D}(Q)$ and there exists $k \in \mathcal{H}$ such that $Q(\phi, \psi) = (k, \psi)_{\mathcal{H}}$ for all $\psi \in \mathcal{D}(Q)$ and then $T\psi = k$.

Using (3.1), it is not difficult to see that if Q is the quadratic form associated to the semibounded self-adjoint operator T, then Q is closed. In fact the closed forms are precisely those which arise from semibounded self-adjoint operators.

Theorem 3.6. *(see [41], Theorem 4.4.2) Let Q be a semibounded form defined on a dense domain $\mathcal{D}(Q) \subseteq \mathcal{H}$. Then Q is closed if and only if there exists a unique semibounded self-adjoint operator T on $\mathcal{H}$ such that Q is the quadratic form associated to T.*

Assume now that $T : \mathcal{D}(T) \subseteq \mathcal{H} \to \mathcal{H}$ is a semibounded symmetric operator, and let Q be the quadratic form defined by

$$\mathcal{D}(Q) = \mathcal{D}(T) \quad Q(\psi, \psi) = (T\phi, \psi)_{\mathcal{H}} \; \forall\phi, \psi \in \mathcal{D}(Q).$$

Then Q is a closable, semibounded quadratic form. If $\hat{Q}$ is the closure of Q, then the associated self-adjoint operator $\hat{T}$ is a self-adjoint extension of T called its Friedrichs extension. The operator $\hat{T}$ is semibounded and in fact

$$\inf \sigma(\hat{T}) = \inf_{0\neq\phi\in\mathcal{D}(T)} \frac{(T\phi, \phi)_{\mathcal{H}}}{||\phi||^2_{\mathcal{H}}}.$$

From the above discussion we deduce that

$$\mathcal{D}(\hat{T}) = \{\phi \in \mathcal{D}(Q) \, : \, \exists k \in \mathcal{H} \text{ such that } Q(\phi, \psi) = (k, \psi)_{\mathcal{H}}, \, \forall\psi \in \mathcal{D}(Q)\},$$
$$\hat{T}\phi = k.$$

Further, $\hat{T}$ is the only self-adjoint extension of T with domain contained in $\mathcal{D}(\hat{Q})$.

We now come to the variational formula and minimax which are useful in determining the eigenvalues of a self-adjoint operator which lie below the essential spectrum.

Let T be a self-adjoint operator on $\mathcal{H}$, and let

$$\lambda_o = \inf_{0 \neq \phi \in \mathcal{D}(T)} \frac{(T\phi, \phi)_{\mathcal{H}}}{||\phi^2||_{\mathcal{H}}}, \qquad \Lambda_o = \sup_{0 \neq \phi \in \mathcal{D}(T)} \frac{(T\phi, \phi)_{\mathcal{H}}}{||\phi^2||_{\mathcal{H}}}.$$

It follows easily from the spectral theorem that $\sigma(T) \subseteq [\lambda_o, \Lambda_o]$ and if they are finite they belong to $\sigma(T)$. Further, T is bounded if and only if λ_o and Λ_o are both finite, and T is semibounded if and only if λ_o, which is called the bottom of the spectrum of T, is finite.

When T is bounded from below, the variational characterization of the bottom of the spectrum may be extended to describe the discrete part of $\sigma(T)$.

Let T be a semibounded self-adjoint operator with domain $\mathcal{D}(T) \subseteq \mathcal{H}$. For every finite-dimensional subspace $\mathcal{L} \subseteq \mathcal{D}(T)$ we define

$$\lambda(\mathcal{L}) = \sup\{(T\phi, \phi)_{\mathcal{H}} : \phi \in \mathcal{L}, ||\phi||^2_{\mathcal{H}} = 1\} \tag{3.2}$$

and define

$$\lambda_n = \inf\{\lambda(\mathcal{L}) : \mathcal{L} \subseteq \mathcal{D}(T), \dim \mathcal{L} = n\}. \tag{3.3}$$

It is clear that λ_n is a non-decreasing sequence, and that λ_1 agrees with λ_o as defined above. In fact we have (see [41], Theorems 4.51, 4.5.2, 4.5.3):

Theorem 3.7 (MiniMax). *Maintaining the notation introduced above, we have the following.*

(1) *If $\sigma_{ess}(T) = \emptyset$, then the numbers λ_n are the eigenvalues of T written in increasing order, and repeated according to multiplicity, and $\lambda_n \to +\infty$ as $n \to +\infty$.*

(2) *If $\sigma_{ess}(T) \neq \emptyset$, then two cases may occur.*

  (i) *There exists $A \leq +\infty$, such that $\lambda_n < A$ for every n and $\lambda_n \to A$ as $n \to +\infty$. Then $A = \inf \sigma_{ess}(T)$, $\sigma(T) \cap (-\infty, A) = \{\lambda_n\}$ and λ_n are eigenvalues of T.*

  (ii) *There exists $A < +\infty$ and N such that $\lambda_N < A$ but $\lambda_n = A$ for every $n > N$. Then $A = \inf \sigma_{ess}(T)$, $\sigma(T) \cap (-\infty, A) = \{\lambda_n\}_{n=1}^N$ and $\lambda_1 \dots \lambda_N$ are eigenvalues of T.*

(3) *$\sigma_{ess}(T)$ is empty if and only if $\lambda_n \to \infty$ as $n \to \infty$.*

(4) *Let Q be the form associated to T with domain $\mathcal{D}(Q)$ and let $\mathcal{D}$ be a core Q. For every n let*

$$\lambda'_n = \inf\{\lambda(\mathcal{L}) : \mathcal{L} \subseteq \mathcal{D}(Q), \dim \mathcal{L} = n\},$$
$$\lambda''_n = \inf\{\lambda(\mathcal{L}) : \mathcal{L} \subseteq \mathcal{D}, \dim \mathcal{L} = n\},$$

then

$$\lambda_n = \lambda_n' = \lambda_n''.$$

In particular, if T is essentially self-adjoint on the domain $\mathcal{D}'$, then we may use $\mathcal{D}'$ instead of $\mathcal{D}$ in the definition of λ_n.

We conclude this review with a definition of the index of a self-adjoint operator. Let T be a self-adjoint operator, P the spectral projection associated to T, and let $\mathcal{W}_- = P((-\infty,0))\mathcal{H} = \{\phi \in \mathcal{D}(T) : (T\phi,\phi)_{\mathcal{H}} < 0\}$ be the space where T is negative definite. We define the *Morse index* of T to be the (possibly infinite) number $i\,(T) = \dim \mathcal{W}_-$.

Note that

$$i\,(T) = \sup\{\dim \mathcal{L} : \mathcal{L} \subseteq \mathcal{D}(T) \text{ and } (T\phi,\phi)_{\mathcal{H}} < 0, \forall \phi \in \mathcal{L}\}$$

and that if $\mathcal{D}$ is a form core for the quadratic form associated to T, in particular, if T is essentially self-adjoint on $\mathcal{D}$, then we may use finite-dimensional subspaces of $\mathcal{D}$ in the definition of $i\,(T)$.

If $i\,(T) < +\infty$ we say that T has finite (Morse) index.

Theorem 3.8. *Let T be a self-adjoint operator. Then:*

(i) *If $i\,(T) < +\infty$, then T is bounded from below, $\inf \sigma_{ess}(T) \geq 0$, and the negative part of the spectrum consists of finitely many eigenvalues $\{\lambda_n\}_{n=1}^N$ repeated according to multiplicity, and $i\,(T) = N$.*

(ii) *If T is bounded from below, and $\inf \sigma_{ess}(T) > 0$, then $i\,(T) < +\infty$.*

(iii) *$i\,(T) < +\infty$ if and only if T is bounded from below, and if having defined*

$$\lambda_n = \inf\{\lambda(\mathcal{L}) : \mathcal{L} \subseteq \mathcal{D}(T), \dim \mathcal{L} = n\}$$

we have $\lambda_n < 0$ only for finitely many $n = 1, 2, \dots, N$. In this case $i\,(T) = N$.

Proof. (i) Assume that $i(T) = N < +\infty$, and let $\mathcal{W}_- = P((-\infty,0))\mathcal{H}$, where P is the spectral projection associated to T, so that $\dim \mathcal{W}_- = N$, $(T\phi,\phi) < 0$ for every $\phi \in \mathcal{W}_-$ and $(T\phi,\phi) \geq 0$ for every $\phi \in \mathcal{W}_-^\perp \cap \mathcal{D}(T)$. Since $\mathcal{W}_-$ is finite-dimensional, and T is positive on $\mathcal{W}_-^\perp \cap \mathcal{D}(T)$, it follows immediately that T is bounded from below. Clearly we must have $\inf \sigma_{ess}(T) \geq 0$ (for otherwise the index would be infinite). Moreover, by Theorem 3.7 (2), if $\lambda_n < \inf \sigma_{ess}(T)$, then λ_n is an eigenvalue of T, and therefore $\lambda_n < 0$ for $n = 1, \dots, N$, while $\lambda_{N+1} \geq 0$. This proves (i) .

(ii) is clear, and (iii) follows from (i) and Theorem 3.7. □

3.2 Schrödinger operators on Riemannian manifolds

Let M be a complete Riemannian manifold, let Δ be its (negative semidefinite) Laplace–Beltrami operator, and assume that $q \in L^\infty_{loc}(M)$. For every open set

$\Omega \in M$ we consider the Schrödinger operator

$$L_\Omega = -\Delta + q$$

originally defined on $C_c^\infty(\Omega)$. In the case where $\Omega = M$ we will simply write L instead of L_M. It follows easily from the divergence theorem that L_Ω is a symmetric operator on $C_c^\infty(\Omega)$, which is associated to the quadratic form

$$Q_\Omega(\phi, \psi) = \int_\Omega (\langle \nabla\phi, \nabla\psi \rangle + q\phi\psi).$$

If L_Ω is bounded from below (this is certainly the case if either q is bounded from below or if Ω is bounded), then denote its Friedrichs extension with the same symbol. According to the discussion of Section 1, we have

$$\inf \sigma(L_\Omega) = \inf\{\frac{\int_\Omega (|\nabla\phi|^2 + q\phi^2)}{\int_\Omega \phi^2} : 0 \neq \psi \in C_c^\infty(\Omega)\}.$$

Note that in the case where $\Omega = M$ and M is geodesically complete, if L is bounded from below, then it is automatically essentially self-adjoint on $C_c^\infty(M)$. See Theorem 3.13 below.

Proposition 3.9. *Assume that Ω is a bounded domain. Then L_Ω has purely discrete spectrum consisting of a non-decreasing sequence of eigenvalues, repeated according to multiplicity,*

$$\lambda_1(L_\Omega) \leq \lambda_2(L_\Omega) \leq \cdots \leq \lambda_n(L_\Omega) \to +\infty \quad \textit{as } n \to +\infty.$$

Further if $\Omega \subseteq \Omega'$, then for every n,

$$\lambda_n(L_{\Omega'}) \leq \lambda_n(L_\Omega),$$

and therefore

$$i(L_\Omega) \leq i(L_{\Omega'}).$$

Indeed, if Ω has a sufficiently smooth boundary, by Rellich's Theorem then $(\lambda + L)^{-1}$ is compact if λ is large enough, and the conclusion follows from Corollary 3.4. In the general case, let $\Omega' \supseteq \Omega$ be a bounded domain with smooth boundary. In the notation of Theorem 3.7 (with $\mathcal{D}' = C_c^\infty(\Omega), C_c^\infty(\Omega')$, respectively), it is clear that

$$\lambda_n(L_{\Omega'}) \leq \lambda_n(L_\Omega) \quad \forall n,$$

and since $\lambda_n(L_{\Omega'}) \to \infty$, the conclusion follows from the variational principle Theorem 3.7.

We now turn our attention to unbounded domains. We begin with a slightly improved version of a result of W.F. Moss and J. Piepenbrink, [114], and D. Fischer-Colbrie and R. Schoen, [54] (see [132]) which will play an important role in the sequel.

Lemma 3.10. *Let $(M, \langle\, ,\, \rangle)$ be a Riemannian manifold, and $\Omega \subset M$ be a domain in M and let $q(x) \in L^{\infty}_{loc}(\Omega)$. The following facts are equivalent:*

(i) *There exists $w \in C^1(\Omega)$, $w > 0$, weak solution of*

$$\Delta w - q(x)w = 0 \quad \text{on } \Omega;$$

(ii) *There exists $\varphi \in H^1_{loc}(\Omega)$, $\varphi > 0$, weak solution of*

$$\Delta \varphi - q(x)\varphi \leq 0 \quad \text{on } \Omega;$$

(iii) $\lambda_1(L_\Omega) \geq 0$.

Proof. We sketch the proof, which is a modification of the original proof in ([54]).

It is trivial that (i) implies (ii) . To prove that (ii) implies (iii) , observe that, by our assumptions, φ satisfies

$$\int \langle \nabla \varphi, \nabla \rho \rangle + q(x)\varphi\rho \geq 0, \quad \forall 0 \leq \rho \in H^1_c(\Omega). \tag{3.4}$$

For every $\epsilon > 0$, set $\psi_\epsilon = \log(\varphi + \epsilon) \in H^1_{loc}$. Then, given $f \in C^\infty_c(\Omega)$ we compute

$$\begin{aligned}
\int \langle \nabla \psi_\epsilon, \nabla (f^2) \rangle &= \int \langle \frac{\nabla \varphi}{\varphi + \epsilon}, \nabla (f^2) \rangle \\
&= \int \langle \nabla \varphi, \nabla \big(\frac{f^2}{\varphi + \epsilon}\big) + \frac{f^2}{(\varphi + \epsilon)^2} \nabla \varphi \rangle \\
&\geq - \int q\varphi \frac{f^2}{\varphi + \epsilon} + \int \frac{|\nabla \varphi|^2}{(\varphi + \epsilon)^2} f^2 = - \int q \frac{\varphi}{\varphi + \epsilon} f^2 + \int |\nabla \psi_\epsilon|^2 f^2,
\end{aligned}$$

where we have used inequality (3.4) applied to the non-negative compactly supported H^1 function $f^2/(\varphi + \epsilon)$. We use the inequality

$$\langle \nabla \psi_\epsilon, \nabla (f^2) \rangle \leq 2|f||\nabla f||\nabla \psi_\epsilon| \leq f^2 |\nabla \psi_\epsilon|^2 + |\nabla f|^2$$

to estimate the left-hand side, and simplify to obtain

$$\int |\nabla f|^2 \geq - \int \frac{\varphi}{\varphi + \epsilon} q f^2,$$

and since $q \in L^\infty_{loc}$, $f^2 \in C^\infty_c$ and $|\varphi/(\varphi + \epsilon)| \leq 1$, we may let $\epsilon \to 0+$ and apply the dominated convergence theorem to conclude that

$$\int |\nabla f|^2 \geq - \int q f^2,$$

as required.

We now come to the implication (iii) $\Rightarrow$ (i) . Let $\{D_n\}$ be an exhaustion of Ω by an increasing sequence of relatively compact domains with smooth boundary. Since $q(x) \in L^\infty_{loc}(\Omega)$, and $\lambda_1(L_{D_n}) > 0$, by domain monotonicity, the Dirichlet problem

$$\begin{cases} (-\Delta + q)v = 0 & \text{in } D_n, \\ v = 1 & \text{on } \partial D_n, \end{cases} \tag{3.5}$$

has a solution v_n which belongs to $C^{0,\alpha}(\overline{D}_n) \cap H^2(D_n)$ for some $0 < \alpha < 1$ (see, e.g., Theorems 8.6, 8.12 and 8.29 in [60]). Moreover, it follows from Theorem 1.1 in [159] (see also [59], Chap VII, Theorem 1.2) that $v_n \in C^{1,\beta}_{loc}(D_n)$ for some $0 < \beta < 1$ independent of n.

We claim that $v_n > 0$ in D_n. By the maximum principle (see [60], p. 35, and note that the result extends to functions in C^1 using a comparison argument modelled, e.g., on [131], Prop. 6.1) it suffices to show that $v_n \geq 0$. Assume by contradiction that $B_n = \{x \in D_n : v_n(x) < 0\} \neq \emptyset$. Then, by the boundary condition, $B_n \subset\subset D_n$ and $v_n^- = \min\{v_n, 0\} \in H^1_c(D_n)$ is a weak solution of the differential inequality $(-\Delta + q)v_n^- \geq 0$. Using the non-zero function $-v_n^-$ as test function, we obtain

$$\int |\nabla v_n^-|^2 + q(x)(v_n^-)^2 \leq 0,$$

contradicting the positivity of $\lambda_1(L_{D_n})$.

Now fix $x_o \in D_0$, and let

$$w_n(x) = \frac{v_n(x)}{v_n(x_o)},$$

so that $w_n \in C^{0,\alpha}(\overline{D}_n) \cap C^{1,\beta}_{loc}(D_n)$. Furthermore, according to Theorem 8.20 in [60] and Theorem 1.1 in [159], for every n there exists a constant C_n such that for every $k > n$,

$$\begin{gathered} C_n^{-1} \leq w_k(x) \leq C_n, \\ |\nabla w_n| \leq C_n, \\ |\nabla w_k(x) - \nabla w_k(y)| \leq C_n d(x,y)^\beta \end{gathered}$$

for every $x, y \in \overline{D}_n$.

The Ascoli-Arzelá theorem and a diagonal argument yield a subsequence $\{w_{n_j}\}$ which converges in $C^1_{loc}(\Omega)$ to a C^1 function w which is a weak solution of $(-\Delta + q)w = 0$. Since $w_n(x_o) = 1$ for every n, $w(x_o) = 1$, and, again by the maximum principle, $w > 0$ on Ω. □

Remark 3.11. As the proof shows, the function w belongs in fact to the space $C^{1,\beta}_{loc}(\Omega)$ for some $0 \leq \beta < 1$. Further, if we assume that $q(x) \in C^{0,\alpha}_{loc}(\Omega)$ for some $0 < \alpha < 1$, then $w \in C^{2,\alpha}_{loc}(\Omega)$ (see, e.g., [9], Theorem 3.55) and it is therefore a classical solution of $\Delta w - q(x)w = 0$ on Ω. Finally, it is easy to see that the equivalences (i) – (iii) extend to the case where Ω is replaced by the exterior $M \setminus K$ of a compact set K, thus extending Proposition 1 in [53].

Our first application of the above lemma is the following theorem that shows that $-\Delta + q$ is bounded from below on M if and only if it bounded from below on some exterior domain (see [13], Prop.1).

Theorem 3.12. *Let M be a Riemannian manifold, then the following are equivalent:*

(i) *L is bounded from below on $C_c^\infty(M)$.*

(ii) *For any relatively compact domain Ω the operator $L_{M\setminus\bar\Omega}$ is bounded from below on $C_c^\infty(M \setminus \bar\Omega)$.*

(iii) *There exists a relatively compact domain Ω such that the operator $L_{M\setminus\bar\Omega}$ is bounded from below on $C_c^\infty(M \setminus \bar\Omega)$.*

Proof. The only implication that requires proof is that (iii) implies (i). By assumption there exists $A \in \mathbb{R}$ such that

$$\int |\nabla\phi|^2 + (q + A)\phi^2 \geq 0, \quad \forall \phi \in C_c^\infty(M \setminus \bar\Omega). \tag{3.6}$$

By replacing q with $q + A$ we may assume that $A = 0$.

Next let Ω', be a relatively compact domain with smooth boundary such that $\Omega \subset\subset \Omega'$, and let $q_1 \in L^\infty_{loc}(M)$ be a function such that

$$q_1 \in C^\infty(\Omega'), \quad q_1 \leq q \text{ on } M, \quad \text{and} \quad q = q_1 \text{ on } M \setminus \tilde\Omega,$$

where $\tilde\Omega$ is a relatively compact domain such that $\Omega' \subset\subset \tilde\Omega$. Note that since $q_1 - q$ is bounded and compactly supported on M, then $-\Delta + q_1 = -\Delta + q + (q_1 - q)$ is bounded from below on $C_c^\infty((M \setminus \bar\Omega)$. Furthermore, since $q_1 - q \leq 0$ on M, then if $-\Delta + q_1$ is bounded from below on $C_c^\infty(M)$, so is $-\Delta + q = -\Delta + q_1 - (q_1 - q)$.

Thus it suffices to prove the assertion under the additional assumption that q is smooth on Ω'.

By (1) and the previous lemma there exists $v \in C^1(M \setminus \bar\Omega)$ such that $v > 0$ and $(-\Delta + q)v = 0$ weakly in $M \setminus \bar\Omega$, and, since $q \in C^\infty(\bar\Omega')$, then $v \in C^\infty(\Omega' \setminus \bar\Omega)$. Choose Ω'' such that $\Omega \subset\subset \Omega'' \subset\subset \Omega'$. By modifying v in $\Omega'' \setminus \bar\Omega$, we may extend v to a function on M, which we still denote v, such that

$$v \in C^1(M) \cap C^\infty(\Omega'), \quad v > 0 \text{ in } M, \quad (-\Delta + q)v = 0 \text{ weakly on } M \setminus \Omega''.$$

Finally, choose a smooth cut-off function σ such that $\sigma = 1$ in a neighborhood of Ω'' and $\operatorname{supp}\sigma \subset\subset \Omega'$.

Now let $\phi \in C_c^\infty(M)$, set $w = \log v$ and compute

$$\begin{aligned}\int \langle \nabla w, \nabla(\phi^2)\rangle + q\phi^2 &= \int \langle \nabla v, \nabla(\frac{\phi^2}{v}) + \frac{\phi^2}{v^2}\nabla v\rangle + q\phi^2 \\ &= \int \phi^2|\nabla w|^2 + \big[\langle \nabla v, \nabla(\frac{\phi^2}{v})\rangle + q\phi^2\big].\end{aligned}$$

Since $\langle \nabla w, \nabla(\phi^2)\rangle \leq \phi^2|\nabla w|^2 + |\nabla\phi|^2$, inserting into the above identity, simplifying and writing $\phi^2 = (1-\sigma)\phi^2 + \sigma\phi^2$ we deduce that

$$\begin{aligned}\int |\nabla\phi|^2 + q\phi^2 &= \int \langle\nabla v, \nabla\big(\frac{\phi^2}{v}\big)\rangle + q\phi^2\\ &= \int \langle\nabla v, \nabla\big(\frac{(1-\sigma)\phi^2}{v}\big)\rangle + qv\frac{(1-\sigma)\phi^2}{v}\\ &\quad+ \int \langle\nabla v, \nabla\big(\frac{\sigma\phi^2}{v}\big)\rangle + qv\frac{\sigma\phi^2}{v}.\end{aligned}$$

Since $\frac{(1-\sigma)\phi^2}{v}$ is C^1, non-negative, and compactly supported in $M\setminus\Omega''$, the first integral on the right-most side vanishes by the assumption that v satisfies $(-\Delta + q)v = 0$ weakly on $M\setminus\Omega''$. On the other hand, since $\frac{\sigma\phi^2}{v}$ is compactly supported in Ω' and v is there smooth, integrating by parts we see that the second integral is equal to

$$\int [-\Delta v + qv]\frac{\sigma\phi^2}{v} \geq -B\int \phi^2$$

where $B = \min(-\Delta v + qv)\sigma/v > -\infty$. Thus

$$\int |\nabla\phi|^2 + q\phi^2 \geq -B\int |\phi|^2$$

as required to show that $-\Delta + q$ is bounded on $C_c^\infty(M)$. □

Our next result shows that if M is complete and $L = -\Delta + q$ is bounded from below on $C_c^\infty(M)$, then L is essentially self-adjoint. This extends a classical result for the Laplace operator due to M.P. Gaffney [56], and later obtained with different methods by P. Chernoff, [37], and R.S. Strichartz [155]. The proof we present, which is taken from [13], Prop. 2 i), is an adaptation of Strichartz's proof.

Theorem 3.13. *Let $(M, \langle\,,\rangle)$ be a complete Riemannian manifold, and assume that $L = -\Delta + q$ is bounded from below on $C_c^\infty(M)$. Then L is essentially self-adjoint on $C_c^\infty(M)$.*

Proof. We may assume, without loss of generality, that L is bounded from below by 1, that is

$$\int |\nabla\phi|^2 + q\phi^2 \geq \int \phi^2, \quad \forall\phi\in C_c^\infty(M). \tag{3.7}$$

Since L is symmetric and densely defined, according to Proposition 3.5 it suffices to show that $ker\, L^* = \{0\}$. Note that the adjoint L^* is defined with domain $\mathcal{D}(L^*) = \{f\in L^2(M) : \exists B>0 : |\int_M f(-\Delta+q)\phi\,dx| \leq B||\phi||_{L^2}\,\forall\phi\in C_c^\infty(M)\}$. In particular, if $f\in\mathcal{D}(L^*)$, then the distribution $(-\Delta+q)f$ is in L^2 and equals L^*f.

Assume therefore that $f\in Ker\,(L^*)$, i.e., $(-\Delta+q)f = 0$. By elliptic regularity, $f\in L^2(M)\cap H^1_{loc}(M)$. Since M is complete, given $o\in M$, there exists a family of smooth cut-off functions $\phi_{r,s}$, $0 < r < s < +\infty$, such that $\phi_{r,s} = 1$ on the ball $B(o,r)$, $\phi_{r,s} = 0$ in $M \setminus B(o,s)$ and $|\nabla \phi_{r,s}| \le C/(s-r)$ for some constant $C > 0$ independent of r, s. Since $f\phi^2_{r,s}$ is compactly supported and belongs to $H^1(M)$, by the definition of the weak equality $(-\Delta + q)f = 0$ we have

$$0 = \int_M \langle \nabla f, \nabla (f\phi^2_{r,s})\rangle + qf^2\phi^2_{r,s} = \int |\nabla (f\phi_{r,s})|^2 + q(f\phi_{r,s})^2 - \int f^2 |\nabla \phi_{r,s}|^2$$
$$\ge \int (f\phi_{r,s})^2 - \frac{C^2}{(s-r)^2}\int f^2,$$

where we have used (3.7) and the properties of $\nabla\phi_{r,s}$. Setting $s = 2r$, and letting $r \to \infty$ we deduce that

$$0 \ge \int f^2,$$

and therefore $f = 0$ in $L^2(M)$, as required. □

Our next task consists in describing the bottom of the essential spectrum of L. We will need the following technical lemma (see [13], p. 317–318).

Lemma 3.14. *Let $(M, \langle\,,\rangle)$ be a complete Riemannian manifold, and assume that L is bounded from below. Let $\lambda \in \sigma_{ess}(L)$ and let K, K' be compact sets in M such that K' is compactly contained in the interior* Int$K$ *of K. Then there exists a characteristic sequence $\{\phi_n\} \subset C^\infty_c(M)$ for L relative to λ, a subsequence $\{\phi'_n\}$ of $\{\phi_n\}$ and $\phi \in L^2(K)$ such that*

(i) $\phi'_n \to \phi$ *in* $L^2(K)$;

(ii) $\omega_n = \phi'_{2n+1} - \phi'_{2n} \to 0$ *in* $L^2(K)$;

(iii) $\nabla\omega_n \to 0$ *in* $L^2(K')$;

(iv) ω_n *is a characteristic sequence for L relative to λ.*

Proof. Let u_n be a characteristic sequence relative to λ. Since L is essentially self-adjoint on $C^\infty_c(M)$, we may find a sequence $\phi_n \in C^\infty_c(M)$ such that $||\phi_n||_2 = 1$ $\forall n$, $||u_n - \phi_n||_2 \le 1/n$ and $||L(u_n - \phi_n)||_2 \le 1/n$. It is clear then that

$$||(L-\lambda)\phi_n||_2 \to 0 \text{ and } (\phi_n, \phi_m)_{L^2} \le 2\max\{1/n, 1/m\}, \tag{3.8}$$

showing that $\{\phi_n\}$ is a characteristic sequence relative to λ.

Let $\chi \in C^\infty_c(M)$ be a cut-off function such that $\chi \equiv 1$ in a neighborhood K_1 of K. Since $\chi^2\phi_n$ is bounded in L^2 and $(L-\lambda)\phi_n \to 0$ in L^2, $(\chi^2\phi_n, (L-\lambda)\phi_n)| \to 0$ as $n \to \infty$. In particular, we may assume that

$$-\int \chi^2\phi_n\Delta\phi_n + \int (q-\lambda)\chi^2\phi^2_n \le 1 \tag{3.9}$$

for every n. On the other hand, integrating by parts,

$$-\int \chi^2\phi_n\Delta\phi_n = \int \langle\nabla(\chi^2\phi_n), \nabla\phi_n\rangle = \int \chi^2|\nabla\phi_n|^2 + 2\chi\phi_n\langle\nabla\chi, \nabla\phi_n\rangle$$
$$\geq \int (\frac{1}{2}\chi^2|\nabla\phi_n|^2 - 2\chi\phi_n|\nabla\chi||\nabla\phi_n| + 2\phi_n^2|\nabla\chi|^2)$$
$$+ \int (\frac{1}{2}\chi^2|\nabla\phi_n|^2 - 2\phi_n^2|\nabla\chi|^2) \geq \int (\frac{1}{2}\chi^2|\nabla\phi_n|^2 - 2\phi_n^2|\nabla\chi|^2)$$

whence, substituting in (3.9) and rearranging, we obtain

$$\frac{1}{2}\int \chi^2|\nabla\phi_n|^2 \leq 2\int \phi_n^2|\nabla\chi|^2 - \int (q-\lambda)\chi^2\phi_n^2 + 1$$
$$\leq 1 + C\int_{\text{supp }\chi} \phi_n^2.$$

Since χ is equal to 1 in the neighborhood K_1 of K, it follows that there exists a constant C' such that

$$\int_{K_1} |\nabla\phi_n|^2 \leq C, \quad \forall n,$$

and therefore ϕ_n is uniformly bounded in $L^2(K_1)$. By Rellich's Theorem a subsequence $\{\phi_n'\}$ of $\{\phi_n\}$ converges to a function ϕ in $L^2(K_1)$. This proves (i) , and therefore (ii) .

Since $(L-\lambda)\phi_n \to 0$ in L^2, $(L-\lambda)\omega_n \to 0$, as $n \to \infty$, and, by (3.8),

$$||\omega_n||_2^2 = 2 - 2(\phi'_{2n+1}, \phi'_{2n}) \geq 2 - \frac{2}{n},$$
$$|(\omega_n, \omega_m)| \leq 4\max\{1/n, 1/m\}, \text{ if } m \neq n,$$

as needed to show that ω_n is a characteristic sequence relative to λ.

Finally, to prove (iii) we argue as in the first part of the proof. We choose a cut-off function χ_1 which is supported in K and equal to 1 in a neighborhood of K', and use the fact that $(L-\lambda)\omega_n \to 0$, $\omega_n \to 0$ in $L^2(K)$ to conclude that there exists a constant C_1 such that

$$\frac{1}{2}\int_{K'} |\nabla\omega_n|^2 \leq \frac{1}{2}\int \chi_1^2|\nabla\omega_n|^2 \leq C_1\int_K \omega_n^2 \to 0 \quad \text{as } n \to \infty. \qquad \square$$

We can now prove the following theorem that relates the bottom of the essential spectrum of L to the bottom of the spectrum of the Dirichlet Schrödinger operator on exterior domains.

Theorem 3.15. *Let $(M, \langle\ ,\ \rangle)$ be a complete Riemannian manifold, and assume that L is bounded from below. For every relatively compact domain Ω, denote by $L_{M\setminus\bar\Omega}$*

the Friedrichs extension of $-\Delta + q$ *originally defined on* $C_c^\infty(M\setminus\bar\Omega)$*. Then*

$$\begin{aligned}\inf \sigma_{ess}(L) &= \sup_{\Omega\subset\subset M} \inf \sigma(L_{M\setminus\bar\Omega}) \\ &= \sup_{\Omega\subset\subset M}\ \inf_{0\neq\phi\in C_c^\infty(M\setminus\bar\Omega)} \frac{\int |\nabla\phi|^2 + q\phi^2}{||\phi||_2^2}.\end{aligned} \tag{3.10}$$

Proof. We begin by noting that the second equality in (3.10) follows from general properties of the Friedrichs extension. Denote by λ_e and δ_e the left- and right-hand side of (3.10), respectively, which are both finite since L and therefore $L_{M\setminus\bar\Omega}$ are bounded from below. Note also that $\lambda_e \in \sigma_{ess}(L)$ since the latter is closed.

We first show that $\lambda_e \le \delta_e$. Indeed, assume by contradiction that $\lambda_e > \delta_e$ and let δ be such that $\delta_e < \delta < \lambda_e$. By definition of δ_e, for every $\Omega \subset\subset M$,

$$\inf_{0\neq\phi\in C_c^\infty(M\setminus\bar\Omega)} \frac{\int |\nabla\phi|^2 + q\phi^2}{||\phi||_2^2} < \delta$$

and therefore there exists a non-zero $\phi_\Omega \in C_c^\infty(M\setminus\bar\Omega)$ such that

$$\int |\nabla\phi_\Omega|^2 + q|\phi_\Omega|^2 < \delta||\phi_\Omega||_2^2.$$

We may choose an exhaustion of M by relatively compact domains Ω_k such that for every k the function $\phi_{\Omega_k} = \phi_k$ is supported in $\Omega_{k+1}\setminus\bar\Omega_k$. Since the functions ϕ_k have disjoint support, they span an infinite-dimensional subspace $\mathcal{E}$ of $C_c^\infty(M)$ where $L \le \delta$ in the sense of quadratic forms. By the minimax principle, L has infinitely many eigenvalues below δ. But this is impossible since $\delta < \lambda_e$ and L is bounded from below.

Next assume by contradiction that $\delta_e > \lambda_e$ and let δ be such that $\lambda_e < \delta < \delta_e$. Using the definition of δ_e we deduce that there exists a relatively open domain Ω such that

$$\inf \sigma(L_{M\setminus\bar\Omega}) \ge \delta > \lambda_e. \tag{3.11}$$

We are going to reach a contradiction, showing that λ_e belongs to $\sigma(L_{M\setminus\bar\Omega})$ (in fact to $\sigma_{ess}(L_{M\setminus\bar\Omega})$.

Let K, K' be compact sets such that $\bar\Omega \subset\subset Int(K') \subset K' \subset\subset Int(K) \subset K$. Since $\lambda_e \in \sigma_{ess}(L)$, we may find a characteristic sequence ω_n for L relative to λ_e with the properties listed in Lemma 3.14. Let θ be a smooth cut-off function such that $\theta = 1$ on a neighborhood of $\bar\Omega$ and supp $\theta \subset K'$. Finally let $\psi_n = (1-\theta)\omega_n$. It is clear that $\psi_n \in C_c^\infty(M\setminus\bar\Omega)$ and therefore it belongs to the domain of $L_{M\setminus\bar\Omega}$. We have

$$\begin{aligned}||(L-\lambda_e)\psi_n||_2 &\le ||(1-\theta)(L-\lambda_e)\omega_n||_2 + 2||\langle\nabla\theta,\nabla\omega_n\rangle||_2 + ||\omega_n\Delta\theta||_2 \\ &\le ||(L-\lambda)\omega_n||_2 + 2C\Big(\int_{K'}|\nabla\omega_n|^2\Big)^{1/2} + C\Big(\int_{K'}\omega_n^2\Big)^{1/2}\end{aligned}$$

and the right-hand side tends to zero as $n \to \infty$ by the properties of ω_n. Furthermore

$$(\psi_n, \psi_m)_{L^2} = \int_M \omega_m \omega_n - \int_M \theta(2-\theta)\omega_n\omega_m,$$

and

$$\Big|\int_M \theta(2-\theta)\omega_n\omega_m\Big| \leq 2\Big(\int_{K'} \omega_m^2\Big)^{1/2}\Big(\int_{K'} \omega_n^2\Big)^{1/2} \to 0 \text{ as } n, m \to \infty,$$

while (see the proof of the lemma)

$$\Big|\int_M \omega_m\omega_n\Big| \leq 4\max\{\frac{1}{n}, \frac{1}{m}\}, \text{ if } m \neq n, \text{ and } \int \omega_n^2 \geq 2 - \frac{2}{n}.$$

Thus ψ_n is a characteristic sequence for $L_{M\setminus\bar{\Omega}}$ relative to λ_e which therefore belongs to $\sigma_{ess}(L_{M\setminus\bar{\Omega}})$, as required to complete the proof. □

Our last task is to relate the index of the operator L with that of L_Ω. We define a "generalized" Morse index by

$$\tilde{i}(L) = \sup\{i(L_\Omega) : \Omega \subset\subset M\}.$$

Recall that since L_Ω has purely discrete spectrum, $i(L_\Omega)$ is the number, counted according to multiplicity, of the negative eigenvalues. Further, since $C_c^\infty(\Omega)$ is a core for the quadratic form Q_Ω associated to L_Ω,

$$i(L_\Omega) = \sup\{\dim \mathcal{L} : \mathcal{L} \subset C_c^\infty(\Omega) \text{ such that } (L_\Omega\phi, \phi)_{L^2} < 0, \forall \phi \in \mathcal{L}\}. \quad (3.12)$$

Similarly, if L is essentially self-adjoint on $C_c^\infty(M)$, then

$$i(L) = \sup\{\dim \mathcal{L} : \mathcal{L} \subset C_c^\infty(M) \text{ such that } (L\phi, \phi)_{L^2} < 0, \forall \phi \in \mathcal{L}\}. \quad (3.13)$$

Then we have the following result of Gulliver, [77], and Fisher–Colbrie, [53].

Lemma 3.16. *Let $(M, \langle\, ,\rangle)$ be a complete Riemannian manifold, and assume that $\tilde{i}(L) < +\infty$. Then there exists a compact set K such that $\lambda_1(L_{M\setminus K}) \geq 0$.*

Proof. Fix a reference point $o \in M$, and let $B_r(o)$ be the geodesic ball of radius r centered at o. Since $B_r(o)$ is almost Euclidean for sufficiently small r, and the bottom of the spectrum of the Dirichlet Laplacian of $\mathbb{R}^n$ on the ball B_R of radius R grows like R^{-2}, as is easily seen by a scaling argument, we deduce that there exists a constant C such that if r is sufficiently small,

$$\int_{B_r(o)} |\nabla\phi|^2 + q\phi^2 \geq \Big(\frac{C}{r^2} - \sup_{B_r(o)} |q|\Big)\int_{B_r(o)} \phi^2,$$

for all $\phi \in C_c^\infty(B_r(o)$, and therefore,

$$\lambda_1(L_{B_r(o)}) > 0 \quad \text{for } r \text{ small enough.}$$

Let

$$r_1 = 2\sup\{r : \lambda_1(L_{B_r(o)}) > 0\}.$$

If $r_1 = +\infty$, then $\lambda_1(M) \geq 0$, and we are done. Otherwise, note that if $r_1 < r < 2r_1$, then the Poincaré inequality on annuli (see the more general Theorem 5.8 of Chapter 5), shows that there exists a constant $C = C(B_{2r_1}(o))$ such that

$$\int_{B_r(o)\setminus\overline{B_{r_1}(o)}} |\nabla\phi|^2 + q\phi^2 \geq \left(\frac{C}{(r-r_1)^2} - \sup_{B_r(o)\setminus\overline{B_{r_1}(o)}} |q|\right) \int_{B_r(o)\setminus\overline{B_{r_1}(o)}} \phi^2,$$

for all $\phi \in C_c^\infty(B_r(o) \setminus \overline{B_{r_1}(o)})$, and therefore,

$$\lambda_1(L_{B_r(o)\setminus\overline{B_{r_1}(o)}}) > 0 \quad \text{for } r \text{ close enough to } r_1.$$

Set

$$r_2 = 2\sup\{r > r_1 : \lambda_1(L_{B_r(o)\setminus\overline{B_{r_1}(o)}}) > 0\}.$$

Again, if $r_2 = +\infty$ we are done. Otherwise $r_1 < r_2 < +\infty$, and, by (strict) domain monotonicity,

$$\lambda_1(L_{B_{r_2}(o)\setminus\overline{B_{r_1}(o)}}) < 0.$$

Iterating the construction we find a sequence $r_1 < r_2 < \cdots < r_k$ such that

$$\lambda_1(L_{B_{r_i}(o)\setminus\overline{B_{r_{i-1}}(o)}}) < 0.$$

For every i let $\phi_i \in C_c^\infty(B_{r_i}(o) \setminus \overline{B_{r_{i-1}}(o)})$ be such that

$$Q_L(\phi_i, \phi_i) = \frac{\int |\nabla\phi_i|^2 + q\phi_i^2}{||\phi_i||^2_{L^2}} < 0.$$

Thus Q is negative definite on the space $\mathcal{W} = \operatorname{span}(\phi_i)$, which is $(k-1)$-dimensional since the functions $\phi_i \in C_c^\infty(B_{r_k}(o))$ have disjoint support, and by definition

$$\tilde{i}(L) \geq i(L_{B_{r_k}}) \geq k-1.$$

It follows that the sequence r_k must be finite, and that

$$\lambda_1(L_{B_r(o)\setminus\overline{B_{r_k}(o)}}) > 0, \quad \forall r > r_k, \quad \text{i.e.,} \quad \lambda_1(L_{M\setminus\overline{B_{r_k}(o)}}) \geq 0.$$

□

Theorem 3.17. *Let $(M, \langle\, ,\, \rangle)$ be a complete Riemannian manifold.*

(i) *If L is essentially self-adjoint on $C_c^\infty(M)$, then $\tilde{i}(L) = i(L)$.*

(ii) *If $\tilde{i}(L) < +\infty$, then L is bounded from below, essentially self-adjoint on $C_c^\infty(M)$ and $i(L) = \tilde{i}(L) < +\infty$.*

Proof. (i) By assumption L is essentially self-adjoint on $C_c^\infty(M)$, and therefore

$$i(L) = \sup\{\dim \mathcal{L} : \mathcal{L} \subset C_c^\infty(M) \text{ such that } (L\phi, \phi)_{L^2} < 0\, \forall \phi \in \mathcal{L}\}.$$

Thus, if $N \le i(L)$ there exists $\mathcal{L} \subset C_c^\infty(M)$ such that $\dim \mathcal{L} = N$ and $(L\phi, \phi) < 0$ for all $\phi \in \mathcal{L}$. Since $\mathcal{L}$ is finite-dimensional, there exists $\Omega \subset\subset M$ such that $\operatorname{supp}\phi \subset \subset \Omega$ for every $\phi \in \mathcal{L}$, and therefore $i(L_\Omega) \ge N$. By definition $\tilde{i}(L) \ge i(L_\Omega) \ge N$, and we conclude that $\tilde{i}(L) \ge i(L)$.

To prove the reverse inequality suppose that $\tilde{i}(L) \ge N$. Again by definition, there exists $\Omega \subset\subset M$ such that $i(L_\Omega) \ge N$. Thus there exist $\mathcal{L} \in C_c^\infty(\Omega)$ such that $\dim \mathcal{L} = N$ and $(L\phi, \phi)_{L^2} < 0$ for every $\phi \in \mathcal{L}$, and clearly $i(L) \ge N$. Thus $i(L) \ge \tilde{i}(L)$.
(ii) In view of (i) it suffices to show that L is bounded from below on $C_c^\infty(M)$.

According to Lemma 3.16, there exists R_0 such that $\lambda_1(L_{M\setminus\overline{B_{R_0}(o)}}) \ge 0$, i.e., for every $\phi \in C_c^\infty(M \setminus \overline{B_{R_0}(o)})$ we have

$$Q(\phi, \phi) = \int |\nabla \phi|^2 + q\phi^2 \ge 0. \tag{3.14}$$

Fix $R > R_o$ suitably large, and let $0 \le \chi \le 1$ be a smooth function such that $\chi = 0$ on $B_R(o)$, $\chi = 1$ on $M \setminus B_{2R}(o)$ and $|\nabla \chi| \le 2/R$. Set $\eta = 1 - (1-\chi)^2$; then $\eta = 0$ on $B_R(o)$, $\eta = 1$ on $M \setminus B_{2R}(o)$ and $\nabla \eta = -2(1-\chi)\nabla\chi$ so that

$$|\nabla \eta|^2 = 4(1-\chi)^2 |\nabla \chi|^2 \le \frac{16}{R^2}(1-\eta) \le \frac{16}{R^2}(1-\eta^2). \tag{3.15}$$

Given $\phi \in C_c^\infty(M)$ we write

$$Q(\phi, \phi) = \int |\nabla \phi|^2 + q\phi^2 = \int (1-\eta^2)(|\nabla\phi|^2 + q\phi^2) + \int \eta^2(|\nabla \phi|^2 + q\phi^2). \tag{3.16}$$

Since

$$\begin{aligned} \eta^2 |\nabla \phi|^2 &= |\nabla(\eta\phi)|^2 - \phi^2 |\nabla \eta|^2 - 2\eta\phi \langle \nabla \phi, \nabla \eta \rangle \\ &\ge |\nabla(\eta\phi)|^2 - \phi^2(1 + |\nabla\eta|^2) - \eta^2 |\nabla \eta|^2 |\nabla \phi|^2, \end{aligned}$$

we may estimate

$$\begin{aligned} \int \eta^2(|\nabla\phi|^2 + q\phi^2) &\ge Q(\eta\phi, \eta\phi) - \int (1 + |\nabla \eta|^2)\phi^2 - \int \eta^2 |\nabla\eta|^2 |\nabla \phi|^2 \\ &\ge -C \int \phi^2 - \frac{16}{R^2} \int (1-\eta^2)|\nabla\phi|^2 \end{aligned}$$

where we have used the fact that, since $\eta\phi \in C_c^\infty(M \setminus \overline{B_{R_0}(o)})$, $Q(\eta\phi, \eta\phi) \ge 0$ by (3.14), and (3.15). Inserting into (3.16) we have

$$Q(\phi, \phi) \ge (1 - \frac{16}{R^2}) \int (1-\eta^2)|\nabla \phi|^2 - (C + \sup_{B_{2R}(o)} |q|) \int \phi^2 \ge -C' \int \phi^2$$

provided $R \ge 4$, with $C' = (C + \sup_{B_{2R}(o)} |q|)$. Thus L is bounded from below on $C_c^\infty(M)$ as required. ∎

As a corollary we have a partial converse of Lemma 3.16.

Corollary 3.18. *Let* $(M, \langle\,,\,\rangle)$ *be a complete Riemannian manifold. Assume that there exists a relatively compact domain* Ω *such that*

$$\lambda_1(L_{M\setminus\overline{\Omega}}) > 0.$$

Then $\tilde{i}(L) < +\infty$.

Proof. Indeed, since $L_{M\setminus\overline{\Omega}}$ is bounded from below, L is bounded from below, and essentially self-adjoint on $C_c^\infty(M)$. Since

$$\inf \sigma_{ess}(L) = \sup_{\Omega \subset\subset M} \lambda_1(L_{M\setminus\overline{\Omega}}) > 0,$$

we conclude that $i(L) = \tilde{i}(L) < +\infty$. □

Observe that the conclusion of Lemma 3.16 is that if $\tilde{i}(L) < +\infty$, then there exists a relatively compact domain Ω such that $\lambda_1(L_{M\setminus\overline{\Omega}}) \geq 0$. In this respect it should be remarked that in general the strict domain monotonicity fails for the bottom of the spectrum of L on external domains, as the case of the Dirichlet Laplacian on external domains of $\mathbb{R}^n$ easily shows.

We also observe that, if we assume that for some $p > 2$ an L^2-Sobolev inequality of the form

$$S_p^{-1}\Big(\int v^{\frac{2p}{p-2}}\Big)^{\frac{p-2}{p}} \leq \int |\nabla v|^2 \tag{3.17}$$

holds for every $v \in C_c^\infty(M)$, and that the negative part of the potential q_- belongs to $L^{\frac{p}{2}}(M)$, then an easy argument shows that $\lambda_1(L_{M\setminus K}) \geq 0$ provided the compact K is large enough. Indeed, for every $v \in C_c^\infty(M \setminus K)$ we have

$$\int q_- v^2 \leq \Big(\int_{M\setminus K} q_-^{\frac{p}{2}}\Big)^{\frac{2}{p}} \Big(\int v^{\frac{2p}{p-2}}\Big)^{\frac{p-2}{p}} \leq S_p^{-1}\Big(\int_{M\setminus K} q_-^{\frac{p}{2}}\Big)^{\frac{2}{p}} \int |\nabla v|^2.$$

Since $q_- \in L^{\frac{p}{2}}$, we may choose K so large that

$$S_p^{-1}\Big(\int_{M\setminus K} q_-^{\frac{p}{2}}\Big)^{\frac{2}{p}} \leq 1,$$

and therefore

$$\int |\nabla v|^2 + qv^2 \geq \int |\nabla v|^2 - q_- v^2 \geq 0,$$

as required. In particular, under the present assumptions, $\inf \sigma_{ess}(L) \geq 0$.

In fact, G.V. Rozenbljum, [141], M. Cwikel, [39], E. Lieb, [109], were able to estimate the index of the operator $-\Delta + q$ on $\mathbb{R}^m$ in terms of a multiple of the $L^{m/2}$ norm of q_-. Later, Li and Yau, [106] improved the constant, expressing it in terms of the Sobolev constant S_p. We state below the result in the form given by Li and Yau.

Theorem 3.19. *Let $(M, \langle\, , \rangle)$ be a complete Riemannian manifold, and assume that the L^2-Sobolev inequality (3.17) holds for some $p > 2$, some constant $S_p > 0$ and every $v \in C_c^\infty(M)$. Assume that $q \in L^\infty_{loc}(M)$ is such that*

$$q_-(x) = \max(-q(x), 0) \in L^{\frac{p}{2}}(M).$$

Then the Schrödinger operator $L = -\Delta + q$ is bounded from below, essentially self-adjoint and has non-negative essential spectrum. Further, there exists a constant $C = C(m, S_p) > 0$ such that

$$i(L) \leq C \|q_-\|_{L^{\frac{p}{2}}(M)}.$$

Chapter 4

Vanishing results

4.1 Formulation of the problem

As we mentioned in the introduction, the aim of this book is to present a unified approach to different geometrical questions such as the study of the constancy of harmonic maps, the topology at infinity of submanifolds, the L^2-cohomology, and the structure and rigidity of Riemannian and Kählerian manifolds (see Sections 6.1, 7.4, 7.5, 7.6, 8.1, and Appendix B).

The common feature of most of these problems lies in the fact that one identifies a suitable function ψ whose vanishing or, more generally, constancy, is the analytic counterpart of the desired geometric conclusion, and, using the peculiarities of the geometric data, one shows that the function ψ satisfies a differential inequality of the form

$$\psi\Delta\psi + a(x)\psi^2 + A|\nabla\psi|^2 \geq 0 \tag{4.1}$$

weakly on M, as well as some suitable non-integrability condition. Typically, ψ represents the length of a section of a suitable vector bundle.

This is reminiscent of Bochner's original method: in the compact case, and under appropriate assumptions on the sign of the function ψ and of the coefficient $a(x)$, one concludes with the aid of the standard maximum principle.

In the non-compact case, one could conclude using a form of the maximum principle at infinity, see for instance [31] and the very recent [131], where, in some cases, one can also relax the boundedness conditions on ψ.

In the general case however, where no sign condition is imposed on $a(x)$ and/or the function ψ is not bounded, this method is not feasible.

The compactness of the ambient manifold is now replaced by the assumption that there exists a positive solution φ of a differential inequality suitably related to (4.1),

$$\Delta\varphi + Ha(x)\varphi \leq 0. \tag{4.2}$$

Combining the two inequalities enables us to rephrase the vanishing of ψ into an appropriate Liouville-type result.

We note that the existence of a positive solution to (4.2) is equivalent to the non-negativity of the bottom of the spectrum of the Schrödinger operator $-\Delta - Ha(x)$ (see Lemma 3.10 above), and one could interpret the condition on its spectrum as a sign condition on $a(x)$ in a suitably integrated sense. We also remark that a somewhat related approach has been used by other authors, see, e.g., P. Berard, [11]. However, he uses the condition on the spectral radius directly, and is therefore forced to restrict consideration to the L^2 case.

The main analytical tool used in proving our geometric results is a Liouville-type theorem for locally Lipschitz solutions of differential inequalities of the type

$$u\mathrm{div}\,(\varphi u) \geq 0$$

on M satisfying suitable non-integrability conditions (see Theorem 4.5 below).

Applying this result to a function u constructed in terms of ψ and φ yields the vanishing result for solutions of (4.1) alluded to above. The fairly weak regularity assumptions imposed on u are indeed necessary in order to treat the geometrical problems at hand.

4.2 Liouville and vanishing results

Theorem 4.1 below is a generalization of a Liouville-type result originally due to by H. Berestycki, L. Caffarelli and L. Nirenberg,[14], in a Euclidean setting. See also Proposition 2.1 in [6]. We remark that in the general case of a Riemannian manifold, the Euclidean technique works as well but does not yield the sharp result we are going to describe. We also note that when $\varphi \equiv 1$ we recover a classical result of Yau's, [168].

Theorem 4.1. *Let $(M, \langle\,,\,\rangle)$ be a complete manifold. Assume that $0 < \varphi \in L^\infty_{loc}(M)$ and $u \in L^\infty_{loc}(M) \cap W^{1,2}_{loc}(M)$ satisfy*

$$u\mathrm{div}\,(\varphi \nabla u) \geq 0, \textit{ weakly on } M. \tag{4.3}$$

If, for some $p > 1$,

$$\left(\int_{\partial B_r} |u|^p \varphi \right)^{-1} \notin L^1(+\infty), \tag{4.4}$$

then u is constant.

Proof. We begin by observing that assumption (4.3) means that, for every $0 \leq \sigma \in C^\infty_c(M)$, we have

$$0 \leq -\int \langle \nabla(\sigma u), \varphi \nabla u \rangle = -\int \{\langle \nabla \sigma, \varphi u \nabla u \rangle + \varphi \sigma |\nabla u|^2\}, \tag{4.5}$$

and it is therefore equivalent to the validity of the differential inequality

$$div\,(\varphi u \nabla u) \geq \varphi |\nabla u|^2 \tag{4.6}$$

in the weak sense on M. Further, by a standard approximation argument, inequality (4.5) holds for every $0 \leq \sigma \in L^\infty(M) \cap W^{1,2}(M)$ compactly supported in M.

Next, let $a(t) \in C^1(\mathbb{R})$ and $b(t) \in C^0(\mathbb{R})$ satisfy

$$\text{(i) } a(u) \geq 0, \qquad \text{(ii) } a(u) + ua'(u) \geq b(u) > 0 \tag{4.7}$$

on M, and, for fixed $\epsilon, t > 0$, let ψ_ϵ be the Lipschitz function defined by

$$\psi_\epsilon(x) = \begin{cases} 1 & \text{if } r(x) \leq t, \\ \dfrac{t+\epsilon - r(x)}{\epsilon} & \text{if } t < r(x) < t+\epsilon, \\ 0 & \text{if } r(x) \geq t+\epsilon. \end{cases}$$

The idea of the proof is to apply the divergence theorem to the vector field $a(u)u\varphi\nabla u$. We use an integrated form of this idea in order to deal with the weak regularity of the functions involved.

For every non-negative compactly supported Lipschitz function ρ, we compute

$$\begin{aligned} -\int \langle \psi_\epsilon a(u)\nabla\rho, \varphi u \nabla u\rangle & \\ &= -\int \langle \nabla(\rho\psi_\epsilon a(u)) - \rho\psi_\epsilon a'(u)\nabla u - \rho a(u)\nabla\psi_\epsilon, \varphi u\nabla u\rangle \\ &\geq \int \rho\psi_\epsilon\varphi|\nabla u|^2[a(u) + a'(u)u] + \rho a(u)\langle\nabla\psi_\epsilon, \varphi u \nabla u\rangle \\ &\geq \int \rho\psi_\epsilon\varphi b(u)|\nabla u|^2 - \frac{1}{\epsilon}\int_{B_{t+\epsilon}\setminus B_t} \rho a(u)\varphi|u||\nabla u|, \end{aligned}$$

where the first inequality follows from (4.6) using as test function $\rho\psi_\epsilon a(u)$, which is non-negative compactly supported and belongs to $L^\infty(M)\cap W^{1,2}(M)$ because of the assumptions imposed on a, u, φ, ψ_ϵ and ρ, while the second inequality is a consequence of (4.7) (ii) , and of the Cauchy–Schwarz inequality.

Choosing ρ in such a way that $\rho \equiv 1$ on $\overline{B}_{t+\epsilon}$ the integral on the leftmost side vanishes, and, applying the Cauchy–Schwarz inequality to the second integral on the right-most side and rearranging, we deduce that

$$\int_{B_t} \varphi b(u)|\nabla u|^2 \leq \left(\frac{1}{\epsilon}\int_{B_{t+\epsilon}\setminus B_t} \frac{a(u)^2}{b(u)}\varphi u^2\right)^{1/2} \left(\frac{1}{\epsilon}\int_{B_{t+\epsilon}\setminus B_t} b(u)\varphi|\nabla u|^2\right)^{1/2}. \tag{4.8}$$

Setting

$$H(t) = \int_{B_t} \varphi b(u)|\nabla u|^2,$$

it follows by the co-area formula (see Theorem 3.2.12 in [51]) that

$$H'(t) = \lim_{\epsilon\to 0+} \frac{1}{\epsilon}\int_{B_{t+\epsilon}\setminus B_t} b(u)\varphi|\nabla u|^2 = \int_{\partial B_t} b(u)\varphi|\nabla u|^2\mathcal{H}^{m-1} \text{ for a.e. } t.$$

Here $\mathcal{H}^{m-1}$ denotes the $(m-1)$-dimensional Hausdorff measure on ∂B_t, which coincides with the Riemannian measure induced on the regular part of ∂B_t (the intersection of ∂B_t with the complement of the cut locus of o, see [51], 3.2.46, or [30], Proposition 3.4).

Since the same conclusion holds for the first integral on the right-hand side of (4.8), letting $\epsilon \to 0+$ in (4.8) and squaring, we conclude that

$$H(t)^2 \leq \left(\int_{\partial B_t} \frac{a(u)^2}{b(u)} \varphi u^2 \right) H'(t) \quad \text{for a.e.} \quad t. \tag{4.9}$$

At this point the proof follows the lines of that of Lemma 1.1 in [138]: assume by contradiction that u is non-constant, so that there exists $R_0 > 0$ such that $|\nabla u|$ does not vanish a.e. in B_{R_o}. Then for each $t > R_0$, $H(t) > 0$, and therefore the RHS of (4.9) is also positive. Integrating the inequality between R and r ($R_0 \leq R < r$) we obtain

$$H(R)^{-1} \geq H(R)^{-1} - H(r)^{-1} \geq \int_R^r \left(\int_{\partial B_t} \varphi \frac{a(u)^2}{b(u)} u^2 \right)^{-1}. \tag{4.10}$$

Now, we consider the sequence of functions defined by

$$a_n(t) = \left(t^2 + \frac{1}{n} \right)^{\frac{p-2}{2}}, \qquad b_n(t) = \min\{p-1, 1\}\, a_n(t)\,, \; \forall n \in \mathbb{N}.$$

Since condition (4.7) holds for every n, so does (4.10), whence, letting $n \to +\infty$ and using the Lebesgue dominated convergence theorem and Fatou's lemma we deduce that there exists $C > 0$ which depends only on p such that

$$\left(\int_{B_R} \varphi |u|^{p-2} |\nabla u|^2 \right)^{-1} \geq C \int_R^r \left(\int_{\partial B_t} \varphi |u|^p \right)^{-1} dt.$$

The required contradiction is now reached by letting $r \to +\infty$ and using assumption (4.4). □

As the above proof shows, the conclusion of the theorem holds if one assumes that $0 < \varphi \in L^2_{loc}(M)$ and $u \in Lip_{loc}(M)$.

We observe that condition (4.4) is implied by $u\varphi^{1/p} \in L^p(M)$. Indeed, if this is the case and we set $f = \int_{\partial B_r} |u|^p \varphi$, then the assumption and the co-area formula show that $f \in L^1(+\infty)$, and by Hölder inequality

$$\int_{r_0}^r f^{-1} \geq (r - r_0)^2 \left(\int_{r_0}^r f \right)^{-1} \to +\infty \text{ as } r \to +\infty.$$

We also note that the conclusion of Theorem 4.1 fails if we assume that $p = 1$ in (4.4). Indeed, taking $\varphi \equiv 1$, (4.3) reduces to $u\Delta u \geq 0$, and P. Li and R. Schoen have constructed in [95] an example of a non-constant, L^1, harmonic, function on a complete manifold. Indeed, let $(M, \langle\,,\rangle$ be a model manifold in the sense of Greene and Wu, namely $M = \mathbb{R}^m$ as a manifold, with the metric given in polar coordinates by

$$\langle\,,\rangle = dr^2 + \sigma(r)^2 d\theta^2,$$

where $d\theta^2$ denotes the standard metric on the unit sphere $\mathbb{S}^{m-1}$, and σ is a smooth odd function on $\mathbb{R}$ which is positive for $r > 0$ and such that $\sigma'(0) = 1$.

Choose a non-negative non-identically zero compactly supported smooth function $a(t)$, and define the non-negative function

$$u(x) = \int_0^{r(x)} \sigma(t)^{-(m-1)} \Big\{ \int_0^t a(s)\sigma(s)^{m-1}\, ds \Big\} dt, \tag{4.11}$$

where $r(x)$ denotes the distance function from 0. It is easily verified that u is smooth, and satisfies

$$\Delta u = a(r(x))$$

on $(\mathbb{R}^m, \langle\,,\rangle)$, and it is therefore a non-constant non-negative subharmonic function. Since u is radial, for ease of notation we will write $u(r)$. If we specify σ to be $\sigma(t) = t$ for $t \in [0, 1]$, and such that

$$\sigma(t) = \big(t(\log t)^\epsilon\big)^{-1/(m-1)} \exp\Big(-\frac{t^2(\log t)^\epsilon}{m-1}\Big),$$

for $t \in [T_o, +\infty)$, for some $\epsilon > 0$, and $T_o > 1$ sufficiently large, then it is easy to check that

$$u(r) \sim C \exp\big(r^2(\log r)^\epsilon\big)$$

and

$$\int_{\partial B_r} u \sim \frac{C'}{r(\log r)^\epsilon}$$

as $r \to +\infty$. Thus, if $\epsilon > 1$, then u is a non-negative, integrable subharmonic function on M. We note that in this case, the manifold $(M, \langle\,,\rangle)$ has finite volume.

Finally, we remark that Theorem 4.1 generalizes [14] (see the proof of Proposition 2.1 therein) in two directions, even in the case where $M = \mathbb{R}^m$. First, in their case $p = 2$; secondly they replace (4.4) by the more stringent condition

$$\int_{B_r} u^2 \varphi \leq C r^2$$

for some constant $C > 0$. To see that the latter implies (4.4) simply note that its validity forces

$$\frac{r}{\int_{B_r} u^2 \varphi} \notin L^1(+\infty)$$

which in turn implies (4.4) (see, e.g., [138], Proposition 1.3). Furthermore, although the approach used in [14] is applicable also in the case of Riemannian manifolds, in this general context, it does not yield a sharp result.

We apply Theorem 4.1 to prove a uniqueness result for harmonic maps which largely improves on previous work in the literature. We recall that a ball $B_R(q)$ in a Riemannian manifold $(N, (\,,\,))$ is said to be regular if it does not intersect the cut locus of q, and, having denoted by $B \geq 0$ an upper bound for the sectional curvature of N on $B_R(q)$, one has $\sqrt{B}R < \pi/2$. Let q_b be the function defined by the formula

$$q_B(t) = \begin{cases} \frac{1}{2}t^2 & \text{if } B = 0, \\ \frac{1}{B}\left(1 - \cos\left(\sqrt{B}t\right)\right) & \text{if } B > 0. \end{cases}$$

Assume that $f, g : (M, \langle\,,\,\rangle) \to B_R(q) \subset N$ are harmonic maps taking values in the regular ball $B_R(q)$ and define functions $\Phi, \psi, \varphi, u : M \to \mathbb{R}$ by setting

$$\begin{aligned} \Phi(x) &= -\log\left(\cos(\sqrt{B}\mathrm{dist}_N(q, f(x)) \cos(\sqrt{B}\mathrm{dist}_N(q, g(x))\right), \\ \varphi(x) &= e^{-\Phi(x)} \quad \text{and} \quad u = \varphi(x)^{-1} q_B\left(\mathrm{dist}_N(f(x), g(x))\right). \end{aligned} \tag{4.12}$$

Clearly, $u \geq 0$ and, since f and g take values in the regular ball $B_R(q)$, there exists a constant $C \geq 1$ such that

$$\begin{aligned} &C^{-1} \leq \varphi \leq 1 \qquad \text{and} \\ &C^{-1}\mathrm{dist}_N(f(x), g(x))^2 \leq u(x) \leq C\mathrm{dist}_N(f(x), g(x))^2 \end{aligned} \tag{4.13}$$

on M. Further, a result of W. Jäger and H. Kaul [86] shows that

$$\mathrm{div}\,(\varphi\nabla u) \geq 0 \quad \text{on } M,$$

and therefore, "a fortiori",

$$u\mathrm{div}\,(\varphi\nabla u) \geq 0 \quad \text{weakly on } M. \tag{4.14}$$

With this preparation we have the following uniqueness result:

Theorem 4.2. *Maintaining the notation introduced above, let $f, g : M \to N$ be harmonic maps taking values in the regular ball $B_R(q) \subset N$, and assume that, for some $p \geq 1$,*

$$\mathrm{dist}_N(f, g)^{2p} \in L^1(M). \tag{4.15}$$

In case $p = 1$, assume also that

$$\int_{\partial B_r} \mathrm{dist}_N(f, g)^2 \leq \frac{C}{r \log^\beta r} \tag{4.16}$$

for some constants $C, \beta > 0$ and for $r(x) >> 1$. If $\mathrm{vol}(M) = +\infty$, then $f \equiv g$.

Proof. As noted above, the functions φ and u satisfy (4.14), and, according to (4.13), the integrability condition (4.15) implies that

$$\varphi u^p = \varphi^{1-p} q_B\left(\mathrm{dist}_N(f(x), g(x))\right)^p \in L^1(M).$$

In particular

$$\left\{\int_{\partial B_r} \varphi u^p\right\}^{-1} \notin L^1(+\infty).$$

If $p > 1$, we can use Theorem 4.1 above to deduce that u is constant, that is, there exists a constant $C_1 \geq 0$, such that

$$q_B\left(\operatorname{dist}_N(f(x), g(x))\right) = C_1\varphi(x).$$

Since $\operatorname{vol}(M) = +\infty$ and φ is bounded away from zero, the integrability condition (4.15) forces $C_1 = 0$ and therefore $\operatorname{dist}_N(f(x), g(x)) \equiv 0$, as required.

The case $p = 1$ is a consequence of the following version of Theorem 4.1 above. □

Theorem 4.3. *Let $(M, \langle\, ,\rangle)$ be a complete manifold. Assume that $0 < \varphi \in L^2_{loc}(M)$ and $u \in Lip_{loc}(M)$ satisfy*

$$\operatorname{div}(\varphi\nabla u) \geq 0 \quad \text{weakly on } M.$$

If $u \geq 0$ and

$$\text{(i)} \int_{\partial B_r} \varphi u \leq \frac{C}{r\log^\beta r}, \qquad \text{(ii)}\ u(x) \leq Ce^{r(x)^2} \tag{4.17}$$

for some constants $C, \beta > 0$ and $r(x) \gg 1$, then u is constant.

Proof. We suppose that u is non-constant to get a contradiction. Thus, we proceed as in the proof of Theorem 4.1 above to arrive at

$$\left\{\int_{B_R} \varphi b(u)\left|\nabla u\right|^2\right\}^{-1} \geq \int_R^r \left\{\int_{\partial B_t} \varphi \frac{a(u)^2}{b(u)}\right\}^{-1} \tag{4.18}$$

for $r > R \geq R_0$ sufficiently large and where the functions $a \in C^1(\mathbb{R})$ and $b \in C^0(\mathbb{R})$ satisfy

$$\text{(i) } a(u) \geq 0; \quad \text{(ii) } a'(u) \geq b(u) > 0 \text{ on } M. \tag{4.19}$$

Now, for every fixed $n \geq 1$, and for every $t \geq 0$, we let

$$a_n(t) = \log^\beta\left(1 + \log\left(1 + t + \frac{1}{n}\right)\right),$$

$$b_n(t) = a_n'(t) = \frac{\beta\log^{\beta-1}\left(1+\log\left(1+t+\frac{1}{n}\right)\right)}{\left(1+\log\left(1+t+\frac{1}{n}\right)\right)\left(1+t+\frac{1}{n}\right)}.$$

It is easy to verify that there exists a constant $\gamma = \gamma(\beta) > 0$ such that, for every $s \geq 0$,

$$\log^{1+\beta}\left(1+\log(1+s)\right) \leq \gamma s\left(1 + \log^{1+\beta}\left(1+\log(1+s)\right)\right)$$

and therefore

$$\frac{a_n(t)^2}{b_n(t)}$$
$$\leq \frac{1}{\beta}\gamma\left(\frac{1}{n}+t\right)\left(1+\log\left(1+t+\frac{1}{n}\right)\right)\left(1+\log^{1+\beta}\left(1+\log\left(1+t+\frac{1}{n}\right)\right)\right).$$

We substitute a_n, b_n into (4.18), let n tend to infinity in the resulting inequality and use the dominated convergence theorem and Fatou's lemma to deduce the existence of a constant $C>0$ such that

$$\left\{\int_{B_R}\frac{\varphi\left|\nabla u\right|^2}{(1+u)\left(1+\log(1+u)\right)\log^{1-\beta}\left(1+\log(1+u)\right)}\right\}^{-1} \tag{4.20}$$
$$\geq C\int_R^r\left\{\int_{\partial B_t}\varphi u\left(1+\log(1+u)\right)\left(1+\log^{1+\beta}\left(1+\log(1+u)\right)\right)\right\}^{-1}.$$

On the other hand, by (4.17),

$$\int_{\partial B_t}\varphi u\left(1+\log(1+u)\right)\left(1+\log^{1+\beta}\left(1+\log(1+u)\right)\right)$$
$$\leq C\int_{\partial B_t}\varphi u t^2\left(\log t\right)^{1+\beta}\leq Ct\log t.$$

By letting $r\to+\infty$, this contradicts (4.20). □

When $(N,(\,,\,))$ is a Cartan–Hadamard manifold, namely, a complete, simply connected manifold of non-positive sectional curvature, the above proof yields the next

Theorem 4.4. *Let $(N,(\,,\,))$ be Cartan–Hadamard and let $f,g:M\to N$ be harmonic maps such that, for some $p\geq 1$,*

$$\operatorname{dist}_N(f(x),g(x))^{2p}\in L^1(M) \tag{4.21}$$

and, for $p=1$, add the conditions

$$\text{(i)}\int_{\partial B_r}\operatorname{dist}_N(f(x),g(x))^2\leq\frac{C}{r\log^\beta r};\quad \text{(ii)}\operatorname{dist}_N(f(x),g(x))^2\leq Ce^{r(x)^2} \tag{4.22}$$

for some constants $\beta,C>0$ and for $r(x)$ large enough. If $\operatorname{vol}(M)=+\infty$, then $f=g$.

As we pointed out after the proof of Theorem 4.1, an L^1-Liouville-type theorem for subharmonic functions is in general false if we do not require some extra assumptions. This explains the role of assumption (4.22) in Theorem 4.4 when $p=1$.

We now come to the following consequence of Theorem 4.1, which will be the main ingredient in the geometric applications of Chapter 6 below.

Theorem 4.5. *Let $(M, \langle\,,\rangle)$ be a complete manifold, $a(x) \in L^\infty_{loc}(M)$ and let $\psi \in Lip_{loc}(M)$ satisfy the differential inequality*

$$\psi\Delta\psi + a(x)\psi^2 + A|\nabla\psi|^2 \geq 0 \qquad \text{weakly on } M \tag{4.23}$$

for some $A \in \mathbb{R}$. Let also $\varphi \in Lip_{loc}(M)$ be a positive solution of

$$\Delta\varphi + Ha(x)\varphi \leq 0, \qquad \text{weakly on } M, \tag{4.24}$$

for some H such that

$$H \geq A + 1, \; H > 0. \tag{4.25}$$

If

$$\left(\int_{\partial B_r} |\psi|^{2(\beta+1)}\right)^{-1} \notin L^1(+\infty) \tag{4.26}$$

for some β such that

$$A \leq \beta \leq H - 1, \; \beta > -1, \tag{4.27}$$

then there exists a constant $C \in \mathbb{R}$ such that

$$C\varphi = |\psi|^H \operatorname{sgn}\psi. \tag{4.28}$$

Further,

(i) *If $H - 1 > A$, then ψ is constant on M, and if in addition, $a(x)$ does not vanish identically, then ψ is identically zero;*

(ii) *If $H - 1 = A$, and ψ does not vanish identically, then φ and therefore $|\psi|^H$ satisfy (4.24) with equality sign.*

Proof. Set, for ease of notation, $\alpha = \frac{\beta+1}{H}$, and let u be the function defined by

$$u = \varphi^{-\alpha}|\psi|^\beta\psi,$$

so that the first assertion in the statement is that u is constant on M.

Noting that the restrictions imposed on β, and Lemma 4.12 in the Appendix at the end of this section, imply that $u \in C^0(M) \cap W^{1,2}_{loc}(M)$. Moreover,

$$\int \varphi^{2\alpha}|u|^2 = \int |\psi|^{2(\beta+1)},$$

so that (4.26) implies that (4.4) holds with $\varphi^{2\alpha}$ in place of φ, and $p = 2$, the constancy of u follows from Theorem 4.1 once we show that the differential inequality

$$u\,div\big(\varphi^{2\alpha}\nabla u\big) \geq 0 \tag{4.29}$$

holds weakly on M. i.e. (see the beginning of the proof of Theorem 4.1), that for every non-negative, compactly supported function $\rho \in L^\infty(M) \cap W^{1,2}(M)$, we have

$$I = \int \left[\langle\varphi^{2\alpha}u\nabla u, \nabla\rho\rangle + \varphi^{2\alpha}|\nabla u|^2\rho\right] \leq 0.$$

Using the definition of u, and Lemma 4.13, we compute

$$\nabla u = -\alpha\varphi^{-\alpha-1}|\psi|^\beta\psi\nabla\varphi + (\beta+1)\varphi^{-\alpha}|\psi|^\beta\nabla\psi$$

whence

$$\begin{aligned} I = (\beta+1)\int\langle\nabla\psi,\psi|\psi|^{2\beta}\nabla\rho\rangle - \alpha\int\varphi^{-1}|\psi|^{2\beta+2}\langle\nabla\varphi,\nabla\rho\rangle \\ + \int\Big[(\beta+1)^2|\psi|^{2\beta}|\nabla\psi|^2\rho + \alpha^2|\psi|^{2\beta+2}\frac{|\nabla\varphi|^2}{\varphi^2}\rho\Big] \\ - 2\alpha(\beta+1)\int|\psi|^{2\beta}\psi\langle\frac{\nabla\varphi}{\varphi},\nabla\psi\rangle. \end{aligned} \tag{4.30}$$

We first consider the first integral on the right-hand side, and assume that $\beta < 0$, the other case being easier. Since

$$\big|(\psi^2+\epsilon)^\beta\psi\nabla\psi\big| \le |\psi|^{2\beta+1}|\nabla\psi| = |\psi|^{1+\beta}|\psi|^\beta|\nabla\psi| \in L^1_{loc}$$

by Lemma 4.13, by the dominated convergence theorem,

$$\begin{aligned} \int|\psi|^{2\beta}\psi\langle\nabla\psi,\nabla\rho\rangle = \lim_{\epsilon\to0+}\int(\psi^2+\epsilon)^\beta\psi\langle\nabla\psi,\nabla\rho\rangle \\ = \lim_{\epsilon\to0+}\Big\{\int\langle\nabla\psi,\nabla\big[\psi(\psi^2+\epsilon)^\beta\rho\big]\rangle - (\psi^2+\epsilon)^\beta\frac{(2\beta+1)\psi^2+\epsilon}{\psi^2+\epsilon}|\nabla\psi|^2\rho\Big\}. \end{aligned} \tag{4.31}$$

According to (4.23), for every non-negative, compactly supported function $\sigma \in W^{1,2}(M)$,

$$\int\langle\nabla\psi,\nabla(\sigma\psi)\rangle \le \int\big(a(x)\psi^2 + A|\nabla\psi|^2\big)\sigma.$$

Applying the above inequality with $\sigma = \rho(\psi^2+\epsilon)^\beta$, and applying the dominated convergence theorem, we deduce that

$$\lim_{\epsilon\to0+}\int(\psi^2+\epsilon)^\beta\frac{(2\beta+1)\psi^2+\epsilon}{\psi^2+\epsilon}|\nabla\psi|^2\rho = (2\beta+1)\int|\psi|^{2\beta}|\nabla\psi|^2\rho,$$

and

$$\begin{aligned} \lim_{\epsilon\to0+}\int\langle\nabla\psi,\nabla\big[\psi(\psi^2+\epsilon)^\beta\rho\big]\rangle \le \lim_{\epsilon\to0+}\int\big[a(x)\psi^2 + A|\nabla\psi|^2\big](\psi^2+\epsilon)^\beta\rho \\ = \int\big[a(x)|\psi|^{2\beta+2} + A|\psi|^{2\beta}|\nabla\psi|^2\big]\rho. \end{aligned}$$

Inserting these expressions into (4.31) we conclude that

$$\int|\psi|^{2\beta}\psi\langle\nabla\psi,\nabla\rho\rangle \le \int\big[a(x)|\psi|^{2\beta+2} + (A-2\beta-1)|\psi|^{2\beta}|\nabla\psi|^2\big]\rho. \tag{4.32}$$

In a similar, but easier way, using (4.24) one verifies that

$$-\int \varphi^{-1}|\psi|^{2\beta+2}\langle\nabla\varphi,\nabla\rho\rangle$$
$$\leq \int \big[-Ha(x)|\psi|^{2\beta+2} - |\psi|^{2\beta+2}\frac{|\nabla\varphi|^2}{\varphi^2} + 2(\beta+1)|\psi|^{2\beta}\psi\langle\frac{\nabla\varphi}{\varphi},\nabla\psi\rangle\big]\rho. \quad (4.33)$$

Substituting (4.32) and (4.33) into (4.30), and recalling the value of α and the condition satisfied by β, we conclude that

$$I \;\leq\; (\beta+1)\int\Big[(A-\beta)|\psi|^{2\beta}|\nabla\psi|^2\rho + \frac{\beta+1-H}{H^2}|\psi|^{2\beta+2}\frac{|\nabla\varphi|^2}{\varphi^2}\rho\Big] \;\leq\; 0,$$

as required to show that (4.29) holds.

In particular, ψ has constant sign, and if we assume that $\psi \not\equiv 0$, multiplying ψ by a suitable constant we may assume that ψ is strictly positive, and

$$\varphi = \psi^H.$$

Inserting this equality into (4.24) we have

$$H\psi^{H-2}\big[\psi\Delta\psi + (H-1)|\nabla\psi|^2 + a(x)\psi^2\big] \leq 0, \quad (4.34)$$

whence, multiplying (4.23) by $H\psi^{H-2}$, and subtracting the resulting inequality from (4.34) we obtain

$$H\big[(H-1)-A\big]\psi^{H-2}|\nabla\psi|^2 \leq 0. \quad (4.35)$$

Thus, if $H-1 > A$, $|\nabla\psi|^2 \equiv 0$, and ψ and therefore φ are constant. It follows from (4.24) that

$$\Delta\varphi + Ha(x)\varphi = Ha(x)\varphi \leq 0, \text{ so that } a(x) \leq 0,$$

while, (4.23) implies that

$$\psi\Delta\psi + a(x)\psi^2 + A|\nabla\psi|^2 = a(x)\psi^2 \geq 0 \text{ so that } a(x) \geq 0,$$

and we conclude that $a(x) \equiv 0$. In particular, if $a(x) \not\equiv 0$, then ψ must vanish identically.

Finally, assume that $A = H-1$, and that ψ does not vanish identically, so that, as noted above, we may assume that ψ is strictly positive, and that $\varphi = \psi^H$. On the other hand, it follows from (4.24) and Lemma 3.10 that there exists a positive C^1 function v satisfying

$$\Delta v + Ha(x)v = 0 \quad \text{weakly on } M. \quad (4.36)$$

Repeating the argument with v in place of φ, we deduce that there exists $\tilde{c} \neq 0$ such that

$$\tilde{c}v = \psi^H = \varphi.$$

Thus φ is a positive multiple of v and we conclude that it also satisfies (4.36). □

We remark that Theorem 4.5 fails if the exponent $2(\beta+1)$ in the integrability condition (4.28) is replaced by $p(\beta+1)$ for some $p > 2$. Indeed, it was shown in [16] that if $a(x)$ and $b(x)$ are non-negative continuous functions on $\mathbb{R}^m$ satisfying

$$a(x) \le \frac{(m-2)^2}{4}|x|^{-2} \quad , \quad a(x) = \frac{(m-2)^2}{4}|x|^{-2} \text{ if } |x| \gg 1$$

and

$$b(x) = \frac{|x|^{(m-2)(\sigma-1)/2}}{(\log|x|)^{\sigma+1}(\log\log|x|)(\log\log\log|x|)^2} \text{ if } |x| \gg 1$$

for some $\sigma > 1$, then the equation

$$\Delta u + a(x)u - b(x)u^\sigma = 0 \tag{4.37}$$

has a family of positive solutions u_α $(\alpha > 0)$ satisfying

$$u_\alpha(0) = \alpha \quad \text{and} \quad u_\alpha(x) \sim |x|^{-(m-2)/2}\log|x| \quad \text{as } |x| \to +\infty.$$

In particular, u_α is a solution of (4.23) with $A = 0$, and

$$\int_{\partial B_r} |u_\alpha|^q \asymp r^{1+(m-2)(2-q)/2}(\log r)^q,$$

so that

$$\left(\int_{\partial B_r} |u_\alpha|^q\right)^{-1} \notin L^1(+\infty)$$

for every $q > 2$.

On the other hand, it is well known that in this case $\lambda_1\left([-\Delta - a(x)]_{\mathbb{R}^m}\right) = 0$, so there exists a positive solution φ of

$$\Delta\varphi + a(x)\varphi = 0 \quad \text{on } \mathbb{R}^m \tag{4.38}$$

(see, e.g., [19], Lemma 3 and subsequent Remark 4). Since in this case $H = 1$, applying Theorem 4.5 we would conclude that

$$c\varphi = u_\alpha$$

for some constant c which is necessarily positive, since both u, $\varphi > 0$. But then u_α would be a solution of (4.38) and this is impossible since it satisfies (4.37) and b is non-zero.

We also note that a minor modification of the above proof yields the following

Theorem 4.6. *Let $a(x)$, $b(x) \in C^0(M)$ and assume that $b(x) \ge 0$. Let $H > 0$, $K > -1$ and $A \in \mathbb{R}$ be constants satisfying*

$$A \le H(K+1) - 1, \tag{4.39}$$

and suppose that there exists a positive $Lip_{loc}(M)$ solution of the differential inequality

$$\Delta\varphi + Ha(x)\varphi \le -K\frac{|\nabla\varphi|^2}{\varphi} \qquad on \quad M. \tag{4.40}$$

Then the differential inequality

$$u\Delta u + a(x)u^2 - b(x)u^{\sigma+1} \ge -A|\nabla u|^2, \qquad \sigma \ge 1, \tag{4.41}$$

has no non-negative $Lip_{loc}(M)$ solutions on M satisfying

$$\operatorname{supp} u \cap \{x \in M \,:\, b(x) > 0\} \neq \emptyset \tag{4.42}$$

and

$$\Big(\int_{\partial B_r} u^{2(\beta+1)}\Big)^{-1} \notin L^1(+\infty), \tag{4.43}$$

for some β satisfying $\beta > -1$, $A \le \beta \le H(K+1) - 1$.

As an immediate corollary of Theorem 4.5 we have

Corollary 4.7. *Let $a(x) \in L^\infty_{loc}(M)$, $A \in \mathbb{R}$, $H \ge A+1$,$H > 0$, and set ${}^HL = -\Delta - Ha(x)$. Assume that $\psi \in Lip_{loc}(M)$ is a changing sign solution of (4.23) satisfying (4.26) for some β such that $\beta > -1$, $A \le \beta \le H-1$. Then $\lambda_1({}^HL_M) < 0$.*

Proof. Assume by contradiction that $\lambda_1({}^HL_M) \ge 0$. By Lemma 3.10 there exists $0 < \varphi \in C^1(M)$ satisfying $\Delta\varphi + Ha(x)\varphi = 0$ on M. By Theorem 4.5, there exists a constant C such that $C\varphi = |\psi|^{H-1}\psi$, and since ψ changes sign, while φ is strictly positive, this yields the required contradiction. □

In the case of Euclidean space, the integrability condition (4.26) follows assuming a suitable upper estimate for ψ, and yields the following (slight) improvement of [14] Theorem 1.7.

Corollary 4.8. *Let $a(x) \in L^\infty_{loc}(\mathbb{R}^m)$, and let $\psi \in Lip_{loc}(\mathbb{R}^m)$ be a changing sign solution of*

$$\psi\Delta\psi + a(x)\psi^2 \ge 0 \qquad on \ \mathbb{R}^m,$$

such that, for some $H \ge 1$,

$$|\psi(x)| = \mathcal{O}\Big(r(x)^{-(m-2)/2H}(\log r(x))^{1/2H}\Big), \qquad as \ r(x) \to +\infty.$$

If ${}^HL = -\Delta - Ha(x)$, then $\lambda_1\big({}^HL_{\mathbb{R}^m}\big) < 0$.

Similar results can be obtained on Riemannian manifolds where $\operatorname{vol}\partial B_r$ satisfies a suitable upper bound. This in turn follows, by the volume comparison theorem, from appropriate lower bounds on the Ricci curvature (see, Section 2.2 and [20], Appendix). We leave the details to the interested reader.

Theorem 4.5 yields also the following generalization of Theorem 2 (and Corollary 2) in [54].

Corollary 4.9. *Let $(M, \langle\,,\rangle)$ be a complete manifold, and let $a(x) \in L^{\infty}_{loc}(M)$ and $A < 0$. Suppose that $\psi \in Lip_{loc}$ is a non-constant weak solution of the differential inequality*

$$\psi\Delta\psi + a(x)\psi^2 + A|\nabla\psi|^2 \geq 0,$$

satisfying

$$\left(\int_{\partial B_r} \psi^2\right)^{-1} \notin L^1(+\infty). \tag{4.44}$$

Then, there exists $H_o \in [0, 1)$ such that, for every $H > H_o$, the differential inequality

$$\Delta\varphi + Ha(x)\varphi \leq 0 \tag{4.45}$$

has no positive, locally Lipschitz weak solution on M, while if $0 \leq H \leq H_o$, such a solution of (4.45) exists.

Proof. Recall that, according to Lemma 3.10, the existence of a positive, locally Lipschitz weak solution of (4.45) is equivalent to

$$\lambda_1\left({}^{H}L\right) \geq 0,$$

where ${}^{H}L = -\Delta - Ha(x)$.

Observe next that if $0 < H_1 \leq H_2$, then, by the variational characterization of the bottom of $\lambda_1\left({}^{H}L\right)$, we have

$$\lambda_1\left({}^{H_1}L\right) \geq \frac{H_1}{H_2}\lambda_1\left({}^{H_2}L\right). \tag{4.46}$$

(see the argument in the proof of Theorem 2 in [54]). Thus, if we denote by S the set of $H \geq 0$ such that (4.45) holds, S is not empty, since $\lambda_1(-\Delta) \geq 0$, and if H_2 is in S, then so is H_1.

An application of Theorem 4.5 with $A < \max\{A, 0\} = 0 = \beta = H - 1$ implies that if $H = 1$, then (4.45) has no positive locally Lipschitz solution, for otherwise ψ would necessarily be constant, against the assumption. Thus $1 \notin S$, and $H_o = \sup S \leq 1$

Now one concludes as in Corollary 2 in [54] showing, by an approximation argument, that S is closed, so that $1 > H_o \in S$. □

To see that Corollary 4.9 implies Theorem 2 and Corollary 2 in [54], it suffices to observe that if $ds^2 = \mu(z)|dz|^2$ is a complete metric on the unit disk D, with Gaussian curvature K, then $\psi = \mu^{-1/2}$ is a non-constant solution of

$$\psi\Delta\psi - K\psi^2 = |\nabla\psi|^2$$

and

$$\int \psi^2 \mathrm{dvol}_{ds^2} = \int \mu^{-1}\mu\, dxdy = \mathrm{vol}_{Eucl}(D) < +\infty.$$

According to the remark after the proof of Theorem 4.1, condition (4.44) holds, and Corollary4.9 implies that there exists $H_o \in [0,1)$ such that equation

$$\Delta\varphi - HK(x)\varphi = 0$$

has no positive solution if $H > H_o$ and has a positive solution if $0 \le H \le H_o$.

Following the above line of investigation, we are naturally led to the next result, which extends some known facts in minimal surfaces theory to minimal hypersurfaces of Euclidean space; see Corollary 4.11 below.

We recall that a minimal hypersurface $f : (M^m, \langle\,,\rangle) \to \mathbb{R}^{m+1}$ is stable if it (locally) minimizes area up to second order or, equivalently, if the bottom of the spectrum $\lambda_1 L_M)$ of the operator $L = -\Delta - |\mathrm{II}|^2$ is non-negative. Here $|\mathrm{II}|$ denotes the length of the second fundamental tensor of the immersion.

We also recall that a Riemannian metric $\widetilde{\langle\,,\rangle}$ on a (generic) manifold M is said to be a pointwise conformal deformation of a metric $\langle\,,\rangle$ if there exists a positive function $\rho \in C^\infty(M)$ such that $\widetilde{\langle\,,\rangle}_x(v,w) = \rho^2(x)\langle\,,\rangle_x(v,w)$, for every $x \in M$ and $v, w \in T_xM$.

Theorem 4.10. *Let $f : (M^m, \langle\,,\rangle) \to \mathbb{R}^{m+1}$ be a complete, stable, minimal hypersurface of dimension $m \ge 2$. Then $\langle\,,\rangle$ cannot be pointwise conformally deformed to a Riemannian metric $\widetilde{\langle\,,\rangle}$ of scalar curvature $\tilde{S}(x) \le 0$ and finite volume.*

Proof. We first consider the case where $m \ge 3$. By contradiction, we assume that there exists a conformal metric $\widetilde{\langle\,,\rangle}$ on M with scalar curvature $\tilde{S}(x) \le 0$ and finite volume $\widetilde{\mathrm{vol}}(M) < +\infty$. Denoting by $S(x)$ the scalar curvature of the original metric, minimality and the Gauss equations imply

$$S(x) = -\left|\mathrm{II}\,(x)\right|^2. \tag{4.47}$$

According to Lemma 3.10, the stability of f is then equivalent to the existence of a positive solution $\varphi \in C^\infty(M)$ of

$$\Delta\varphi - S(x)\varphi = 0 \qquad \text{on } M. \tag{4.48}$$

Setting

$$H = \frac{4\,(m-1)}{m-2} > 1; \qquad a(x) = -\frac{1}{H}S(x),$$

we can rewrite (4.48) in the form

$$\Delta\varphi + Ha(x)\varphi = 0 \qquad \text{on } M.$$

Now, let

$$\widetilde{\langle\,,\rangle} = \psi^{\frac{4}{m-2}}\langle\,,\rangle. \tag{4.49}$$

Then the smooth positive function ψ is a solution of the Yamabe equation, and, since $\tilde{S}(x) \leq 0$, we deduce that

$$\Delta\psi + a(x)\psi = -\frac{1}{H}\tilde{S}(x)\psi^{\frac{m+2}{m-2}} \geq 0, \text{ on } M. \tag{4.50}$$

Since

$$\int_M \psi^{\frac{2m}{m-2}} d\text{vol} = \widetilde{\text{vol}}(M) < +\infty$$

we have

$$\frac{1}{\int_{\partial B_r(o)} \psi^{2(\beta+1)}} \notin L^1(+\infty)$$

where

$$\beta = \frac{2}{m-2}$$

satisfies

$$0 < \beta < H - 1.$$

Applying Theorem 4.5, case 1, with $A = 0$ we therefore conclude that ψ, and therefore φ, is a positive constant and $S(x) \equiv 0$. According to (4.47) and (4.49) we deduce that $f(M)$ is an affine hyperplane and hence $\left(M, \widetilde{\langle\, , \rangle}\right)$ is homothetic to $(\mathbb{R}^m, can)$. But this clearly contradicts the assumption that $\widetilde{\text{vol}}(M) < +\infty$.

The case $m = 2$ is completely similar. This time, we replace (4.49) with

$$\widetilde{\langle\, , \rangle} = \psi^2 \langle\, , \rangle$$

and, instead of (4.50), we use the corresponding Yamabe equation

$$\psi\Delta\psi - S(x)\psi^2 = -\tilde{S}(x)\psi^4 + |\nabla\psi|^2.$$

Thus, ψ satisfies

$$\psi\Delta\psi - S(x)\psi^2 \geq |\nabla\psi|^2.$$

Since

$$\int_M \psi^2 d\text{vol} = \widetilde{\text{vol}}(M) < +\infty$$

we have

$$\frac{1}{\int_{\partial B_r(o)} \psi^2} \notin L^1(+\infty).$$

On the other hand, the stability assumption implies the existence of a positive, smooth solution φ of (4.48). Therefore we can apply Theorem 4.5, case 1, with the choices $\beta = 0$, $a(x) = -S(x)$, $H = 1$, $A = -1$. Reasoning as above, we reach the desired contradiction. □

Using a classical universal covering argument, together with the Riemann-Köbe uniformization theorem, we easily recover Corollary 4 in [54]:

Corollary 4.11. *Let $f : (M, \langle\,,\rangle) \to \mathbb{R}^3$ be a 2-dimensional, complete, stable, minimal surface. Then $f(M)$ is parabolic, and hence is an affine plane.*

Proof. Let $\pi : (\bar{M}, \bar{\langle\,,\rangle}) \to (M, \langle\,,\rangle)$ be the Riemannian universal covering of M. Then, $\bar{f} = f \circ \pi : (\bar{M}, \bar{\langle\,,\rangle}) \to \mathbb{R}^3$ defines a complete, minimal surface. Moreover $\bar{f}$ is stable because any positive solution φ of (4.48) on M lifts to a positive solution $\bar{\varphi} = \varphi \circ \pi$ of $\bar{\Delta}\bar{\varphi} - \bar{S}(y)\bar{\varphi} = 0$ on $\bar{M}$. Here the bar-quantities refer to the covering metric $\bar{\langle\,,\rangle}$. Since there are no compact minimal surfaces in the Euclidean space, the Uniformization Theorem implies that $(\bar{M}, \bar{\langle\,,\rangle})$ is conformally diffeomorphic to either $\mathbb{R}^2$ or the open unit disc $D_1 \subset \mathbb{R}^2$. In view of Theorem 4.10 the second possibility cannot occur so that M must be parabolic. To conclude that f is totally geodesic, simply note that, by (4.48), φ is a positive superharmonic function. Therefore φ must be constant and $S(x) = -\left|\mathrm{II}\right|^2 \equiv 0$. □

4.3 Appendix: Chain rule under weak regularity

This section provides the technical justification for the distributional computations needed in the proofs of Theorems 4.5 and 5.16 and Lemma 5.17 in the next section. First, we present a regularity result.

Lemma 4.12. *Let $a(x) \in L^\infty_{loc}(M)$ and $A \in \mathbb{R}$. Let $\psi \in Lip_{loc}(M)$ be a weak solution of*

$$\psi\Delta\psi + a(x)\psi^2 + A\left|\nabla\psi\right|^2 \geq 0 \text{ on } M.$$

Then

$$|\psi|^{p-1}\psi \in W^{1,2}_{loc}(M) \tag{4.51}$$

provided

$$\begin{cases} p \geq 1 & \text{if } A \geq 1, \\ p > \max\left\{0, \frac{A+1}{2}\right\} & \text{if } A < 1 \end{cases}$$

and, furthermore,

$$\nabla\left(\left(\psi^2 + \varepsilon\right)^{(p-1)/2}\psi\right) \stackrel{L^2}{\rightharpoonup} \nabla\left(|\psi|^{p-1}\psi\right) \quad \text{as } \varepsilon \to 0+. \tag{4.52}$$

Proof. We treat only the case $p < 1$, the other case being easier. Consider the family of functions $\left(\psi^2 + \varepsilon\right)^{(p-1)/2}\psi$ and note that, as $\varepsilon \to 0+$,

$$\left(\psi^2 + \varepsilon\right)^{(p-1)/2}\psi \to |\psi|^{p-1}\psi \quad \text{in } L^2_{loc}.$$

We are going to use the fact that if a sequence $\{f_n\}$ is uniformly bounded in $W^{1,2}_{loc}$ and converges to f strongly in L^2_{loc}, then the limit function f is in $W^{1,2}_{loc}$ and ∇f_n converges to ∇f weakly in L^2_{loc} (see [55], Lemma 6.2, page 16). Since

$$|\nabla((\psi^2+\varepsilon)^{(p-1)/2}\psi)| = \left(\psi^2+\varepsilon\right)^{(p-1)/2}\frac{p\psi^2+\varepsilon}{\psi^2+\varepsilon}|\nabla\psi| \leq \left(\psi^2+\varepsilon\right)^{(p-1)/2}|\nabla\psi|$$

it suffices to show that the right-hand side is uniformly bounded in L^2_{loc} as $\varepsilon \to 0+$.

By assumption, for any $0 \le \rho \in Lip_c(M)$, we have

$$-\int \langle \nabla\psi, \nabla(\rho\psi)\rangle \ge -\int a(x)\,\psi^2\rho - A\int |\nabla\psi|^2\,\rho,$$

that is,

$$-\int \psi\,\langle \nabla\psi, \nabla\rho\rangle \ge -\int a(x)\,\psi^2\rho + (-A+1)\int |\nabla\psi|^2\,\rho. \tag{4.53}$$

Fix $\varepsilon > 0$ and choose

$$\rho = \left(\psi^2+\varepsilon\right)^{p-1}\phi^2$$

where $0 \le \phi \in C_c^\infty(M)$. Then,

$$\nabla\rho = 2(p-1)\phi^2\left(\psi^2+\varepsilon\right)^{p-2}\psi\nabla\psi + 2\phi\left(\psi^2+\varepsilon\right)^{p-1}\nabla\phi,$$

so that, using the Cauchy–Schwarz and Young inequalities and the fact that $p-1 < 0$, we estimate

$$\begin{aligned}
&LHS \text{ of } (4.53)\\
&= -2\int \phi\left(\psi^2+\varepsilon\right)^{p-1}\psi\langle\nabla\psi,\nabla\phi\rangle - 2(p-1)\int \phi^2\left(\psi^2+\varepsilon\right)^{p-2}\psi^2|\nabla\psi|^2\\
&\le 2\int \phi\left(\psi^2+\varepsilon\right)^{p-1/2}|\nabla\psi|\,|\nabla\phi| - 2(p-1)\int \phi^2\left(\psi^2+\varepsilon\right)^{p-1}|\nabla\psi|^2\\
&\le \frac{4}{\eta}\int \left(\psi^2+\varepsilon\right)^{p}|\nabla\phi|^2 - (2p-2-\eta)\int \phi^2\left(\psi^2+\varepsilon\right)^{p-1}|\nabla\psi|^2.
\end{aligned}$$

Moreover

$$\begin{aligned}
RHS \text{ of } (4.53) &= -\int a(x)\psi^2\left(\psi^2+\varepsilon\right)^{p-1}\phi^2 + (-A+1)\int \phi^2\left(\psi^2+\varepsilon\right)^{p-1}|\nabla\psi|^2\\
&\ge -\int |a(x)|\left(\psi^2+\varepsilon\right)^{p}\phi^2 + (-A+1)\int \phi^2\left(\psi^2+\varepsilon\right)^{p-1}|\nabla\psi|^2,
\end{aligned}$$

for $\eta > 0$. Combining the two inequalities and rearranging we obtain

$$\begin{aligned}
&(2p-A-1-\eta)\int \phi^2\left(\psi^2+\varepsilon\right)^{p-1}|\nabla\psi|^2\\
&\qquad\le \frac{4}{\eta}\int \left(\psi^2+\varepsilon\right)^{p}|\nabla\phi|^2 + \int |a(x)|\left(\psi^2+\varepsilon\right)^{p}\phi^2.\\
&\qquad\qquad\le \int \max\{1,|\psi|^{2p}\}\left(\frac{4}{\eta}|\nabla\phi|^2 + |a(x)|\phi^2\right).
\end{aligned}$$

Since $2p-A-1 > 0$, we may choose $\eta > 0$ small enough that the $(2p-A-1-\eta) > 0$, and conclude that the left-hand side is uniformly bounded in L^2_{loc} as $\varepsilon \to 0+$, as required to conclude the proof. □

Next we prove that, in the above assumptions, one can use the ordinary chain rule to compute the weak gradient of $|\psi^{p-1}|\psi$ even if $p < 1$. Note that, in this situation, the function $x \longmapsto |x|^{p-1}x$ is not Lipschitz so that standard results in the literature do not apply directly.

Lemma 4.13. *Let $0 < p_o$ (< 1) and assume that $\psi \in Lip_{loc}(M)$ satisfies (4.51) and (4.52), for every $p > p_o$. Then for every such p,*

$$|\psi|^{p-1}\nabla\psi \in L^2_{loc}(M) \tag{4.54}$$

and

$$\nabla\left(|\psi|^{p-1}\psi\right) = p|\psi|^{p-1}\nabla\psi, \text{ a.e. on } M, \tag{4.55}$$

where the LHS is understood in the sense of distribution and the RHS is defined almost everywhere, and is equal to 0 where the $\nabla\psi$ vanishes.

Proof. Let $p_o < p'$ (< 1) be any real number, and $\Omega \subset\subset M$ a fixed domain. Using $\nabla\psi \in L^2(\Omega)$ as a test function in (4.52) we have

$$\int_\Omega \langle\nabla\big((\psi^2+\varepsilon)^{(p'-1)/2}\psi\big),\nabla\psi\rangle \to \int_\Omega \langle\nabla(|\psi|^{p'-1}\psi),\nabla\psi\rangle, \text{ as } \varepsilon\to 0+. \tag{4.56}$$

Since

$$p'\left(\psi^2+\varepsilon\right)^{(p'-1)/2}|\nabla\psi|^2 \le \langle\nabla\big((\psi^2+\varepsilon)^{(p'-1)/2}\psi\big),\nabla\psi\rangle,$$

it follows from (4.56) and the monotone convergence theorem that

$$\lim_{\varepsilon\to 0}\left(\psi^2+\varepsilon\right)^{(p'-1)/2}|\nabla\psi|^2 = \begin{cases} 0 & \text{if } \nabla\psi = 0,\\ |\psi|^{p'-1}|\nabla\psi|^2 & \text{if } \nabla\psi\neq 0,\ \psi\neq 0,\\ +\infty & \text{if } \nabla\psi\neq 0,\ \psi = 0\end{cases}$$

is integrable on Ω. In particular, the set where $\psi = 0$ and $\nabla\psi \neq 0$ has measure zero, showing that the vector field $|\psi|^{(p'-1)/2}\nabla\psi$ is defined almost everywhere and in $L^2(\Omega)$. Therefore we may use this vector field in (4.52) and, arguing as above, show that $|\psi|^{3(p'-1)/4}\nabla\psi \in L^2(\Omega)$. Iterating, we deduce that, for every n,

$$|\psi|^{(2^n-1)(p'-1)/2^n}\nabla\psi \in L^2(\Omega). \tag{4.57}$$

Now, given $p > p_o$, let $p' = 2^n(p-1)/(2^n-1)+1$ so that $(2^n-1)(p'-1)/2^n = p-1$. Choosing n large enough that $p' > p_o$, shows that (4.54) holds.

Finally, to prove (4.55), let $\rho \in C^\infty_c(M)$. By (4.52),

$$\int\langle\nabla\big((\psi^2+\varepsilon)^{(p-1)/2}\psi\big),\nabla\rho\rangle \to \int\langle\nabla(|\psi|^{p-1}\psi),\nabla\rho\rangle, \text{ as } \varepsilon\to 0+. \tag{4.58}$$

On the other hand,

$$\nabla\big((\psi^2+\varepsilon)^{(p-1)/2}\psi\big) = (\psi^2+\varepsilon)^{(p-1)/2}\frac{p\psi^2+\varepsilon}{\psi^2+\varepsilon}\nabla\psi \to p|\psi|^{p-1}\nabla\psi$$

pointwise a.e., and its absolute value is bounded above by $p|\psi|^{p-1}|\nabla\psi|$ which is in L^2_{loc} by (4.54). Therefore we may apply the dominated convergence theorem to the left-hand side of (4.58) to obtain

$$\int \langle p|\psi|^{p-1}\nabla\psi, \nabla\rho\rangle = \int \langle \nabla(|\psi|^{p-1}\psi), \nabla\rho\rangle$$

as required. □

Chapter 5

A finite-dimensionality result

As briefly mentioned at the beginning of the previous chapter, typical geometric applications of Theorem 4.5 are obtained by applying it when the function ψ is the norm of the section of a suitable vector bundle. In appropriate circumstances, the theorem guarantees that certain vector subspaces of such sections are trivial, the main geometric assumption being the existence of a positive solution φ of the differential inequality

$$\Delta\varphi + Ha(x)\varphi \leq 0 \quad \text{weakly on } M, \tag{5.1}$$

where $a(x)$ is a lower bound for the relevant curvature term. According to Lemma 3.10 this amounts to requiring that the bottom of the spectrum of the Schrödinger operator $-\Delta - Ha(x)$ is non-negative.

We now consider the case where it is only assumed that (5.1) has a positive solution outside a compact set, which is related to the finiteness of the Morse index of the operator (see Lemma 3.16, Corollary 3.18 and Lemma 3.10).

Theorem 5.1. *Let $(M, \langle\,,\rangle)$ be a connected, complete, m-dimensional Riemannian manifold and E a Riemannian (Hermitian) vector bundle of rank l over M. The space of its smooth sections is denoted by $\Gamma(E)$. Having fixed*

$$a(x) \in C^0(M)$$

and constants $A \in \mathbb{R}$, p and H satisfying

$$H \geq p \geq A + 1, \quad p > 0, \tag{5.2}$$

let $V = V(a, A, p, H) \subset \Gamma(E)$ be any vector space with the following property:

(P) *Every $\xi \in V$ has the unique continuation property, i.e., ξ is the null section whenever it vanishes on some domain; furthermore the locally-Lipschitz function $\psi = |\xi|$ satisfies*

$$\begin{cases} \psi\Delta\psi + a(x)\psi^2 + A|\nabla\psi|^2 \geq 0 & \text{weakly on } M \\ \int_{B_r} \psi^{2p} = o(r^2) & \text{as } r \to +\infty. \end{cases} \tag{5.3}$$

If there exists a solution $0 < \varphi \in Lip_{loc}$ of the differential inequality

$$\Delta\varphi + Ha(x)\varphi \leq 0 \tag{5.4}$$

weakly outside a compact set $K \subset M$*, then*

$$\dim V < +\infty. \tag{5.5}$$

The spaces of harmonic functions, and more generally, harmonic forms on a Riemannian manifold are the most typical examples of spaces of sections for which the conditions of the theorem hold.

This situation can be generalized to the following setting. Let E be a Riemannian (Hermitian) vector bundle of rank l over M with a compatible connection D and let Δ_E be a differential operator acting on the space of smooth sections $\Gamma(E)$ of the form

$$\Delta_E = \Delta_B + \mathfrak{R} \tag{5.6}$$

where $\Delta_B = -\mathrm{Trace}\left(D^2\right)$ is the rough Laplacian, and $\mathfrak{R}$ is a smooth symmetric endomorphism of E. A smooth section $\xi \in \Gamma(E)$ is called Δ_E-harmonic if $\Delta_E \xi = 0$. We define the vector spaces

$$\mathcal{H}(E) = \{\xi \in \Gamma(E) : \Delta_E \xi = 0\}$$

and

$$L^{2p}\mathcal{H}(E) = \left\{\xi \in \mathcal{H}(E) : |\xi| \in L^{2p}(M)\right\}.$$

Note that Δ_E-harmonic sections satisfy the (strong) unique continuation property. In local coordinates, the condition $\Delta_B \xi + \mathfrak{R}\xi = 0$ becomes a system of l elliptic differential equations satisfying the structural assumptions of Aronszajn–Cordes, see Appendix A below. Indeed, we have

Proposition 5.2. *Let E be a rank l vector bundle over the connected manifold $(M, \langle\, ,\rangle)$ and let Δ_E be a differential operator acting on sections of E and satisfying (5.6). Let $\xi \in \mathcal{H}(E)$ be a Δ_E-harmonic section of E. If ξ has a zero of infinite order at some point $p \in M$, then ξ vanishes identically on M.*

Proof. Notation is as in Appendix A. By assumption $\Delta_E \xi = 0$ so that, according to (5.6),

$$\Delta_B \xi + \mathcal{R}\xi = 0. \tag{5.7}$$

where $\Delta_B = -trD^2$. Let $\left\{\xi^A\right\} \subset \Gamma(E)$ be a smooth, local orthonormal frame of E. Writing

$$\xi = \sum_{1 \le A \le l} u_A \xi^A$$

and setting

$$u = (u_A) \in \mathbb{R}^l,$$

straightforward computations give the local expression of (5.7):

$$\{\Delta u_A + F_A(u, \nabla u)\}\, \xi^A = 0, \tag{5.8}$$

where $F = (F_A)$ is a suitable function satisfying

$$|F(u, \nabla u)| \leq A|u| + B|\nabla u|$$

for some constants $A, B > 0$. Therefore, the local (vector-valued) function $u = (u_A)$ satisfies the assumptions of Theorem A.5, proving that harmonic sections locally enjoy the unique continuation property. A standard connectedness argument now completes the proof. Indeed, let

$$\Omega = \{x \in M : \xi = 0 \text{ in a neighborhood of } x\}.$$

Clearly, Ω is open. Furthermore, $\Omega \neq \emptyset$ because, according to Theorem A.5, ξ vanishes in a sufficiently small neighborhood of p. If we show that Ω is closed, then $\Omega = M$ and we are done. But this is obvious because, if $x \in \Omega'$ is a limit point of Ω, then, according to Remark A.3 and the local unique continuation property, we have $\xi = 0$ near x, proving that $x \in \Omega$. □

From (5.6) we deduce the Bochner-Weitzenböck formula, $\forall \xi \in \mathcal{H}(E)$,

$$-\frac{1}{2}\Delta|\xi|^2 = \langle \Delta_B \xi, \xi \rangle - |D\xi|^2 = -\langle \mathfrak{R}\xi, \xi \rangle - |D\xi|^2$$

which in turn implies that the differential inequality

$$|\xi|\Delta|\xi| - \langle \mathfrak{R}\xi, \xi \rangle = |D\xi|^2 - |\nabla|\xi||^2 \geq 0 \tag{5.9}$$

holds in the sense of distributions. The last inequality in (5.9) is known as (the first) " Kato inequality". We recall for completeness that, in case there exists a constant $k > 0$ such that

$$|D\xi|^2 - |\nabla|\xi||^2 \geq k|\nabla|\xi||^2, \tag{5.10}$$

one says that a " refined Kato inequality" holds.

If we let

$$\mathfrak{R}_-(x) = \sup_{|\xi|=1} \langle -\mathfrak{R}\xi, \xi \rangle,$$

then, from (5.9), we obtain

$$|\xi|(\Delta|\xi| + \mathfrak{R}_-(x)|\xi|) \geq 0.$$

and we are naturally led to consider the Schrödinger operator

$${}^H L = -\Delta - H\mathfrak{R}_-(x)$$

with $H > 0$ a real number. Accordingly, from Theorem 5.1 we immediately deduce the following

Corollary 5.3. *Maintaining the notation introduced above, assume that, for some $H \geq 1$,*

$$\operatorname{Ind}\left({}^{H}L\right) < +\infty.$$

Then,

$$\dim L^{2p}\mathcal{H}\left(E\right) < +\infty$$

for every $1 \leq p \leq H$.

As examples of bundles where the above considerations apply, we mention the space of spinors and of the exterior differential q-forms. In these settings the role of the operator Δ_E is played by the Dirac and the Hodge–De Rham Laplacians, respectively. Both operators can be written in terms of the rough Laplacian via a Bochner-type formula. In the spinorial case the endomorphism $\mathfrak{R}$ is given by the formula

$$\langle \mathfrak{R}\left(x\right)v, v\rangle = \frac{{}^{M}\mathrm{Scal}\left(x\right)}{4}\left|v\right|^{2}, \quad ,\forall v \in E_x,$$

where ${}^{M}\mathrm{Scal}$ denotes the scalar curvature of M (see [165]). For differential 1-forms $\mathfrak{R}$ is given by

$$\langle \mathfrak{R}\left(x\right)v, v\rangle = {}^{M}\mathrm{Ric}\left(x\right)\left(v, v\right)$$

where ${}^{M}\mathrm{Ric}$ is the Ricci tensor of M. In the case of differential k-forms on a locally, conformally flat manifold M of even dimension $m = 2k \geq 4$, one has (see [17])

$$\langle \mathfrak{R}\left(x\right)v, v\rangle = \frac{k!k\ {}^{M}\mathrm{Scal}\left(x\right)}{2\left(2k-1\right)}\left|v\right|^{2}.$$

The expression of $\mathfrak{R}$ for the exterior bundle $\Lambda^{q}\left(T^{*}M\right), q \geq 2$, on a general manifold is quite complicated but can be estimated in terms of the curvature operator $\mathcal{R}$ of M (the linear extension to $\Lambda^{2}TM$ of the $(2,2)$-Riemann curvature tensor) by (see [58])

$$\langle \mathfrak{R}\left(x\right)v, v\rangle \geq C\lambda\left(x\right)\left|v\right|^{2}$$

where $C = C(m,q) > 0$ is a constant depending on m and q and

$$\lambda\left(x\right) = \min_{V \in \Lambda^{2}T_{x}M} < \mathcal{R}_{x}(V), V > .$$

For future use we record the following particular case of Corollary 5.3.

Corollary 5.4. *Let $(M, \langle\,,\rangle)$ be a complete Riemannian manifold whose Ricci curvature satisfies*

$$\mathrm{Ric} \geq -a(x),$$

for some non-negative continuous function $a(x)$. Let ${}^{H}L = -\Delta - Ha(x)$, and assume that, for some $H \geq 1$,

$$\operatorname{Ind}\left({}^{H}L\right) < +\infty.$$

Then,

$$\dim L^{2p}\mathcal{H}(E) < +\infty$$

for every $1 \leq p \leq H$.

Remark 5.5. In the special case $p = 1$, finiteness results have been widely investigated by many authors under different assumptions. We limit ourselves to quote [104], [105] by P. Li and J. Wang, where Morse index assumptions are used in a way similar to ours, and [28] by G. Carron where quantitative-dimensional estimates are obtained assuming that the underlying manifold supports a global Sobolev inequality. We note explicitly that Theorem 5.1 on the one hand allows us to deal with integrability exponents $p \neq 1$, and, on the other hand, it enables us to avoid the request that $\psi = |\xi|$ be a solution of the more stringent inequality

$$\psi\Delta\psi + a(x)\psi^2 \geq k|\nabla\psi|^2 \tag{5.11}$$

for some constant $k > 0$. Often, in geometrical contexts, the validity of (5.11) depends on a refined Kato inequality. This is the case, for instance, in the Li-Wang papers cited above. The greater generality we achieve enables us to deal with the space of harmonic, $2p$ -integrable q-forms instead of restricting ourselves to the closed and co-closed ones for which a refined Kato inequality does hold.

5.1 Peter Li's lemma

The proof of Theorem 5.1 is based on the following version of a classical result due to Li (see [94] Lemma 11).

Lemma 5.6. *Let* $(E, \langle\,,\rangle_E)$ *be a Riemannian vector bundle of rank* l *over a compact Riemannian manifold* $\left(\bar\Omega, \langle\,,\rangle\right)$ *with (possibly empty) boundary* $\partial\bar\Omega$. *Let* $L^2\Gamma(E)$ *be the vector space of continuous sections of* E *endowed with the* L^2*-scalar product*

$$(\xi_1, \xi_2) = \int_{\bar\Omega} \langle\xi_1, \xi_2\rangle_E .$$

Let T *be a vector subspace of* $L^2\Gamma(E)$ *of positive finite dimension. Then, there exists a (non-zero) section* $\bar\xi \in T$ *such that, for any* $p > 0$,

$$(\dim T)^{\min\{1,p\}} \int_{\bar\Omega} |\bar\xi|_E^{2p} \leq \min\{l, \dim T\}^{\min\{1,p\}} \operatorname{vol}\left(\bar\Omega\right) \sup_{\bar\Omega} |\bar\xi|_E^{2p} . \tag{5.12}$$

Proof. It suffices to consider the case $p = 1$. The general case then follows by noting that, if $p > 1$, then

$$\begin{aligned}\dim T \int_\Omega |\bar\xi|_E^{2p} &= \dim T \int_\Omega |\bar\xi|_E^{2(p-1)} |\bar\xi|_E^2 \leq \left(\dim T \int_\Omega |\bar\xi|_E^2\right) \sup_\Omega |\bar\xi|_E^{2(p-1)} \\ &\leq \left\{\min\{l, \dim T\} \operatorname{vol}(\Omega) \sup_\Omega |\bar\xi|_E^2\right\} \sup_\Omega |\bar\xi|_E^{2(p-1)} .\end{aligned}$$

On the other hand, if $p < 1$, we can use Hölder inequality to obtain

$$\begin{aligned}\dim T \int_\Omega \left|\bar{\xi}\right|^{2p} &\leq \dim T \left(\int_\Omega \left|\bar{\xi}\right|^2\right)^p \mathrm{vol}\,(\Omega)^{1-p} \\ &= (\dim T)^{1-p}\left(\dim T \int_\Omega \left|\bar{\xi}\right|^2\right)^p \mathrm{vol}\,(\Omega)^{1-p} \\ &\leq (\dim T)^{1-p}\left(\mathrm{vol}\,(\Omega)\min\{l, \dim T\}\sup_\Omega \left|\bar{\xi}\right|^2\right)^p \mathrm{vol}\,(\Omega)^{1-p}\end{aligned}$$

which implies

$$(\dim T)^p \int_\Omega \left|\bar{\xi}\right|^{2p} \leq \mathrm{vol}\,(\Omega)\min\{l, \dim T\}^p \sup_\Omega \left|\bar{\xi}\right|^{2p}.$$

Thus, let $p = 1$ and consider the function $F(x) : \bar{\Omega} \to \mathbb{R}$ defined by

$$F(x) = \sum_{i=1}^{\dim T} |f_i(x)|_E^2 \tag{5.13}$$

where $\{f_i(x)\}$ is an orthonormal basis of $T \subset L^2\Gamma(E)$. Since T is finite-dimensional, different orthonormal bases in T are related by constant coefficients unitary matrices, and therefore F is independent of the chosen orthonormal basis. Let $x_0 \in \bar{\Omega}$ be the point of absolute maximum of F. Then

$$\dim T = \int_{\bar{\Omega}} F(x)\, d\mathrm{vol}\,(x) \leq F(x_0)\,\mathrm{vol}\left(\bar{\Omega}\right). \tag{5.14}$$

We shall show that there exists $\bar{\xi} \in T$ such that $\left(\bar{\xi}, \bar{\xi}\right) = 1$ and

$$F(x_0) \leq \min\{l, \dim T\} \sup_{\bar{\Omega}} \left|\bar{\xi}\right|_E^2. \tag{5.15}$$

Toward this end we consider the evaluation map $\varepsilon_{x_0} : T \to E_{x_0}$

$$\varepsilon_{x_0}\xi = \xi(x_0).$$

Since T is finite-dimensional, ε_{x_0} is continuous with norm

$$\sup_{(\xi,\xi)=1} |\varepsilon_{x_0}\xi|_E = \left|\varepsilon_{x_0}\bar{\xi}\right|_E \tag{5.16}$$

for some $\bar{\xi}$ in the unit sphere of T. Note that the kernel $S = \mathrm{Ker}\,(\varepsilon_{x_0})$ consists of those sections in T vanishing at x_0 and we have the orthogonal decomposition

$$T = S^\perp \oplus S$$

with $\dim S^{\perp} \leq \min\{l, \dim T\}$. Let $\{\xi_j\}$ be an orthonormal basis of T adapted to this decomposition, namely, $\xi_1, \ldots, \xi_{\dim S^{\perp}}$ is a basis of $S^{\perp}$ and $\xi_{\dim S^{\perp}+1}, \ldots,$ $\xi_{\dim T}$ is a basis of S. Since F is independent of the choice of $\{f_j\}$, we have

$$F(x) = \sum_{i=1}^{\dim T} |\xi_i(x)|_E^2 .$$

Whence, using (5.16), we get

$$\begin{aligned} F(x_0) &= \sum_{i=1}^{\dim S^{\perp}} |\xi_i(x_0)|_E^2 = \sum_{i=1}^{\dim S^{\perp}} |\varepsilon_{x_0}\xi_i|_E^2 \leq \dim S^{\perp} \sup_{\substack{\xi \in T \\ (\xi,\xi)=1}} |\varepsilon_{x_0}\xi|_E^2 \\ &\leq \min\{l, \dim T\} \sup_{\bar{\Omega}} \left|\bar{\xi}\right|_E^2 \end{aligned}$$

proving (5.15). □

In view of Peter Li's lemma, the strategy of the proof of Theorem 5.1 consists in showing that there are a geodesic ball $B_{\bar{R}} \subset M$ and a constant $C > 0$, such that the following a-priori local estimate holds true:

$$\sup_{B_{\bar{R}}} |\xi|^{2p} \leq C \int_{B_{\bar{R}}} |\xi|^{2p}$$

for every $\xi \in V$. This is obtained in Lemma 5.17 below combining a local, weak Harnack inequality for solutions of (5.3), see Theorem 5.16 below, and the annuli-estimate technique of Li and Wang, [104], [105]. As in the previous section, to derive this estimate we use a technique based on the interaction of the differential inequalities (5.3) and (5.4), which is the crucial ingredient to extend the result to situations where a refined Kato inequality does not hold (see Remark 5.5 above).

The proof of the local weak Harnack inequality is based on the following idea (see, e.g., [140] pp. 486-487): if we have a local Sobolev inequality, then the weak Harnack inequality for subsolutions of a suitable class of PDEs is a formal consequence, via the Moser iteration procedure, of a Caccioppoli-type inequality. Schematically:

$$\text{Sobolev} + \text{Caccioppoli} \quad \underset{\substack{\uparrow \\ \text{Moser Iteration}}}{\Longrightarrow} \quad \text{Weak Harnack.} \tag{5.17}$$

As we shall see, there is no obstruction against a manifold supporting a local Sobolev inequality. Therefore our first task will be to obtain a Caccioppoli inequality.

To obtain the estimate on annuli we will also need a suitable Poincaré inequality on annuli valid for functions vanishing only on one boundary component.

We begin by collecting the results we are going to use.

5.2 Poincaré-type inequalities

The following result is due to Li and Schoen, see Corollary 1.1 in [95] and Remark 5.10 below. It is a Riemannian version of the classical L^p-Poincaré inequality ($p \geq 1$) for functions with zero boundary conditions. Note that the case $p = 2$ gives a lower estimate for the first Dirichlet eigenvalue of $-\Delta$. The case $p = 1$ will be used below to obtain a local Sobolev inequality.

Theorem 5.7. *Let R be such that $5R \leq \operatorname{diam}(M)$, and $B_{5R}(o)$ is a relatively compact geodesic ball in $(M, \langle \, , \rangle)$. Let $B \geq 0$ be such that*

$$^M\mathrm{Ric} \geq -(m-1)B^2 \quad \text{on } B_{5R}(o).$$

Then, for every $p \geq 1$,

$$\int_{B_R(o)} |u|^p \leq C_p \int_{B_R(o)} |\nabla u|^p \quad \forall u \in C_0^\infty(B_R(o))$$

with

$$C_p = \left(\frac{pR \exp(2m(1+BR))}{1+BR} \right)^p .$$

Proof. Fix $x_1 \in \partial B_{2R}(o)$ and let $\rho(y) = \operatorname{dist}_{(M,\langle \, , \rangle)}(y, x_1)$. By the Laplacian comparison theorem, and the assumption on the Ricci curvature, it follows that, on $B_{3R}(o)$,

$$\Delta\rho \leq (m-1)B\coth(Br) \leq \frac{m-1}{\rho} + (m-1)B \tag{5.18}$$

in the weak sense, that is, for every $0 \leq \phi \in C_c^\infty(B_{3R}(o))$,

$$\int \rho\Delta\phi \leq \int \Big[\frac{m-1}{\rho} + (m-1)B\Big]\phi.$$

Letting $\alpha > 0$ be a constant to be chosen later, we compute

$$\Delta e^{-\alpha\rho} = \alpha e^{-\alpha\rho}(-\Delta\rho + \alpha) \tag{5.19}$$

in the weak sense. Since $B_R(o) \subset B_{3R}(x_1) \setminus B_R(x_1)$, we deduce that

$$\Delta e^{-\alpha\rho} \geq \alpha e^{-\alpha\rho}\Big[\alpha - \frac{m-1}{R} - (m-1)B\Big]$$

on $B_R(o)$, and choosing

$$\alpha = m(\frac{1}{R} + B),$$

we obtain

$$\Delta e^{-\alpha\rho} \geq \alpha e^{-3\alpha R}(\frac{1}{R} + B) \quad \text{on } B_R(o). \tag{5.20}$$

Let $\psi \in C_c^\infty(B_R(o))$, $\psi \geq 0$ and apply the divergence theorem to obtain

$$\int_{B_R(o)} \psi \Delta e^{-\alpha\rho} = -\int_{B_R(o)} \langle \nabla\psi, \nabla e^{-\alpha\rho}\rangle,$$

whence, using (5.20), the Schwarz inequality, and the fact that $\rho \geq R$ on $B_R(o)$,

$$e^{-3R\alpha}(\frac{1}{R} + B)\int_{B_R(o)} \psi \leq \int_{B_R(o)} |\nabla\psi| e^{-\alpha\rho} \leq \int_{B_R(o)} |\nabla\psi| e^{-\alpha R}$$

and rearranging

$$\int_{B_R(o)} \psi \leq \frac{R}{1+BR} e^{2\alpha R} \int_{B_R(o)} |\nabla\psi|.$$

Now let $u \in C_c^\infty(B_R(o))$. Applying Kato's and Hölder's inequality, and inserting the value of α we get

$$\begin{aligned}
\int_{B_R(o)} |u|^p &\leq \frac{R}{1+BR} e^{2m(1+BR)} \int_{B_R(o)} |\nabla(|u|^p)| \\
&= \frac{R}{1+BR} e^{2m(1+BR)} \int_{B_R(o)} p|u|^{p-1} |\nabla|u|| \\
&\leq \frac{pR}{1+BR} e^{2m(1+BR)} \int_{B_R(o)} |u|^{p-1} |\nabla u| \\
&\leq \frac{pR}{1+BR} e^{2m(1+BR)} \left\{\int_{B_R(o)} |u|^p\right\}^{1-1/p} \left\{\int_{B_R(o)} |\nabla u|^p\right\}^{1/p},
\end{aligned}$$

and the required conclusion follows by simplifying. □

There are several variants of this result, each requiring some special condition on the family of functions under consideration. For instance, one can replace the zero-boundary condition with the zero-mean value condition, namely each function u has to satisfy $\int_B u = 0$. The corresponding inequality is usually known under the name of Neumann–Poincaré and it yields a lower bound for the first (non-zero) Neumann eigenvalue.

We shall employ a version of the Poincaré inequality on annuli valid for functions vanishing only on one of the boundary components. We state it in the form we will need. In a Euclidean setting, due to the special structure of the manifold, different techniques have been used by L. Hedberg, to obtain more general and refined statements of these type of inequalities , see Lemma 2.1 in [81] and Theorem 4.1 in [82].

Theorem 5.8. *Let $B_{\bar{R}}(o)$ be a relatively compact geodesic ball in the Riemannian manifold $(M, \langle\,,\rangle)$. Assume that*

$$\mathrm{Ric} \geq -(m-1)\, B^2 \ \text{on } B_{\bar{R}}(o)$$

for some $B > 0$. *Fix* $p > 1$ *and* $0 < R_1 < R_2 < \bar{R}$, *then for every* $u \in C^0\left(\overline{B_{R_2} \setminus B_{R_1}}\right) \cap W^{1,p}(B_{R_2} \setminus B_{R_1})$ *such that*

$$u = 0 \quad \text{on } \partial B_{R_1}$$

we have

$$\int_{B_{R_2} \setminus B_{R_1}} |u|^p \leq C^p \int_{B_{R_2} \setminus B_{R_1}} |\nabla u|^p$$

with

$$C = p(R_2 - R_1)\left(\frac{\sinh(BR_2)}{\sinh(BR_1)}\right)^{(m-1)}.$$

The proof of the theorem is based on the next lemma. For the sake of completeness we recall that, for any $p > 1$, the p-Laplacian of a function $u \in W^{1,p}_{loc}$ is defined by the expression

$$\Delta_p u = \operatorname{div}\left(|\nabla u|^{p-2} \nabla u\right)$$

where the divergence is understood in the weak sense.

Lemma 5.9. *Let* $A = \Omega_2 \setminus \overline{\Omega}_1 \subset (M, \langle\, , \rangle)$ *be an annular domain with compact boundary* $\partial\Omega_1 \cup \partial\Omega_2$. *Let* $p > 1$ *and* $0 \leq \phi \in Lip\left(\bar{A}\right)$ *be a non-zero solution of the problem*

$$\begin{cases} \Delta_p \phi \geq 0 & \text{weakly on } A, \\ \phi = 0 & \text{on } \partial\Omega_2. \end{cases} \tag{5.21}$$

Suppose also that

$$|\nabla \phi| > 0 \quad \text{on } \bar{A}. \tag{5.22}$$

Then, there exists an explicit constant

$$C = \frac{\inf_{\bar{A}} |\nabla \phi|^p}{p^p \sup_{\bar{A}} |\phi|^p} > 0$$

such that

$$C \int_A |u|^p \leq \int_A |\nabla u|^p \tag{5.23}$$

for every $u \in C^0\left(\overline{A}\right) \cap W^{1,p}(A)$ *satisfying*

$$u = 0 \quad \text{on } \partial\Omega_1. \tag{5.24}$$

Proof. By assumption,

$$-\int \left\langle |\nabla \phi|^{p-2} \nabla \phi, \nabla \rho \right\rangle \geq 0, \tag{5.25}$$

for every $0 \leq \rho \in C_0^\infty(A)$. As a matter of fact, since $\phi \in Lip(\bar{A}) \subset W^{1,p}(A)$, we have the validity of (5.25) for every $0 \leq \rho \in W_0^{1,p}(A)$. Note that $\phi |u|^p \in W_0^{1,p}(A)$. Indeed $\phi |u|^p$ lies in $W^{1,p}(A)$ and vanishes continuously on ∂A. Therefore, we can choose $\rho = \phi |u|^p$ in (5.25). Using Schwarz and Hölder inequalities we get

$$\begin{aligned} 0 &\leq -\int \left\langle |\nabla \phi|^{p-2} \nabla \phi, \nabla (\phi |u|^p) \right\rangle \qquad (5.26) \\ &= -\int |\nabla \phi|^p |u|^p - \int p |u|^{p-1} \phi |\nabla \phi|^{p-2} \langle \nabla \phi, \nabla |u| \rangle \\ &\leq -\int |\nabla \phi|^p |u|^p + p \int \phi |u|^{p-1} |\nabla |u|| |\nabla \phi|^{p-1} \\ &\leq -\int |u|^p |\nabla \phi|^p + p \left(\int |\phi|^p |\nabla u|^p \right)^{1/p} \left(\int |u|^p |\nabla \phi|^p \right)^{(p-1)/p} \end{aligned}$$

proving the Caccioppoli-type inequality

$$\int |u|^p |\nabla \phi|^p \leq p^p \int |\phi|^p |\nabla u|^p, \qquad (5.27)$$

which, recalling the properties of ϕ, easily yields (5.23). □

Proof of Theorem 5.8. We simply have to choose the test function ϕ in Lemma 5.9. One observes that, in case of model manifolds, the p-equilibrium potential of the condenser $E = \left(B_{R_2}, \overline{B_{R_1}}\right)$ is suitable for the purpose. The general case can be obtained by a model-comparison argument as follows.

Set $r(x) = \text{dist}_M(x, o)$ and define $\phi(x) = \alpha(r(x))$ where

$$\alpha(r) = c \int_r^{R_2} \frac{dt}{\left[B^{-1} \sinh(Bt)\right]^{\frac{m-1}{p-1}}}$$

with

$$c = \left(\int_{R_1}^{R_2} \frac{dt}{\left[B^{-1} \sinh(Bt)\right]^{\frac{m-1}{p-1}}} \right)^{-1}.$$

Then $\phi \geq 0$, $\phi = 0$ on ∂B_{R_2}, $\phi = 1$ on ∂B_{R_1} and

$$|\nabla \phi(x)| = c[B^{-1} \sinh(r(x))]^{\frac{1-m}{p-1}} > 0 \quad \text{on } \overline{B_{R_2} \setminus B_{R_1}}.$$

Moreover , since $\alpha' \leq 0$, using the Laplacian comparison theorem, we obtain, pointwise in the complement of $cut(o)$,

$$\begin{aligned} \Delta_p \phi &= (p-1)(-\alpha')^{p-2} \alpha'' - (-\alpha')^{p-1} \Delta r(x) \\ &\geq (p-1)(-\alpha')^{p-2} \alpha'' - (-\alpha')^{p-1} (m-1) \coth r(x) = 0. \end{aligned}$$

As usual, this latter extends weakly on all of the annulus. Therefore, Lemma 5.9 applies to obtain the desired inequality. □

Remark 5.10. Obviously, Theorem 5.8 implies the validity of Theorem 5.7 for $p > 1$. Indeed, having fixed a geodesic ball $B_r(o)$, we can choose an annular domain $B_{R_2}(z) \setminus B_{R_1}(z)$ such that

$$B_r(o) \subset B_{R_2}(z) \setminus B_{R_1}(z).$$

Since every $u \in C_0^\infty(B_r(o))$ satisfies $u = 0$ on the interior boundary component $\partial B_{R_1}(z)$, Theorem 5.8 applies and the asserted result follows. We point out that the corresponding Poincaré constant is different from that of Lemma 5.9, but this is irrelevant in local applications. Similarly, this yields the result for annular regions $\Omega_2 \setminus \overline{\Omega}_1$. Finally, we remark that Theorem 5.8 could be established using computations in polar coordinates.

5.3 Local Sobolev inequality

It is well known that for $m \geq 2$, the standard Euclidean space $(\mathbb{R}^m, \mathrm{can})$ enjoys the L^1-Sobolev inequality

$$S_1^{-1}\left(\int |v|^{\frac{m}{m-1}}\right)^{\frac{m-1}{m}} \leq \int |\nabla v| \tag{5.28}$$

for every $v \in C_0^\infty(\mathbb{R}^m)$, where

$$S^{-1} = m\omega_m^{1/m}$$

with ω_m the volume of the unit ball of $\mathbb{R}^m$. The situation may change drastically if we consider a general complete Riemannian manifold. This depends on the isoperimetric properties of the manifold. In fact, a classical result by Federer and Fleming asserts that (5.28) is equivalent to

$$S_1^{-1}\mathrm{vol}(\Omega)^{\frac{m-1}{m}} \leq \mathrm{vol}(\partial\Omega)$$

for every relatively compact domain Ω with smooth boundary. In particular, integrating the above inequality yields a uniform lower bound for the volumes of geodesic balls of prescribed radius, namely

$$\inf_{x\in M} \mathrm{vol} B_r(x) \geq S_1^{-1} r^m > 0,$$

and the underlying manifold has at least Euclidean volume growth.

Note however that obstructions of these types have a global nature, and indeed, locally, the usual Euclidean Sobolev inequality is valid on every Riemannian manifold.

Theorem 5.11. *Let $B_{2R}(o)$ be a relatively compact geodesic ball in the Riemannian manifold $(M, \langle\, ,\rangle)$ of dimension $m = \dim M \geq 2$. Assume that $2R \leq \mathrm{diam}(M)$.*

Then, for each domain $\Omega \subset\subset B_R(o)$ there exists a constant $S_1 = S_1(\Omega, m) > 0$ such that the L^1-Sobolev inequality

$$S_1^{-1}\left(\int |v|^{\frac{m}{m-1}}\right)^{\frac{m-1}{m}} \leq \int |\nabla v| \tag{5.29}$$

holds for every $v \in C_0^\infty(\Omega)$.

In particular, the Sobolev inequalities

$$S_2^{-1}\left(\int v^{\frac{mq}{m-1}}\right)^{\frac{m-1}{mq}} \leq \left\{\int |\nabla v|^2\right\}^{1/2}, \quad \forall v \in C_0^\infty(\Omega) \tag{5.30}$$

hold for any choice of

$$q \in \begin{cases} \left[1, \frac{2(m-1)}{m-2}\right] & \text{if } m \geq 3, \\ [1, +\infty) & \text{if } m = 2, \end{cases} \tag{5.31}$$

and for some appropriate constant $S_2 = S_2(\Omega, m, q) > 0$.

Proof of Theorem 5.11. We supply a proof for the sake of completeness. Note, first of all, that (5.30) follows applying (5.29) to the function

$$v = |u|^q$$

for every $u \in C_0^\infty(\Omega)$. Indeed, assume that $q > 1$ satisfies (5.31). Then,

$$S_1^{-1}\left(\int |u|^{\frac{mq}{m-1}}\right)^{\frac{m-1}{m}} \leq q\int |u|^{q-1}|\nabla u|,$$

whence, applying Hölder inequality with conjugate exponents

$$\frac{mq}{(m-1)(q-1)}, \qquad \frac{mq}{m+q-1},$$

and simplifying we obtain

$$S_1^{-1}\left(\int |u|^{\frac{mq}{m-1}}\right)^{\frac{m-1}{mq}} \leq q\left(\int |\nabla u|^{\frac{mq}{m+q-1}}\right)^{\frac{m+q-1}{mq}}.$$

Set

$$p = \frac{2(m+q-1)}{mq},$$

and note that $p \geq 1$ by the assumptions on q. If $p = 1$ the above inequality is (5.30) and we are done. Otherwise, the desired conclusion is obtained by applying Hölder inequality with exponents p and $p/(p-1)$ to the right-hand side.

We now come to the proof of (5.29). For each $\bar{x} \in \Omega$ there exists a small geodesic ball $B_\varepsilon(\bar{x}) \subset\subset B_R(o)$ such that, within $B_\varepsilon(\bar{x})$, the metric $\langle\,,\rangle$ is quasi Euclidean, i.e., there exists a positive constant $C > 0$ such that

$$C^{-1}\mathrm{can}_{\mathbb{R}^m} \leq \langle\,,\rangle \leq C\,\mathrm{can}_{\mathbb{R}^m}$$

in the sense of quadratic forms. As a consequence both the length of the gradient, and the Riemannian volume element are controlled from above and below by their Euclidean counterparts, i.e.,

$$C_1^{-1}|^{\mathbb{R}^m}\nabla v|_{\mathbb{R}^m} \leq |\nabla v| \leq C_1|^{\mathbb{R}^m}\nabla v|_{\mathbb{R}^m}$$

and,

$$C_2^{-1}d\mathrm{vol}_{\mathbb{R}^m} \leq d\mathrm{vol} \leq C_2 d\mathrm{vol}_{\mathbb{R}^m},$$

for appropriate constants $C_1, C_2 > 0$, and it follows that (5.29) holds on $B_\varepsilon(\bar{x})$ for some choice of $S_1(B_\varepsilon(\bar{x})) > 0$. To conclude, using Theorem 5.7 and a standard partition of unity argument, we paste these inequalities together. Let $\left\{B_{\varepsilon_j}\right\}_{j=1}^n$ be a finite covering of the compact $\bar{\Omega}$ by small geodesic balls contained in $B_R(o)$. The L^1-Sobolev constant relative to B_{ε_j} is denoted by S_1^j. Choose a partition of unitity $\{\varphi_j\}_{j=1}^n$ subordinated to this covering and satisfying $\mathrm{supp}(\varphi_j) \subset B_{\varepsilon_j}$.

Let $v \in C_0^\infty(\Omega)$. Then

$$\begin{aligned}
\left(\int_\Omega |v|^{\frac{m}{m-1}}\right)^{\frac{m-1}{m}} &= \left(\int_\Omega |\textstyle\sum \varphi_j v|^{\frac{m}{m-1}}\right)^{\frac{m-1}{m}} \\
&\leq \sum \left(\int_\Omega |\varphi_j v|^{\frac{m}{m-1}}\right)^{\frac{m-1}{m}} \qquad \text{(Sobolev on } B_{\varepsilon_j}) \\
&\leq \sum S_1^j \int |\nabla(\varphi_j v)| \\
&\leq \sum S_1^j \left\{\int_\Omega \varphi_j |\nabla v| + \int_\Omega |v|\,|\nabla \varphi_j|\right\} \\
&\leq \sum S_1^j \left\{\int_\Omega |\nabla v| + \max_{\bar{\Omega}} |\nabla \varphi_j| \int_\Omega |v|\right\} \qquad \text{(Poincaré)} \\
&\leq S_1 \int_\Omega |\nabla v|
\end{aligned}$$

where we have set

$$S_1 = \sum S_1^j \left(1 + C_3 \max_{\bar{\Omega}} |\nabla \varphi_j|\right)$$

with $C_3 > 0$ the Poincaré constant of Theorem 5.7. □

5.4 L^2 Caccioppoli-type inequality

It is well known that if v is a non-negative subharmonic function v on a domain $\Omega \subset\subset M$, then there exists an absolute constant $C > 0$, such that, for every $\eta \in C_0^\infty(\Omega)$,

$$C \int_\Omega |\nabla v|^2 \eta^2 \leq \int_\Omega v^2 |\nabla \eta|^2 . \tag{5.32}$$

Indeed, in the proof of Lemma 5.9 we showed that (5.32) holds with $C = 1/4$. This integral inequality is usually referred to as the Caccioppoli inequality for subharmonic functions on Ω. The exponent 2 entering in the gradient terms is intimately related with the structure of the Laplace operator. This is well understood if we consider, for instance, the p-Laplacian or other divergence-form non-linear operators modelled on the p-Laplacian; see [140]. Generalizing the above situation, we say that an L^μ-Caccioppoli-type inequality is valid if

$$C \int_\Omega \eta^\mu v^q |\nabla v|^\mu \leq \int_\Omega v^{q+\mu} |\nabla \eta|^\mu \tag{5.33}$$

holds with $q \geq 0$ and $\mu > 1$. We shall come back to (5.33) in the next section. Here, we limit ourselves to proving the following simple variant of (5.32).

Lemma 5.12. *Let $(M, \langle\, ,\rangle)$ be a Riemannian manifold and let Ω be a relatively compact domain in M. Let $0 < w \in C^0(\bar{\Omega})$ and $v \in Lip_{loc}(\Omega)$ satisfy the differential inequality*

$$v\mathrm{div}\,(w\nabla v) \geq 0 \tag{5.34}$$

weakly on Ω. Then, for any fixed $q \geq 0$,

$$D_q \int_\Omega w|v|^q |\nabla v|^2 \eta^2 \leq \int_\Omega w|v|^{q+2} |\nabla \eta|^2 \qquad \forall \eta \in C_0^\infty(\Omega), \tag{5.35}$$

where

$$D_q = \frac{(1+q)^2}{4}.$$

Proof. Inequality (5.34) means that, for each $0 \leq \rho \in Lip_0(\Omega)$,

$$-\int \langle w\nabla v, \nabla(v\rho)\rangle = -\int w\rho |\nabla v|^2 - \int wv \langle \nabla v, \nabla \rho\rangle \geq 0.$$

We choose

$$\rho = (v^2+\delta)^{q/2}\eta^2$$

with $\eta \in C_0^\infty(\Omega)$, and use the Schwarz and the Young inequality to obtain

$$\begin{aligned} 0 &\geq \int w(v^2+\delta)^{q/2}\eta^2 (1 + \frac{qv^2}{v^2+\delta}) |\nabla v|^2 - \int 2w(v^2+\delta)^{q/2}|v||\eta||\nabla v||\nabla \eta| \\ &= \int w(v^2+\delta)^{q/2}\eta^2 (1 + \frac{qv^2}{v^2+\delta} - \varepsilon) |\nabla v|^2 - \frac{1}{\varepsilon}\int w(v^2+\delta)^{q/2} v^2 |\nabla \eta|^2. \end{aligned}$$

Letting $\delta \to 0$, and using the dominated convergence theorem, we deduce that

$$\varepsilon(1+q-\varepsilon)\int w|v|^q\eta^2|\nabla v|^2 \le \int w|v|^{q+2}v^2|\nabla\eta|^2,$$

and the conclusion follows by optimizing with respect to $\varepsilon \in (0, q+1]$. □

5.5 The Moser iteration procedure

The Moser iteration is a by now standard procedure that enables one to get an L^∞-L^q inequality starting from an L^p-L^q one. The process is explained in the next

Lemma 5.13. *Let $\gamma, \mu > 1$ and $R > 0$ be fixed real numbers. Suppose that v is a non-negative locally bounded function such that, for every $0 < r_1 < r_2 \le R_o$ and for every $q \ge 1$, the integral inequality*

$$\|v^q\|_{L^{\mu\gamma}(B_{r_1})} \le \frac{C}{r_2 - r_1}\|v^q\|_{L^\mu(B_{r_2})} \tag{5.36}$$

holds for some constant $C = C(R_o, \mu) > 0$. Then, having fixed $q_0 > 0$, and $0 < \bar{R} < R \le R_o$, the following L^μ-weak Harnack inequality holds:

$$\|v^{q_0}\|_{L^\infty(B_{\bar{R}})} \le \left(\frac{C'}{R-\bar{R}}\right)^{\frac{\gamma}{\gamma-1}}\|v^{q_0}\|_{L^\mu(B_R)}. \tag{5.37}$$

Proof. We fix $q_0 \ge 1$, $0 < \bar{R} < R <\le R_o$ and, for every $j = 0, 1, 2, \ldots$, we set

$$q_j = q_0\gamma^j; \quad R_j = \bar{R} + \left(R - \bar{R}\right)2^{-j}.$$

An application of (5.36) with $q = q_j$, $r_2 = R_j$ and $r_1 = R_{j+1}$ gives

$$\left\|v^{q_0\gamma^j}\right\|_{L^{\mu\gamma}(B_{R_{j+1}})} \le \frac{2^{j+1}C}{(R-\bar{R})}\left\|v^{q_0\gamma^j}\right\|_{L^\mu(B_{R_j})}$$

which, in turn, can be written in the form

$$\|v^{q_0}\|_{L^{\mu\gamma^{j+1}}(B_{R_{j+1}})} \le \left\{\frac{2^{j+1}C}{(R-\bar{R})}\right\}^{\gamma^{-j}}\|v^{q_0}\|_{L^{\mu\gamma^j}(B_{R_j})}. \tag{5.38}$$

Iterating (5.38) n-times and noting that $\{R_n\} \searrow \bar{R}$ we deduce

$$\begin{aligned}\|v^{q_0}\|_{L^{\mu\gamma^{n+1}}(B_{\bar{R}})} &\le \|v^{q_0}\|_{L^{\mu\gamma^{n+1}}(B_{R_{n+1}})}\\ &\le \prod_{j=0}^{n}\left\{\frac{2^{j+1}C}{(R-\bar{R})}\right\}^{\gamma^{-j}}\|v^{q_0}\|_{L^\mu(B_R)}\\ &= 2^{\sum_{j=0}^{n}\frac{j+1}{\gamma^j}}\left\{\frac{C}{(R-\bar{R})}\right\}^{\sum_{j=0}^{n}\gamma^{-j}}\|v^{q_0}\|_{L^\mu(B_R)}.\end{aligned}$$

The validity of (5.37) with $q_0 \geq 1$ follows letting $n \to +\infty$, and recalling that for a space X of finite measure and for any $f \in L^\infty(X)$,

$$\|f\|_{L^p(X)} \to \|f\|_{L^\infty(X)} \text{ as } p \to +\infty.$$

At this point one uses a standard argument (see, e.g., [95]) to prove that (5.36) holds for every $q_0 > 0$. We outline the argument for completeness. Clearly it suffices to consider the case $q_0 < 1$.

W use the above result to conclude that for every $0 < R \leq R_o$, and $\epsilon, \theta > 0$ such that $\epsilon + \theta \leq 1$ we have

$$\begin{aligned}\sup_{B_{\theta R}} v &\leq \left(\frac{C'}{\epsilon R}\right)^{\frac{\gamma}{\gamma-1}} \int_{B_{(\theta+\epsilon)R}} v \\ &\leq \left(\frac{C'}{\epsilon R}\right)^{\frac{\gamma}{\gamma-1}} \Big[\sup_{B_{(\theta+\epsilon)R}} v\Big]^{1-q_0} \int_{B_R} v^{q_0}.\end{aligned}$$

Setting $M(\theta) = \sup_{B_{\theta r}} v$ we rewrite the last inequality in the form

$$M(\theta) \leq \left(\frac{C'}{\epsilon R}\right)^{\frac{\gamma}{\gamma-1}} M(\theta+\epsilon)^\lambda \int_{B_R} v^{q_0},$$

where $\lambda = 1 - q_0$. Setting $\theta_0 = 1 - \tau$, $0 < \tau < 1$, $\theta_i = \theta_{i-1} + 2^{-i}\tau$ and $\epsilon_i = 2^{-i}\tau$, we get

$$M(\theta_{i-1}) \leq \left(\frac{C'}{2^{-i}\tau R}\right)^{\frac{\gamma}{\gamma-1}} M(\theta_i)^\lambda \int_{B_R} v^{q_0},$$

whence, iterating,

$$M(\theta_0) \leq \left\{\left(\frac{C'}{\tau R}\right)^{\frac{\gamma}{\gamma-1}}\right\}^{\sum_{i=0}^{n-1}\lambda^i} \times 2^{\frac{\gamma}{\gamma-1}\sum_{i=0}^{n-1}(i+1)\lambda^i} M(\theta_n)^{\lambda^n} \left(\int_{B_R} v^{q_0}\right)^{\sum_{i=0}^{n-1}\lambda^i}.$$

Letting $n \to +\infty$, and noting that

$$\theta_n = \theta_0 + \tau \sum_1^n 2^{-i} \to 1, \quad \lambda^n \to 0$$

and

$$\sum_0^\infty \lambda^i = \frac{1}{1-\lambda} = 1/q_0 \quad \sum_0^\infty (i+1)\lambda^i = \frac{d}{d\lambda}\sum_0^\infty \lambda^i = (1/q_0)^2,$$

we conclude that

$$M(1-\tau) \leq \left(\frac{2^{1/q_0}C'}{\tau R}\right)^{\frac{\gamma}{(\gamma-1)q_0}} \left(\int_{B_R} v^{q_0}\right)^{1/q_0},$$

that is, letting $\bar{R} = (1-\tau)R$, recalling the definition of $M(\theta)$,

$$\sup_{B_{\bar{R}}} v^{q_0} \leq \left(\frac{2^{1/q_0}C'}{R-\bar{R}}\right)^{\frac{\gamma}{(\gamma-1)}} \int_{B_R} v^{q_0}$$

as required to conclude the proof. □

Obviously, a natural problem is now to understand when an integral inequality like that in (5.36) is satisfied. This is addressed in the next

Lemma 5.14. *Let $\mu, \gamma > 1$ and $R > 0$ be fixed real numbers. Assume that the L^μ-Sobolev inequality*

$$S^{-1} \|\eta\|_{L^{\mu\gamma}(B_R)} \leq \|\nabla \eta\|_{L^\mu(B_R)} \tag{5.39}$$

holds for all $\eta \in C_0^\infty(B_R)$ and for some constant $S = S(R, \gamma, \mu) > 0$. Let $v \geq 0$ be a locally Lipschitz function for which for every $\tilde{q} > 0$ the L^μ-Caccioppoli-type inequality

$$\int_{B_R} v^{\tilde{q}} \eta^\mu |\nabla v|^\mu \leq \frac{C^\mu}{(\tilde{q}+1)^\mu} \int_{B_R} v^{\tilde{q}+\mu} |\nabla \eta|^\mu, \quad \forall \eta \in C_0^\infty(B_R) \tag{5.40}$$

is valid for some constant $C = C(R, \mu) > 0$ independent of $\tilde{q}$. Then for every $q \geq 1$ and for every $0 < r_1 < r_2 \leq R$, we have

$$\|v^q\|_{L^{\mu\gamma}(B_{r_1})} \leq \frac{2S(1+C)}{r_2 - r_1} \|v^q\|_{L^\mu(B_{r_2})}. \tag{5.41}$$

Proof. By density, the Sobolev inequality (5.39) holds for every function in $W_0^{1,\mu}$. Fix $0 < r_1 < r_2 \leq R$ and let $\eta = \eta_{r_1, r_2} \in C^\infty(M)$ be a cut-off function satisfying

$$\text{(i) } 0 \leq \eta \leq 1, \ \text{(ii) } \eta \equiv 0 \text{ on } M \setminus B_{r_2}, \ \text{(iii) } \eta \equiv 1 \text{ on } B_{r_1}, \ \text{(iv) } |\nabla \eta| \leq \frac{2}{r_2 - r_1}.$$

Assume first that $q > 1$. We set $\tilde{q} = q - 1$ and estimate

$$\begin{aligned}
\|v^q\|_{L^{\mu\gamma}(B_{r_1})} &\leq \left\|\eta v^{\tilde{q}+1}\right\|_{L^{\mu\gamma}(B_{r_2})} \text{ (Sobolev)} \\
&\leq S \left\|\nabla\left(\eta v^{\tilde{q}+1}\right)\right\|_{L^\mu(B_{r_2})} \\
&= S \left\|v^{\tilde{q}+1} \nabla \eta + (\tilde{q}+1) \eta v^{\tilde{q}} \nabla v\right\|_{L^\mu(B_{r_2})} \text{ (Minkowski)} \\
&\leq S \left\|v^{\tilde{q}+1} \nabla \eta\right\|_{L^\mu(B_{r_2})} + S(\tilde{q}+1) \left\|\eta v^{\tilde{q}} \nabla v\right\|_{L^\mu(B_{r_2})} \text{ (Caccioppoli)} \\
&\leq S \left\|v^{\tilde{q}+1} \nabla \eta\right\|_{L^\mu(B_{r_2})} + SC \frac{\tilde{q}+1}{\mu\tilde{q}+1} \left\|v^{\tilde{q}+1} \nabla \eta\right\|_{L^\mu(B_{r_2})} \\
&\leq \frac{2S(1+C)}{r_2 - r_1} \|v^q\|_{L^\mu(B_{r_2})}
\end{aligned}$$

as required. The case $q = 1$ follows by dominated convergence. □

As an application of the previous results we deduce the validity of an L^q-weak Harnack inequality for weak solutions v of $v \operatorname{div}(w \nabla v) \geq 0$.

Proposition 5.15. *Let B_R be a relatively compact geodesic ball in the Riemannian manifold $(M, \langle\, , \rangle)$ of dimension $m \geq 2$. Let also $0 < w \in C^0\left(\bar{B}_R\right)$. Having fixed $0 < \bar{R} < R$ and $q \geq 1$, there exists a constant $C = C(R, q, w) > 0$ such that*

$$\sup_{B_{\bar{R}}} |v|^{2q} \leq C \int_{B_R} |v|^{2q}$$

for every $v \in Lip_{loc}(B_R)$ satisfying the differential inequality

$$v \operatorname{div}(w \nabla v) \geq 0$$

weakly on B_R.

Proof. As noted in in Theorem 5.11 above, a local L^2-Sobolev inequality is always satisfied. On the other hand, by Lemma 5.12 and Kato's inequality $\left|\nabla |v|\right| \leq |\nabla v|$, the function $|v|$ satisfies an L^2-Caccioppoli inequality. Therefore, according to Lemma 5.14, we can apply the Moser iteration procedure of Lemma 5.13 with $q_o = q$ to obtain the desired conclusion. □

5.6 A weak Harnack inequality

In this section we prove an L^p-version of the Harnack inequality for (weak) solutions of differential inequalities of the type

$$u\Delta u + a(x) u^2 + A |\nabla u|^2 \geq 0. \tag{5.42}$$

Theorem 5.16. *Let $B_{R+1}(o)$ be a relatively compact geodesic ball in a Riemannian manifold $(M, \langle\, , \rangle)$ of dimension $m \geq 2$ and let*

$$a \in C^0\left(B_{R+2}(o)\right), \quad A \in \mathbb{R}, \quad \text{and } p \geq A + 1, \quad p > 0. \tag{5.43}$$

Then, there exists a constant $C > 0$ depending on the above data and the geometry of $B_{R+1}(o)$, such that

$$\sup_{B_R(o)} u^{2p} \leq C \int_{B_{R+1}(o)} u^{2p} \tag{5.44}$$

for every locally Lipschitz, weak solution u of (5.42).

Proof. We shall show that, for every $x \in \bar{B}_R(o)$, there exists $\varepsilon > 0$ and a constant $C' > 0$ independent of u such that

$$\sup_{B_\varepsilon(x)} |u|^{2p} \leq C' \int_{B_{2\varepsilon}(x)} |u|^{2p}. \tag{5.45}$$

Since $\bar{B}_R(o)$ is compact, the desired inequality will follow from (5.45) using a covering argument.

Let us consider the Schrödinger operator

$$L = -\Delta - pa(x)$$

on $L^2(B_{3\varepsilon}(x))$. Since the first Dirichlet eigenvalue of $-\Delta$ on $B_r(x)$ grows like r^{-2} as $r \to 0+$, we can choose $\varepsilon > 0$ so small that

$$\lambda_1(L_{B_{3\varepsilon}(x)}) > 0.$$

Let w be the corresponding, positive, first eigenfunction, i.e., a solution of the eigenvalue problem

$$\begin{cases} \Delta w + pa(x)\, w = -\lambda_1(L_{B_{3\varepsilon}(x)})w \leq 0 & \text{on } B_{3\varepsilon}(x), \\ w > 0 & \text{on } B3\varepsilon(x), \\ w \equiv 0 & \text{on } \partial B_{3\varepsilon}(x). \end{cases} \tag{5.46}$$

The regularity theory for elliptic equations implies that $w \in C^1(B_{3\varepsilon}(x))$. Combining u and w, we define a new function

$$v = w^{-1}|u|^{p-1}u.$$

According to the first part of the proof of Theorem 4.5 (with $H = p$, $\beta = p-1$ and $\alpha = 1$) the function v satisfies the differential inequality

$$v\mathrm{div}\left(w^2 \nabla v\right) \geq 0$$

weakly on $B_{3\varepsilon}(x)$. Therefore Proposition 5.15 applies with $q = 1$ and gives

$$\sup_{B_\varepsilon(x)} |v|^2 \leq C \int_{B_{2\varepsilon}(x)} |v|^2,$$

for some constant $C > 0$ depending on $w|_{\bar{B}_{2\varepsilon}(x)}$ and the geometry of $B_{2\varepsilon}(x)$. Thus (5.45) holds with

$$C' = \left(\frac{\sup_{B_{2\varepsilon}} w}{\inf_{B_\varepsilon} w}\right)^2 C.$$

□

5.7 Proof of the abstract finiteness theorem

The weak Harnack inequality of Theorem 5.16 differs from the estimate

$$\sup_{B_R(x)} |\xi|^{2p} \leq C \int_{B_R(x)} |\xi|^{2p} \tag{5.47}$$

needed to apply Li's Lemma in that on both sides of (5.47) there is the same ball $B_R(x)$. Accordingly, one is led to search for an integral, a-priori estimate on annuli of the type

$$\int_{B_{R+1}(x)\setminus B_R(x)} |\xi|^{2p} \leq C' \int_{B_R(x)} |\xi|^{2p}.$$

This is the core of the proof. Very recently, [104] and [105], Peter Li and J. Wang have developed a technique to solve the problem in the L^2-setting. We point out that the L^2 assumption is crucial for their argument to work in that they need the presence of a refined Kato inequality (see Remark 5.5 above). However L^{2p}-harmonic forms, in general, do not have this property. We are able to circumvent the problem, using once again the reduction procedure which is based on a combination of the two basic differential inequalities (5.3) and (5.4) in a new one of the type $v\mathrm{div}\,(w\nabla v) \geq 0$.

Lemma 5.17. *Keeping notation and assumptions of Theorem* 5.1, *having fixed an origin* $o \in M$, *there exist* $\bar{R} > 0$ *and a constant* $C > 0$ *depending on* p, H *and the geometry of* $B_{\bar{R}}(o)$ *such that*

$$\sup_{B_{\bar{R}}(o)} |\xi|^{2p} \leq C \int_{B_{\bar{R}}(o)} |\xi|^{2p} \tag{5.48}$$

for $\xi \in V$.

Proof. From now on, we assume that all the geodesic balls under consideration are centered at the point $o \in M$ and so, to simplify the notation, we omit it from the writing.

We set $u = |\xi|$ and choose $R_0 > 0$ so large that $K \subset B_{R_0}$. We shall show that (5.48) is met with $\bar{R} = R_0 + 1$. To this end, let us note that by Theorem 5.16 there exists a constant $D > 0$ independent of u such that

$$\sup_{B_{R_0+1}} u^{2p} \leq D \int_{B_{R_0+2}} u^{2p} = D \left(\int_{B_{R_0+2}\setminus B_{R_0+1}} + \int_{B_{R_0+1}} \right) u^{2p}.$$

The goal is to prove that

$$\int_{B_{R_0+2}\setminus B_{R_0+1}} u^{2p} \leq E \int_{B_{R_0+1}} u^{2p} \tag{5.49}$$

for some constant $E > 0$ independent of u. We set

$$\alpha = \frac{p}{H}$$

and consider the function

$$v = \varphi^{-\alpha} u^p \text{ on } M \setminus B_{R_0}.$$

As above, the first part of the proof of Theorem 4.5 shows that

$$v\mathrm{div}\left(\varphi^{2\alpha} \nabla v\right) \geq 0 \tag{5.50}$$

weakly on $M \setminus B_{R_0}$ Moreover, since

$$\int_{B_{R_0+2}\setminus B_{R_0+1}} u^{2p} = \int_{B_{R_0+2}\setminus B_{R_0+1}} \varphi^{2\alpha}\left(\varphi^{-\alpha}u^p\right)^2 \leq \left(\sup_{B_{R_0+2}\setminus B_{R_0+1}} \varphi^{2\alpha}\right)\int_{B_{R_0+2}\setminus B_{R_0+1}} v^2$$

the desired inequality (5.49) will follow once we prove

$$\int_{B_{R_0+2}\setminus B_{R_0+1}} v^2 \leq E\int_{B_{R_0+1}} u^{2p}. \tag{5.51}$$

Towards this end, we fix a sequence $\{R_k\} \nearrow +\infty$, and we consider the family of compactly supported, Lipschitz functions $\{\phi_k\}$ defined by

$$\phi_k(x) = \begin{cases} 0 & \text{on } B_{R_0}, \\ r(x) - R_0 & \text{on } B_{R_0+1} \setminus B_{R_0}, \\ 1 & \text{on } B_{R_0+2} \setminus B_{R_0+1}, \\ \dfrac{R_k - r(x)}{R_k - R_0 - 2} & \text{on } B_{R_k} \setminus B_{R_0+2}, \\ 0 & \text{on } M \setminus B_{R_k}. \end{cases}$$

Furthermore, we set

$$\phi_\infty = \begin{cases} 0 & \text{on } B_{R_0}, \\ r(x) - R_0 & \text{on } B_{R_0+1} - B_{R_0}, \\ 1 & \text{on } M - B_{R_0+1}. \end{cases}$$

According to (5.50) we can apply Lemma 5.12 with $q = 0$ to obtain

$$\begin{aligned} D_0 &\int_{B_{R_0+2}\setminus B_{R_0}} \phi_\infty^2 |\nabla v|^2 \\ &\leq D_0 \sup_{B_{R_0+2}\setminus B_{R_0}} \varphi^{-2\alpha} \int_{B_{R_0+2}\setminus B_{R_0}} \varphi^{2\alpha}\phi_\infty^2 |\nabla v|^2 \\ &\leq D_0 \sup_{B_{R_0+2}\setminus B_{R_0}} \varphi^{-2\alpha} \int_{M\setminus B_{R_0}} \varphi^{2\alpha}\phi_k^2 |\nabla v|^2 \\ &\leq \sup_{B_{R_0+2}\setminus B_{R_0}} \varphi^{-2\alpha} \int_{M\setminus B_{R_0}} \varphi^{2\alpha} v^2 |\nabla \phi_k|^2 \\ &\leq \sup_{B_{R_0+2}\setminus B_{R_0}} \varphi^{-2\alpha} \left\{\int_{B_{R_0+1}\setminus B_{R_0}} \varphi^{2\alpha}v^2 + \int_{B_{R_k}\setminus B_{R_0+2}} \varphi^{2\alpha}v^2|\nabla\phi_k|^2\right\} \\ &\leq \sup_{B_{R_0+2}\setminus B_{R_0}} \varphi^{-2\alpha} \left\{\int_{B_{R_0+1}\setminus B_{R_0}} u^{2p} + \frac{1}{(R_k-R_0-2)^2}\int_{B_{R_k}\setminus B_{R_0}} u^{2p}\right\}. \end{aligned}$$

Letting $k \to +\infty$ we deduce

$$\int_{B_{R_0+2}\setminus B_{R_0}} \phi_\infty^2 |\nabla v|^2 \le \tilde{D} \int_{B_{R_0+1}\setminus B_{R_0}} u^{2p} \tag{5.52}$$

where we have set

$$\tilde{D} = \frac{1}{D_0 \sup_{B_{R_0+2}\setminus B_{R_0}} \varphi^{-2\alpha}} > 0.$$

On the other hand, Theorem 5.8 above implies that there exists a constant $C_1 > 0$ depending on the geometry of B_{R_0+2} such that the following Poincaré-type inequality holds:

$$C_1 \int_{B_{R_0+2}\setminus B_{R_0}} f^2 \le \int_{B_{R_0+2}\setminus B_{R_0}} |\nabla f|^2$$

for every $f \in H^1(B_{R_0+2})$ with

$$f|_{\bar{B}_{R_0}} = 0.$$

Applying this inequality to the function $\phi_\infty v$, and using the Schwarz and Young inequalities, we get

$$\begin{aligned}
C_1 &\int_{B_{R_0+2}\setminus B_{R_0}} \phi_\infty^2 v^2 \\
&\le \int_{B_{R_0+2}\setminus B_{R_0}} |\nabla(\phi_\infty v)|^2 \\
&= \int_{B_{R_0+2}\setminus B_{R_0}} \phi_\infty^2 |\nabla v|^2 + v^2 |\nabla \phi_\infty|^2 + 2v\phi \langle \nabla v, \nabla \phi_\infty \rangle \\
&\le \int_{B_{R_0+2}\setminus B_{R_0}} \left\{ 2\phi_\infty^2 |\nabla u|^2 + 2v^2 |\nabla \phi_\infty|^2 \right\} \\
&\le 2 \int_{B_{R_0+2}\setminus B_{R_0}} \phi_\infty^2 |\nabla v|^2 + 2 \int_{B_{R_0+1}\setminus B_{R_0}} v^2 \\
&\le 2 \int_{B_{R_0+2}\setminus B_{R_0}} \phi_\infty^2 |\nabla v|^2 + 2 \sup_{B_{R_0+1}\setminus B_{R_0}} \varphi^{-2\alpha} \int_{B_{R_0+1}\setminus B_{R_0}} u^{2p},
\end{aligned}$$

whence, using (5.52), we conclude that

$$\begin{aligned}
C_1 \int_{B_{R_0+2}\setminus B_{R_0+1}} v^2 &\le C_1 \int_{B_{R_0+2}\setminus B_{R_0}} \phi_\infty^2 v^2 \\
&\le 2\left(\tilde{D} + \sup_{B_{R_0+1}\setminus B_{R_0}} \varphi^{-2\alpha} \right) \int_{B_{R_0+1}\setminus B_{R_0}} u^{2p} \\
&\le 2\left(\tilde{D} + \sup_{B_{R_0+1}\setminus B_{R_0}} \varphi^{-2\alpha} \right) \int_{B_{R_0+1}} u^{2p},
\end{aligned}$$

as required to prove (5.51). □

We are now in a position to conclude the proof of Theorem 5.1.

Proof of Theorem 5.1. Let $B_{\bar{R}} \subset M$ and $C > 0$ be as in Lemma 5.17. From the unique continuation property we have that the restriction map

$$\xi \longmapsto \xi|_{B_{\bar{R}}}$$

defines an injective homomorphism of V into $L^2\Gamma\left(E|\, B_{\bar{R}}\right)$, the space of continuous sections of E on $B_{\bar{R}}$ endowed with the L^2-inner product.

Let T be any finite-dimensional subspace of V. We have to prove that $t = \dim T$ is bounded from above by an absolute constant. To this end we apply Peter Li's lemma to deduce that there exists $\bar{\xi} \in T$ such that

$$t \int_{B_{\bar{R}}} \left|\bar{\xi}\right|^{2p} \leq \operatorname{vol}\left(B_{\bar{R}}\right) \min\left\{l, t\right\} \sup_{B_{\bar{R}}} \left|\bar{\xi}\right|^{2p}.$$

On the other hand, using $u = \left|\bar{\xi}\right|$ in (5.48) of Lemma 5.17, we see that

$$\sup_{B_{\bar{R}}} \left|\bar{\xi}\right|^{2p} \leq C \int_{B_{\bar{R}}} \left|\bar{\xi}\right|^{2p}.$$

As a consequence

$$t \leq \operatorname{vol}\left(B_{\bar{R}}\right) \min\left\{l, t\right\} C$$

which in turn implies

$$t = \dim T \leq l \max\left\{C \operatorname{vol}\left(B_{\bar{R}}\right), 1\right\}.$$ □

Chapter 6

Applications to harmonic maps

In this section we show the usefulness of Theorem 4.5 by deriving a number of results on harmonic maps. We begin by establishing a Liouville-type theorem which compares with classical work by Schoen and Yau, [146]. Direct inspection shows that our result, emphasizing the role of a suitable Schrödinger operator related to the Ricci curvature of the domain manifold, unifies in a single statement the situations considered in [146]; see Remark 6.22 below. We also give a version of this result in case the domain manifold is Kähler and see how this allows weaker integrability conditions on the energy density of the map. From this, we derive a number of geometric conclusions. We then provide a sharp upper estimate on the growth of the energy of a harmonic map. We close the section with a Schwarz-type lemma for harmonic maps with bounded dilation, and some applications to the fundamental group which extend results by Schoen and Yau and Lemaire ([93]).

6.1 Harmonic maps of finite L^p-energy

6.1.1 A vanishing theorem

In [146], using harmonic maps techniques, Schoen and Yau studied the fundamental group of a manifold of non-negative Ricci curvature and of a stable minimal hypersurface immersed into non-positively curved ambient spaces. Basic tools in their investigation are represented by vanishing-type theorems for finite-energy harmonic maps. This section aims to unify and extend their results in the following

Theorem 6.1. *Let $(M, \langle\, ,\rangle)$ be a complete manifold whose Ricci tensor satisfies*

$$ {}^{M}\mathrm{Ric} \geq -\rho(x)\,, \quad \text{on } M \tag{6.1}$$

for some continuous function $\rho(x)$. Having fixed $H > \frac{m-2}{m-1}$, set ${}^{H}L = -\Delta - H\rho(x)$ and assume that

$$\lambda_1({}^{H}L_M) \geq 0. \tag{6.2}$$

Let $(N, (\, ,\,))$ be a manifold of non-positive sectional curvature ${}^{N}\mathrm{Riem} \leq 0$. Then, any harmonic map $f : M \to N$ with energy density satisfying

$$|df|^2 \in L^\gamma(M) \tag{6.3}$$

for some $\frac{m-2}{m-1} \leq \gamma \leq H$, is constant.

Proof. We recall the Weitzenböck–Bochner formula of Theorem 1.2:

$$\frac{1}{2}\Delta |df|^2 = |Ddf|^2 + \sum_i \left(df \left({}^M\mathrm{Ric}\,(e_i,\cdot)^\# \right), df\,(e_i) \right) - \sum_{i,j} \left({}^N\mathrm{Riem}\,(df\,(e_i), df\,(e_j))\, df\,(e_j), df\,(e_i) \right). \tag{6.4}$$

The curvature assumptions imply that

$$\Delta |df|^2 \geq 2\,|Ddf|^2 - 2\rho\,(x)\,|df|^2 \text{ on } M.$$

Therefore, the non-negative function $u = |df| \in Lip_{loc}(M)$ satisfies

$$u\Delta u + \rho\,(x)\,u^2 \geq |Ddf|^2 - |du|^2,$$

pointwise on the open set

$$\Omega = \{x \in M : u\,(x) \neq 0\}$$

and weakly on all of M. Whence, recalling the refined Kato inequality

$$|Ddf|^2 - |d\,|df||^2 \geq \frac{1}{m-1}\,|d\,|df||^2$$

of Proposition 1.3 above, we see that u satisfies

$$u\Delta u + \rho\,(x)\,u^2 \geq \frac{1}{(m-1)}\,|du|^2$$

weakly on M. Moreover, the condition $u \in L^{2\gamma}(M)$ implies

$$\frac{1}{\int_{\partial B_r} u^{2\gamma}} \notin L^1(+\infty).$$

On the other hand, since $\lambda_1(L_M) \geq 0$, there exists a positive function $\varphi \in C^1(M)$ satisfying

$$\Delta\varphi + H\rho\,(x)\,\varphi = 0. \tag{6.5}$$

Applying Theorem 4.5, case (i), with the choices $A = -1/(m-1)$, $\beta = \gamma - 1$ we conclude that u and φ are non-negative constants. Suppose by contradiction that $u \equiv c > 0$. Then, the integrability condition (6.3) forces $\mathrm{vol}\,(M) < +\infty$. On the other hand, since φ is constant, (6.5) forces $\rho\,(x) \equiv 0$, that is, M has non-negative Ricci curvature. In particular $\mathrm{vol}\,(M) = +\infty$ which gives the desired contradiction. □

We now show how Theorem 6.1 improves in the case of pluriharmonic maps defined on a Kähler manifold by enlarging the range of admissible γ's. The improvement relies on the Weitzenböck–Bochner formula for pluriharmonic maps.

Theorem 6.2. *Let $(M, \langle\, , \rangle\, , J_M)$ be a complete Kähler manifold with Ricci tensor satisfying* (6.1), *and assume that* (6.2) *holds for $H > 0$. Let $(N, (\, ,))$ be a Riemannian manifold with non-positive Hermitian curvature. Then any pluriharmonic map $f : M \to N$ with energy density satisfying* (6.3) *with $0 < \gamma \leq H$, is constant.*

Proof. We set $u = |df|$. Since N has non-positive Hermitian curvature, and the Ricci tensor of M satisfies (6.1), an application of Corollary 1.25 shows that u satisfies the differential inequality

$$u\Delta u + \rho_+ (x)\, u^2 \geq |\nabla u|^2 \tag{6.6}$$

pointwise on $\Omega = \{x \in M : u(x) \neq 0\}$ and weakly on all of M. To complete the proof we apply Theorem 4.5 with $A = -1$ and $\beta = \gamma - 1$. □

Corollary 6.3. *Let $(M, \langle\, , \rangle\, , J_M)$ be a complete Kähler manifold with Ricci tensor satisfying* (6.1) *and assume that* (6.2) *holds for $H > 0$. Let $(N, (\, ,))$ be a Riemannian, locally symmetric space whose irreducible local factors are all of non-compact or Euclidean type. Then any pluriharmonic map $f : M \to N$ with energy density satisfying* (6.3) *with $0 < \gamma \leq H$, is constant.*

Proof. Indeed, by Theorem 1.19, $(N, (\, ,))$ has non-positive Hermitian curvature. □

Remark 6.4. The same conclusion holds if M is as above, and $f : M \to N$ is a holomorphic map into a Hermitian manifold N with non-positive holomorphic bisectional curvature. Indeed, according to Corollary 1.29, $u = |df|$ satisfies inequality (6.6).

Remark 6.5. In the above proofs the term involving the curvature tensor of the target manifold N was dealt with in a rather crude way and one may expect better results from a more careful analysis. However, this term is not easy to handle and in general requires extra assumptions; see, e.g., Section 6.2 below. Particularly favorable instances occur when f is a holomorphic map from a Kähler manifold M into a Hermitian manifold N. This situation is considered in Chapter 8 below. In this case, the Weitzenböck–Bochner formula for the energy density of f contains a non-linear term of the form $|df|^4$ which arises from the curvature tensor of N. In Theorem 8.11, starting from this observation, adapting a result by Li and Yau, and using "a-priori" estimates for solutions of Yamabe-type equations, we are able to extend the range of γ to

$$0 < \gamma \leq H + \sqrt{H(H-2)}, \quad (H \geq 2).$$

We stress that this provides the only instance where the integrability exponent is allowed to be greater than H.

6.1.2 Convergent harmonic maps

Under appropriate circumstances we can guarantee that $|df|^2 \in L^\gamma(M)$ for some γ in the admissible range. Towards this end, we recall the following estimate ([139]).

Lemma 6.6. *Let $f : (M, \langle\, ,\rangle) \to (N, (\, ,))$ be a harmonic map such that $u(x) \to q \in N$ as $r(x) \to +\infty$. Let $\eta(r)$ be a non-increasing, positive function such that $\eta(r) \to 0$ as $r \to +\infty$ and*

$$\operatorname{dist}_N(f(x), q)^2 \le C\eta(r(x)) \tag{6.7}$$

for $r(x)$ sufficiently large and for some constant $C > 0$. Then, either

$$|df|^2 \in L^1(M) \tag{6.8}$$

or both the following conditions are satisfied:

$$\frac{1}{\operatorname{vol}(\partial B_r)} \in L^1(+\infty) \tag{6.9}$$

and there exists $R > 0$ such that, for every $r \ge R$,

$$\int_{B_r} |df|^2 \le C\eta(r)\left\{\int_r^{+\infty} \frac{dt}{\operatorname{vol}(\partial B_t)}\right\}^{-1}. \tag{6.10}$$

Proof. Let $\rho(y) = \operatorname{dist}_N(f(y), q)$. Then ρ^2 is smooth in the geodesic ball $B_T(q)$ for $T > 0$ sufficiently small. Furthermore, since

$${}^N\mathrm{Hess}\left(\rho^2\right) = 2\rho\, {}^N\mathrm{Hess}(\rho) + 2d\rho \otimes d\rho$$

by the Hessian comparison theorem we can also suppose that ρ^2 is strictly convex on $B_T(q)$. Next, we choose $R_1 > 0$ so large that

$$\overline{f\left(M - \overline{B_{R_1}}\right)} \subset B_T(q)$$

and

$$\sup_{\partial B_r} \rho^2(f) \le \eta(r), \quad r \ge R_1. \tag{6.11}$$

Let

$$\Lambda = \inf_{B_T(q)} \lambda(y) > 0$$

where $\lambda(y)$is the smallest eigenvalue of $\mathrm{Hess}\left(\rho^2\right)(y)$. Then, by the composition law

$${}^M\Delta(v \circ f) = dv(\tau(f)) + \sum_{i=1}^m {}^N\mathrm{Hess}(v)(df(e_i), df(e_i))$$

valid for $v : N \to \mathbb{R}$ and $\{e_i\}_{i=1,\dots,m=\dim M}$ a local orthonormal frame on M, we deduce

$$^M\Delta(\rho^2(f)) \geq \Lambda \left|df\right|^2 \tag{6.12}$$

on $M \setminus B_{R_1}$. Let $r > R_1$. Applying the divergence theorem on $B_r \setminus B_{R_1}$, with the aid of (6.11), (6.12), Gauss' lemma, and the Cauchy–Schwarz inequality, we obtain

$$\begin{aligned}\Lambda \int_{B_r \setminus B_{R_1}} \left|df\right|^2 &\leq \int_{B_r \setminus B_{R_1}} {}^M\Delta(\rho^2 \circ f) \\ &\leq 2 \int_{\partial(B_r \setminus B_{R_1})} \rho(f)|df| \leq C + \left(\eta(r)\mathrm{vol}\,(\partial B_r) \int_{\partial B_r} |df|^2\right)^{1/2}.\end{aligned}$$

Letting

$$\gamma(r) = \int_{B_r \setminus B_{R_1}} \left|df\right|^2$$

rearranging and using the co-area formula, the above inequality becomes

$$\gamma(r) \leq C\left\{1 + \eta(r)^{\frac{1}{2}} \gamma'(r)^{\frac{1}{2}} \mathrm{vol}\,(\partial B_r)^{\frac{1}{2}}\right\} \tag{6.13}$$

for $r > R_1$ and some constant $C > 0$. To complete the proof, assume that $\left|df\right|^2 \notin L^1(M)$. Then $\gamma(r) \to +\infty$ as $r \to +\infty$ and therefore so does the RHS of (6.13). We deduce that there exist $R_2 \geq R_1$ and $C_1 > 0$ sufficiently large that

$$\gamma(r)^2 \leq C_1 \eta(r) \mathrm{vol}\,(\partial B_r) \gamma'(r),$$

for every $r \geq R_2$. Let $R_2 \leq r \leq t$. Integrating over (r, t) and taking into account the monotonicity of η, we have

$$\frac{1}{\gamma(r)} - \frac{1}{\gamma(t)} \geq \frac{C}{\eta(r)} \int_r^t \frac{ds}{\mathrm{vol}\,(\partial B_s)}.$$

Whence, letting $t \to +\infty$ we conclude the validity of both (6.9) and (6.10). □

In a similar way we can also prove

Lemma 6.7. *Let $f : (M, \langle\, ,\rangle) \to (N, (\, ,\,))$ be a harmonic map such that, for some $R_0 > 0$, $\overline{f(M \setminus \overline{B_{R_0}})}$ is compact and contained in a domain $D \subset N$ supporting a strictly convex function. Then, either $|df|^2 \in L^1(M)$ or $\mathrm{vol}(\partial B_r)^{-1} \in L^1(+\infty)$ and there exists $R > 0$ such that, for every $r \geq R$,*

$$\int_{B_r} \left|df\right|^2 \leq C\left\{\int_r^{+\infty} \frac{dt}{\mathrm{vol}\,(\partial B_r)}\right\}^{-1}, \tag{6.14}$$

for some $C > 0$.

We are now ready to show the validity of

Theorem 6.8. *Let $(M, \langle\, ,\rangle)$ be a complete Riemannian manifold such that*

$$\mathrm{Ric} \geq -\rho\,(x) \quad on\ M$$

for some function $\rho \in C^0\,(M)$ and assume that the Schrödinger operator $L = -\Delta - \rho_+\,(x)$ satisfies

$$\lambda_1(L_M) \geq 0.$$

Suppose that $\mathrm{vol}(\partial B_r)$ is at most of polynomial growth in r, as $r \to +\infty$. Let $(N, (\ ,\ ))$ be a complete manifold with non-positive sectional curvature and let $q \in N$. Assume that

$$\mathrm{vol}\,(\partial B_r) = O\left(r^{\alpha+3}\right), \ as\ r \to +\infty, \tag{6.15}$$

for some $\alpha \geq 0$. Then any harmonic map $f : (M, \langle\, ,\rangle) \to (N, (\ ,\ ))$ such that

$$\mathrm{dist}_N\,(f\,(x)\,, q) = \begin{cases} O(r(x)^{-\alpha}) & if\ \alpha > 0, \\ o(1) & if\ \alpha = 0, \end{cases} \quad as\ \ r(x) \to +\infty, \tag{6.16}$$

is constant.

Proof. The proof is similar to that of of Theorem 6.1. The only delicate point in order to apply Theorem 4.5 is to show that

$$\frac{1}{\int_{\partial B_r} |df|^2} \notin L^1\,(+\infty)\,. \tag{6.17}$$

Note that if $\mathrm{vol}\,(\partial B_r)^{-1} \notin L^1\,(+\infty)$, then, by Lemma 6.6, $|df|^2 \in L^1\,(M)$ and (6.17) holds true. Therefore, suppose $\mathrm{vol}\,(\partial B_r)^{-1} \in L^1\,(+\infty)$. From (6.15) we deduce

$$\int_r^{+\infty} \frac{dt}{\mathrm{vol}\,(\partial B_r)} \geq C \int_t^{+\infty} \frac{dt}{r^{\alpha+3}} = C' \frac{1}{r^{\alpha+2}}$$

which implies

$$\left\{\int_r^{+\infty} \frac{dt}{\mathrm{vol}\,(\partial B_r)}\right\}^{-1} \leq C'' r^{\alpha+2}. \tag{6.18}$$

First, we consider the case where $\alpha > 0$. From estimate (6.10) of Lemma 6.6 we deduce

$$\int_{B_r} |df|^2 \leq Cr^2 \tag{6.19}$$

for $r \geq R > 0$ sufficiently large. It follows that

$$\frac{r}{\int_{B_r} |df|^2} \notin L^1\,(+\infty)$$

and, "a-fortiori" (6.17) is true. The case where $\alpha = 0$ is similar. □

We end this section showing that the growth estimates obtained in (6.10) and (6.14) are rather sharp despite the simplicity of their proofs. To this end we consider rotationally symmetric maps between models, referring to [139] for the details and proofs that we omit here.

Let η, $\zeta : [0,+\infty) \to [0,+\infty)$ be smooth functions such that $\eta(t)$, $\zeta(t) > 0$ for $t > 0$, and

$$\eta'(0) = \zeta'(0) = 1, \quad \text{and} \quad \eta^{(2k)}(0) = \zeta^{(2k)}(0) = 0 \ \forall k = 0, 1, \ldots,$$

and let $M(\eta)$ and $M(\zeta)$ be the Riemannian manifolds obtained by endowing $\mathbb{R}^m$ with the metrics defined in polar coordinates by

$$\langle\,,\rangle_\eta = dr^2 + \eta^2(r)d\theta^2 \qquad \text{and} \qquad \langle\,,\rangle_\zeta = dr^2 + \zeta^2(r)d\theta^2,$$

where $d\theta^2$ is the canonical metric on S^{m-1}.

Next we consider rotationally symmetric maps $F_\rho : M(\zeta) \to M(\eta)$ defined in polar coordinates by

$$F_\rho : (r,\theta) \mapsto \big(\rho(r),\theta\big) \qquad \rho(0) = 0.$$

A computation shows that, up to a constant factor, the energy of F_ρ on B_r is given by

$$E_r(F_\rho) = \int_0^r \left[(\rho')^2 + \frac{m-1}{\zeta^2}\eta^2(\rho)\right](s)\zeta^{m-1}(s)\,ds, \tag{6.20}$$

and therefore F_ρ is a (smooth) harmonic map if and only if ρ is a non-negative solution of

$$\begin{cases} \rho''(r) + (m-1)\frac{\zeta'}{\zeta}\rho'(r) - \frac{m-1}{\zeta^2}\eta(\rho(r))\eta'(\rho(r)) = 0 & r > 0, \\ \lim_{r\to 0^+}\rho(r) = 0. \end{cases} \tag{6.21}$$

Assuming that $m \geq 3$, we choose $\eta(r) = r$, so that $M(\eta)$ is the standard Euclidean space, and $\zeta(r)$ such that, for some $\delta > 1$ and sufficiently large $R_o > 0$,

$$\zeta(t) = t(\log t)^\delta \qquad \forall t \geq R_o. \tag{6.22}$$

According to [139] Theorem 4.1 there exists $L > 0$ such that, for every ρ_∞ in $(0, L]$, (6.21) has a solution ρ satisfying

$$\rho'(r) \geq 0, \qquad \lim_{r\to+\infty}\rho(r) = \rho_\infty. \tag{6.23}$$

It follows that for every $R_1 < L$ there exists a harmonic map $F_\rho : M(\zeta) \to M(\eta)$ with the property that, for suitably large R_2,

$$F_\rho\Big(M(\zeta) \setminus \overline{B_{R_2}}\Big) \subset B^{M(\eta)}_{R_1},$$

where $B_{R_1}^{M(\eta)}$ denotes the ball of radius R_1 centered at 0 in $M(\eta)$. In particular, integrating (6.21) between 1 and r we obtain

$$\rho'(r) = C\zeta(r)^{-(m-1)} + (m-1)\zeta(r)^{-(m-1)} \int_1^r \eta(\rho)\eta'(\rho)\zeta(s)^{m-3}\, ds.$$

Using (6.22) we see that the integral on the right-hand side has the same order of magnitude as

$$\int_1^r \zeta^{m-3}(s)\, ds \asymp r^{m-2}\big(\log r\big)^{(m-3)\delta} \qquad \text{as } r \to +\infty,$$

and therefore

$$\rho'(r) \asymp r^{-1}\big(\log r\big)^{-2\delta} \qquad \text{as } r \to +\infty.$$

Inserting this result into (6.20) we conclude that

$$\begin{aligned} E_r(F_\rho) &\asymp \int_1^r \big[(\rho')^2 + (m-1)\zeta^{-2}\eta^2(\rho)\big](s)\zeta^{m-1}(s)\, ds \\ &\asymp \int_1^r \zeta^{m-3}(s)\, ds \asymp r^{m-2}\big(\log r\big)^{(m-3)\delta} \quad \text{as } r \to +\infty. \end{aligned}$$

On the other hand, we have $\operatorname{vol}\partial B_t = c_m t^{m-1}(\log t)^{\delta(m-1)}$ for $t \geq R_o$, so that

$$\int_r^{+\infty} \frac{dt}{\operatorname{vol}(\partial B_r)} = \int_r^{+\infty} \frac{dt}{t^{m-1}(\log t)^{\delta(m-1)}} \asymp \frac{1}{r^{m-2}(\log r)^{\delta(m-1)}}$$

showing that the estimate provided by Lemma 6.7 is sharp up to a power of $\log r$.

6.1.3 Further remarks on convergent harmonic maps

In this section we take a closer look at Lemma 6.6 above.

We first note that the assumption that $\overline{f\left(M \setminus \overline{B_{R_1}}\right)} \subset B_T(q)$, $0 < T << 1$, cannot be replaced by $\overline{f(M)} \subset B_T(q)$, for otherwise the result is trivial. In fact, harmonic maps having small images and limiting value at infinity are necessarily constant. This is true regardless both of the geometry of the domain manifold and of the rate of convergence at infinity. More precisely we have the following

Proposition 6.9. *Let $(M, \langle\,,\rangle_M)$, $(N, \langle\,,\rangle_N)$ be non-compact Riemannian manifolds and let $f : M \to N$ be a harmonic map. Suppose that $\overline{f(M)}$ is contained in a regular geodesic ball $B_R^N(y_0)$ in N, and that*

$$\lim_{x\to\infty} f(x) = y_0 \in N.$$

Then $f \equiv y_0$.

Proof. To see this, let $v : B_R^N(y_0) \to R$ be the smooth function defined by

$$v(y) = 1 - \cos\left(\sqrt{k_+}\, \mathrm{dist}_N(y, y_0)\right),$$

where $0 < k_+ < \left(\frac{\pi}{2R}\right)^2$ denotes an upper bound for the sectional curvatures of $B_R^N(y_0)$. Then, according to the Hessian comparison theorem, v is convex. Moreover, it is non-negative and vanishes exactly at y_0. Since f is a harmonic map with $\overline{f(M)} \subset B_R^N(y_0)$, the composition $u = v \circ f : M \to R$ is a well-defined, non-negative, subharmonic function. It suffices to show that $u(x) \equiv 0$. To this end, suppose the contrary and assume that $u(x_0) > 0$ for some $x_0 \in M$. Since, by assumption,

$$\lim_{x \to \infty} u(x) = 0$$

the non-negative function u must attain its positive absolute maximum. By the maximum principle u is a positive constant, contradicting the limit condition. □

As a matter of fact, the limit assumption can be relaxed provided we require some further property on the harmonic map f. For instance, one can consider the situation of rotationally symmetric harmonic maps between model manifolds considered at the end of the previous section. Clearly, if $f = F_\rho : M(\zeta) \to M(\eta)$ is the rotationally symmetric map induced by ρ, then

$$f\left(\partial B_r^{M(\zeta)}\right) = \partial B_{\rho(r)}^{M(\eta)}$$

where $B_R^{\cdot}$ denotes the geodesic ball of radius R centered at the pole of the model manifold at hand. In view of this latter property one can introduce the following

Definition 6.10. A smooth map $f : (M^m, \langle\,,\rangle_M, x_0) \to (N^n, \langle\,,\rangle_N, y_0)$ between pointed, complete Riemannian manifolds of dimensions m and n is said to be weakly rotationally symmetric if, for every $r > 0$ there is $\bar{r} > 0$ such that

$$f\left(\partial B_r^M(x_0)\right) \subseteq \partial B_{\bar{r}}^N(y_0).$$

We have the following version of the Liouville property proved above.

Proposition 6.11. *Let $f : (M, \langle\,,\rangle_M, x_0) \to (N, \langle\,,\rangle_N, y_0)$ be a harmonic map which is weakly rotationally symmetric. Suppose that $\overline{f(M)}$ is contained in a regular geodesic ball $B_R^N(y_0)$ of N. If*

$$\lim_{n \to +\infty} f(x_n) = y_0$$

along some sequence $\{x_n\} \to \infty$, then $f \equiv y_0$.

Proof. Indeed, let $v(y)$ and $u(x) = v \circ f(x)$ be the functions defined above. Again, we have to show that $u(x) \equiv 0$. By assumption, we find a sequence of real numbers $\{r_n\} \nearrow +\infty$ with the property that, for each n,

$$u(x) < \frac{1}{n} \quad \text{on } \partial B_{r_n}^M(x).$$

Since f is subharmonic, the maximum principle yields that, for each n,

$$(0 \leq) \ u(x) < \frac{1}{n} \text{ on } B^M_{r_n}(x).$$

Whence, letting $n \to +\infty$, we conclude $u(x) \equiv 0$ as desired. □

6.2 Harmonic maps of bounded dilations and a Schwarz-type lemma

In this section, we compare the above results based on L^γ energy growth assumptions with other Liouville-type theorems. Towards this end, we recall some definitions.

Definition 6.12. Given a smooth map $f : (M, \langle\, , \rangle) \to (N, (\, ,))$ between Riemannian manifolds of dimension m and n respectively, let

$$\lambda_1(x) \geq \lambda_2(x) \geq \cdots \geq \lambda_s(x) \geq \lambda_{s+1}(x) \geq \cdots \geq \lambda_{\min\{n,m\}} \geq 0$$

be the eigenvalues of the quadratic form $f^*_x(\, ,)$. We say that f has (first) dilation of order at most k and constant $T \geq 1$ if, for every $x \in M$, either $d_x f = 0$ or

$$\lambda_1(x) \leq T\left(1 + r(x)^2\right)^{\frac{k}{2}} \lambda_2(x).$$

We will denote with $\mathcal{H}_{k,T}$ the set of all smooth, harmonic maps f with dilation of order at most k and constant T, and with $L^{2p}\mathcal{H}_{k,T}$ the subset of maps $f \in \mathcal{H}_{k,T}$ with energy density satisfying $|df|^2 \in L^{2p}$.

Remark 6.13. If f has bounded dilation of order at most k on M, then f cannot have rank 1 at any point of M, for otherwise $\lambda_2(x) = 0$.

Remark 6.14. Obviously, when M is compact, $\mathcal{H}_{k,T} = L^{2p}\mathcal{H}_{k,T}$. Furthermore, the dilation condition can always be reduced to the case $k = 0$ for a sufficiently large $T > 0$. Therefore, in the compact realm, one simply considers the set of harmonic maps of uniformly bounded dilation and write $\mathcal{H}_T$.

A special class of harmonic maps of (uniformly) bounded dilation is represented by (anti-) holomorphic maps $f : M \to N$ from a Kähler manifold $(M, \langle\, , \rangle, J_M)$ to a Hermitian manifold $(N, (\, ,), J_N)$. Indeed, let $X \in T_x M$ be an eigenvector corresponding to the eigenvalue λ of $f^*(\, ,)$. Then, assuming for instance that f is holomorphic, we have

$$\begin{aligned}\lambda &= f^* \langle\, , \rangle_N (X, X) = \langle d_x f(X), d_x f(X) \rangle_N = \langle J_N d_x f(X), J_N d_x f(X) \rangle_N \\ &= \langle d_x f(J_M X), d_x f(J_M X) \rangle_N = f^* \langle\, , \rangle_N (J_M X, J_M X).\end{aligned}$$

Therefore, also $J_M X$ is an eigenvector of λ, proving that the eigenvalues of $f^* \langle\, , \rangle_N$ have (geometric hence algebraic) multiplicity at least 2. It follows that f has bounded dilation of order at most $k = 0$ and constant $T = 1$.

Harmonic maps of controlled dilations enjoy the Schwarz-type property described in the next theorem, which extends to this situation seminal results obtained by Ahlfors, [2] and Yau, [169].

Theorem 6.15. *Let $(M, \langle\, , \rangle)$ and $(N, (\, ,))$ be complete, Riemannian manifolds of dimensions m and n, respectively. Assume that the Ricci curvature of M satisfies*

$$ {}^M\mathrm{Ric} \geq -(m-1)\, B^2 \left(1 + r(x)^2\right)^{-\frac{\gamma}{2}} \quad \text{on } M, \tag{6.24} $$

for some $0 \leq \gamma < 2$ and some constant $B > 0$. Suppose also that the sectional curvature of N satisfies

$$ {}^N\mathrm{Riem}_z \leq -K(z) \quad \text{on } N, $$

for some function $K(z)$ on N. Let $f \in \mathcal{H}_{k,T}$ be such that

$$ K(f(x)) \leq -C^2 \left(1 + r(x)^2\right)^{\frac{k-\gamma}{2}} \quad \text{on } M $$

for some $C > 0$. Then,

$$ \sup_M |df|^2 \leq \frac{(m-1)\, B^2 T \min\{m,n\}^2}{2C^2}. \tag{6.25} $$

Proof. To simplify the notation, set

$$ a(x) = (m-1)\, B^2 \left(1 + r(x)^2\right)^{-\frac{\gamma}{2}}. $$

Let $\ell \leq \min\{m,n\}$ be the rank of df at x, and choose a local orthonormal frame $\{e_i\}$ of M which diagonalizes the quadratic form $f^* \langle\, , \rangle_N$ at x. Thus

$$ |df|^2 = \sum_{j=1,\dots,\ell} f^* \langle e_j, e_j \rangle = \sum_{j=1,\dots,\ell} \lambda_j $$

and

$$ {}^N\mathrm{Riem}\left(df(e_i), df(e_j), df(e_i), df(e_j)\right) \leq -C^2 \left(1 + r(x)^2\right)^{\frac{k-\gamma}{2}} \lambda_i \lambda_j, $$

where $\{\lambda_i\}$, $i = 1, \dots, \ell$ are the non-zero eigenvalues of $f^* \langle\, , \rangle_N$ arranged in decreasing order. From the Bochner–Weitzenböck formula for harmonic maps, we have

$$ \Delta |df|^2 \geq 2\, |Ddf|^2 - 2a(x)\, |df|^2 + 4C^2 \left(1 + r(x)^2\right)^{\frac{k-\gamma}{2}} \sum_{i<j=1}^{\ell} \lambda_i \lambda_j. \tag{6.26} $$

On the other hand, using the dilation assumption on f we obtain

$$
\begin{aligned}
|df|^4(x) &= \Big(\sum_{1\le i\le \ell} \lambda_i(x)\Big)^2 \le \ell^2\lambda_1(x)^2 \\
&\le \min\{m,n\}^2 T\big(1+r(x)^2\big)^{\frac{k}{2}}\lambda_1(x)\lambda_2(x) \\
&\le \min\{m,n\}^2 T\big(1+r(x)^2\big)^{\frac{k}{2}} \sum_{1\le i<j\le \ell} \lambda_i(x)\lambda_j(x).
\end{aligned}
$$

We therefore conclude that the smooth function $u = |df|^2$ satisfies the differential inequality

$$\Delta u + a(x)\, u \ge b(x)\, u^2 \text{ on } M$$

with

$$a(x) = \frac{2(m-1)B^2}{\left(1+r(x)^2\right)^{\frac{\gamma}{2}}}, \quad b(x) = \frac{4C^2}{T\min\{m,n\}^2}\frac{1}{\left(1+r(x)^2\right)^{\frac{\gamma}{2}}}.$$

Note that

$$\frac{a(x)}{b(x)} = \frac{(m-1)B^2 T\min\{m,n\}^2}{2C^2}.$$

Furthermore, according to (2.30), we have

$$\log\big(\mathrm{vol}(B_r^M)\big) \asymp r^{1-\gamma/2}$$

so that

$$\liminf_{r\to+\infty} \frac{\log \mathrm{vol}\left(B_r^M\right)}{r^{2-\gamma}} = 0.$$

Therefore, to conclude the validity of (6.25), we can invoke the a-priori estimates contained in the next theorem that we quote from [130]. □

Theorem 6.16. *Let $(M, \langle\, ,\rangle)$ be a complete manifold and let $a(x), b(x) \in C^0(M)$. Set $a_+(x) = \max\{a(x), 0\}$. Assume*

$$\sup_M a_+(x) < +\infty$$

and

$$b(x) > 0 \text{ on } M, \qquad b(x) \ge \frac{B}{r(x)^\mu}$$

for $r(x) >> 1$ and some $0 \le \mu < 2$. Suppose furthermore

$$\frac{a_+(x)}{b(x)} \le E \text{ on } M$$

for some $E \ge 0$. Let $u \in C^1(M)$ be a non-negative, weak solution of

$$\Delta u + a(x)\, u \ge b(x)\, u^\sigma \text{ on } M$$

with $\sigma > 1$. *If*

$$\liminf_{r\to+\infty} \frac{\log \operatorname{vol}\left(B_r^M\right)}{r^{2-\mu}} < +\infty,$$

then

$$u(x) \leq E^{\frac{1}{\sigma-1}} \text{ on } M.$$

Corollary 6.17. *Let* $(M, \langle\,,\rangle, J_M)$ *be a complete Kähler manifold such that*

$$^{M}\mathrm{Ric} \geq -(m-1)B^2\left(1+r(x)^2\right)^{-\frac{\gamma}{2}} \text{ on } M \tag{6.27}$$

for some constants $B > 0$ *and* $0 \leq \gamma < 2$. *Let* $(N, (\,,), J_N)$ *be a Hermitian manifold with holomorphic bisectional curvature bounded above by* $k(z)$, $z \in N$. *Let* $f : M \to N$ *be a holomorphic map such that*

$$k(f(x)) \leq -C^2\left(1+r(x)^2\right)^{-\frac{\gamma}{2}} \quad \text{on } M$$

for some $C > 0$. *Then*

$$\sup_M |df|^2 \leq \frac{(m-1)B^2}{C^2}.$$

Proof. According to Corollary 1.29, the function $u = |df|^2$ satisfies

$$\Delta u + 2\rho(x)u \geq -2k(f(x))u^2 + \frac{|\nabla u|^2}{u}$$

pointwise on $\Omega = \{x \in M : u(x) > 0\}$ and weakly on M. Here, $-\rho(x)$ denotes the function on the RHS of (6.27). Now the result follows easily. □

Remark 6.18. Using a different technique, which relies on a-priori estimates valid under Ricci curvature assumptions, one may obtain the conclusions of Theorem 6.15 and Corollary 6.17 replacing the assumption $0 \leq \gamma < 2$ with $\gamma \leq 2$; see [135].

Remark 6.19. Note that in the assumption of Corollary 6.17 the holomorphic function f is distance-decreasing provided

$$\frac{(m-1)B^2}{C^2} < 1$$

and this is an exact analog of the classical Schwarz lemma.

It is also possible to obtain versions of the Schwarz lemma for volumes. We describe an instance of this in the holomorphic setting, extending results first obtained by Chern, [36], and, for generic complete Kähler domains, by Yau, [169].

Theorem 6.20. *Let* $(M, \langle\,,\rangle, J_M)$ *be a complete Kähler manifold of complex dimension* m *and scalar curvature* $s(x)$, *and let* $(N, (\,,), J_N)$ *be a Hermitian manifold of the same dimension and with Ricci curvature bounded from above by* $R(z)$,

$z \in N$. Denote by dvol_M and dvol_N the volume elements of M and N respectively. Suppose that $f : M \to N$ is holomorphic and that

$$\text{(i) } R\left(f\left(x\right)\right) < 0 \text{ on } M; \quad \text{(ii) } R\left(f\left(x\right)\right) \leq -\frac{1}{r\left(x\right)^{\mu}} \text{ on } M \setminus B_{R_0}$$

for some $R_0 > 0$ and some $0 \leq \mu < 2$. Assume also that

$$\text{(i) } s_- \in L^{\infty}\left(M\right); \quad \text{(ii) } \frac{s_-\left(x\right)}{R\left(f\left(x\right)\right)} \geq -2mE \text{ on } M$$

for some $E \geq 0$. If

$$\liminf_{r \to +\infty} \frac{\log \mathrm{vol}\left(B_r\right)}{r^{2-\mu}} < +\infty,$$

then

$$\left\{\frac{f^* \mathrm{dvol}_N}{\mathrm{dvol}_M}\right\}^{\frac{1}{m}} \leq E.$$

In particular, if $E < 1$, then f is volume-decreasing.

Proof. We set $u = |df|^2 = \mathrm{tr} f^*\left(,\right)$ and

$$v = \left\{\frac{f^* \mathrm{dvol}_N}{\mathrm{dvol}_M}\right\}^4 = \left(\det f^*\left(,\right)\right)^2.$$

A computation due to Chern, [36], shows that the function v satisfies the differential inequality

$$\Delta v \geq -2R\left(f\left(x\right)\right) uv + 2s\left(x\right) v + \frac{\left|\nabla v\right|^2}{v}.$$

The arithmetic-geometric means inequality implies that

$$v^{\frac{1}{4m}} = \left(\det f^*\left(,\right)\right)^{1/2m} \leq \frac{1}{2m} \mathrm{tr} f^*\left(,\right) = \frac{1}{2m} u,$$

whence, letting $w = v^{\frac{1}{4m}}$, and inserting into the above inequality, we obtain

$$\Delta w \geq -R\left(f\left(x\right)\right) w^2 + \frac{1}{2m} s\left(x\right) w + \frac{\left|\nabla w\right|^2}{w}.$$

The result now follows from Theorem 6.16. □

For applications of the type of Schwarz's lemma described above see, e.g., Ph. Griffiths' book, [70], and the references therein.

6.3 Fundamental group and harmonic maps

As mentioned at the beginning of Section 6.1.1, Schoen and Yau have used harmonic map techniques, and most notably the Liouville-type Theorem 6.1, to study the fundamental group of manifolds of non-negative Ricci curvature and of stable minimal hypersurfaces immersed into non-positively curved ambient spaces (see [146] and chapter XII in [148]). We generalize and unify their topological results in the following obstruction theorem.

Theorem 6.21. *Let $(M, \langle\,,\rangle)$ be a complete, m-dimensional manifold whose Ricci tensor satisfies*

$$^{M}\mathrm{Ric} \geq -\rho(x), \quad \textit{on } M \tag{6.28}$$

and assume that the Schrödinger operator $L = -\Delta - \rho$ satisfies

$$\lambda_1(L_M) \geq 0. \tag{6.29}$$

If D is any compact domain in M with smooth, simply connected boundary, then there is no non-trivial homomorphism of $\pi_1(D)$ into the fundamental group of a compact manifold with non-positive sectional curvature.

Remark 6.22. If $(M, \langle\,,\rangle)$ has non-negative Ricci curvature, we can obviously choose $\rho(x) \equiv 0$ in (6.28) so that condition (6.29) is automatically satisfied. Similarly, suppose that $(M, \langle\,,\rangle)$ is isometrically immersed as a complete, stable, minimal hypersurface into a space $\bar{M}$ of non-negative sectional curvature. Then, according to the Gauss equations, $^{M}\mathrm{Ric} \geq -|\mathrm{II}|^2$, where $|\mathrm{II}|$ denotes the length of the second fundamental tensor of the immersion. Moreover, the stability assumption amounts to the fact that the operator $L = -\Delta - |\mathrm{II}|^2$ satisfies (6.29). These are the geometric situations considered in [146].

A further situation where the topological conclusion of Theorem 6.21 holds true is when $(M, \langle\,,\rangle)$ is a complete, m-dimensional manifold satisfying both the following properties:

(a) M supports the (global) Sobolev inequality

$$S(\alpha)^{-1}\left(\int_M v^{\frac{2}{1-\alpha}}\right)^{1-\alpha} \leq \int_M |\nabla v|^2, \ \forall v \in C_c^\infty(M)$$

where $0 < \alpha < 1$, and $S(\alpha) > 0$ is a constant.

(b) The Ricci tensor of M satisfies

$$^{M}\mathrm{Ric} \geq -\rho(x), \ \text{on } M$$

with

$$\|\rho_+(x)\|_{L^{\frac{1}{\alpha}}(M)} \leq S(\alpha)^{-1}.$$

This readily follows from Lemma 7.33 in Section 7.4. It should be noted that, in the above assumptions, we also have some information on the topology at infinity of the manifold at hand. See Section 7.1.

The proof of Theorem 6.21 relies on the following generalized version of a result by Schoen and Yau, [146].

Theorem 6.23. *Let* $(M, \langle\, ,\rangle)$ *be a complete Riemannian manifolds whose Ricci tensor satisfies* (6.28) *and* (6.29). *Let* $(N, (\, ,\,))$ *be a compact manifold of non-positive sectional curvature* ${}^N\mathrm{Riem} \leq 0$. *Then, any smooth map* $f : M \to N$ *of finite energy* $|df| \in L^2(M)$ *is homotopic to a constant on each compact set of* M.

Proof. Arguing as in [146], we see that the smooth map $f : M \to N$ is homotopic, on each compact set, to a harmonic map $g : M \to N$ of finite energy $|dg| \in L^2(M)$. To reach the desired conclusion it suffices to show that g is constant, and this follows at once from Theorem 6.1. □

We now come to the proof of the main theorem of the section. The arguments are detailed from [146].

Proof of Theorem 6.21. Let $D \subset M$ be a compact domain with smooth, simply connected boundary and let $\sigma \in \mathrm{Hom}(\pi_1(D), \pi_1(N))$, where $(N, (\, ,\,))$ is a compact manifold of non-positive sectional curvature. We have to show that σ is trivial.

By the Cartan–Hadamard theorem, the universal cover of N is contractible, so that all the higher homotopy groups $\pi_j(N)$ $(j \geq 2)$ of N vanish. Thus N is aspherical, or equivalently, is a $K(\pi, 1)$ space. According to the theory of aspherical spaces, see, e.g., [163], the set $[D; N]$ of (free) homotopy classes of maps from D to N is in one-to-one correspondence with the set of conjugacy classes of homomorphisms $\mathrm{Hom}(\pi_1(D), \pi_1(N))$. The correspondence is simply given by $f \longmapsto [f_\#]$, where $f_\#$ is the homomorphism induced by f. Accordingly, we find a continuous map $f : D \to N$ such that

$$\sigma = \alpha \circ f_\#,$$

for some $\alpha \in \mathrm{Aut}(\pi_1(N))$.

Since ∂D is simply connected, the restriction $f|_{\partial D} : \partial D \to N$ is homotopic to a constant. This follows again by the above mentioned property of aspherical spaces. Let $H : \partial D \times [0, \varepsilon] \to M$ be a (continuous) homotopy with $H(\cdot, 0) = f$, $H(\cdot, \varepsilon) = \mathrm{const}$. Suppose also $\varepsilon > 0$ is small enough. Using the normal exponential map $\exp^\perp$ relative to ∂D, see, e.g., [30], the homotopy map H enables us to extend the definition of f to a smooth domain D', $D \subset\subset D'$, along the directions normal to ∂D. Namely, for any $x \in M$ with $\mathrm{dist}_M(x, \partial D) = t \leq \varepsilon$, there are uniquely determined $x' \in \partial D$ and $\nu \in (T_{x'}\partial D)^\perp \subset T_{x'}M$ with $|\nu| = 1$ such that $x = \exp_{x'}(t\nu)$. Thus, one defines

$$f(x) = H(x', t).$$

By construction, the extended continuous map $f : D' \to N$ is constant on $\partial D'$ and therefore can be further extended by the same constant on $M \setminus D'$.

Now, we use the Whitney approximation procedure, see, e.g., [92], to get a smooth map $\bar{f}$ on M which is homotopic to f. In fact, we can take $\bar{f}$ in the same homotopy class of f, relative to the closed set $M \setminus D''$, where $D' \subset\subset D'' \subset\subset M$. This is because f is constant, hence smooth, on $M \setminus D''$. Clearly, we have

$$\sigma = \alpha' \circ (\bar{f}|_D)_\#$$

for some $\alpha' \in \mathrm{Aut}(\pi_1(N))$. On the other hand, $\bar{f}$ is constant on $M \setminus D''$, hence it has finite energy. Applying Theorem 6.23 we deduce that $\bar{f}$ is homotopic to a constant on each compact subset of M. Whence, it follows that $(\bar{f}|_D)_\#$, and therefore σ, are trivial homomorphisms. □

6.4 A generalization of a finiteness theorem of Lemaire

In the previous section, following Schoen and Yau, we used vanishing results for harmonic maps in order to obtain information on the fundamental group of a manifold whose Ricci curvature is non-negative in some integral sense. In this section we take a different point of view. The purpose is to prove finiteness theorems for harmonic maps of bounded dilation into a negatively curved manifold, on the assumption that the domain manifold has finitely generated fundamental group. The results we consider here generalize to non-compact settings previous work by Lemaire, [93]. Maintaining the notation introduced in Definition 6.12, Lemaire's result states

Theorem 6.24. *Suppose that $(M, \langle\,,\rangle)$ and $(N, (\,,\,))$ are compact manifolds. Suppose also that the sectional curvatures of N are strictly negative. Then, for every $T > 0$, the set $\mathcal{H}_T$ is finite.*

We provide a somewhat direct and qualitative proof of the result, where the compactness of the target manifold plays no role.

Proof. Since $(M, \langle\,,\rangle)$ is compact and the sectional curvatures of the complete manifold $(N, (\,,\,))$ are negative and bounded away from zero, we see that $\mathcal{H}_T$ is a compact subset of $C^0(M, N)$ endowed with the uniform topology. Indeed, the Schwarz-type lemma of Theorem 6.15 asserts that

$$\sup_{f \in \mathcal{H}_T} \sup_M |df| = C < +\infty.$$

In particular

$$\sup_{f \in \mathcal{H}_T} \mathrm{diam} f(M) \leq C \mathrm{diam}(M) < +\infty. \tag{6.30}$$

By Ascoli's theorem, $\mathcal{H}_T$ is relatively compact in $C^0(M, N)$. Elliptic estimates now show that, for every sequence $\{f_n\} \subseteq \mathcal{H}_T$ converging to f, there is a subsequence whose higher derivatives converge locally uniformly on M. It follows that the limit function f is in fact a smooth harmonic map; see also [148], Chapter IX. Finally, f has dilation bounded by T. Indeed, $f_n^* \langle , \rangle_N \to f^* \langle , \rangle_N$ as $n \to +\infty$, pointwise on M and in the sense of quadratic forms. Set λ_j^n and λ_j for the j-th eigenvalues of $f_n^* \langle , \rangle_N$ and $f^* \langle , \rangle_N$, respectively. Then, from the variational characterization, we deduce $\lambda_j^n \to \lambda_j$ as $n \to +\infty$, proving that

$$\lambda_1 (x) \leq T\lambda_2 (x) \ \text{ on } M.$$

As a consequence, the relatively compact set $\mathcal{H}_T$ is, in fact, compact.

Accordingly, we have only to show that $\mathcal{H}_T$ is discrete. To this end, we adapt an argument of [148], Chapter XI. Let $R > 0$ be so large that, for every $f \in \mathcal{H}_T$,

$$f (M) \subset B_R^N.$$

Such a radius exists because of (6.30). Next, set

$$i = \operatorname{inj} \left(\overline{B_R^N} \right) > 0.$$

For any chosen $f \in \mathcal{H}_T$, we consider the ball $\mathcal{B}_i (f)$ in $C^0 (M, N)$, centered at f and of radius i:

$$\mathcal{B}_i (f) = \left\{ g \in C^0 (M, N) : \sup_M \operatorname{dist}_N (f (x), g (x)) < i \right\}.$$

We claim that every harmonic map $g \in \mathcal{B}_i (f)$ is homotopic to f. Indeed, consider the function $\rho : M \to \mathbb{R}$ given by $\rho (x) = \operatorname{dist}_N (f (x), g (x))$. Then $\rho^2 (x)$ is smooth and subharmonic on M. By the maximum principle, ρ^2 must be constant. We can therefore construct a (geodesic) homotopy between f and g in the following way: for every $x \in M$, let $\gamma_x : [0, \rho] \to N$ be the unique geodesic segment, of length $\rho (< i)$, satisfying $\gamma_x (0) = f (x)$ and $\gamma_x (\rho) = g (x)$. The desired homotopy map is defined by setting $H (x, t) = \gamma_x (t)$ and this proves the claim.

To conclude, recall that every free homotopy class in $[M; N]$ contains at most one harmonic map of bounded dilation. Therefore, if $f \in \mathcal{H}_T$, then $\mathcal{H}_T \cap \mathcal{B}_i (f) = \{f\}$, showing that $\mathcal{H}_T$ is discrete □

We are now going to generalize Lemaire's theorem to the case where the domain manifold is non-compact.

Theorem 6.25. *Let $(M, \langle , \rangle)$ be a geodesically complete, parabolic manifold satisfying, for some $0 \leq k < 2$,*

$$^M\mathrm{Ric} \geq -(m-1) B^2 (1 + r (x))^{-k}, \ \textit{on } M.$$

Assume further that the fundamental group $\pi_1(M)$ of M is finitely generated. Let $(N, (\, , \,))$ be a compact manifold with strictly negative sectional curvatures. Then, $L^2\mathcal{H}_{k,T}$ is a finite set.

The proof of Theorem 6.25 can be obtained by adapting Lemaire's original argument to the present setting. The crucial ingredients of the proof are the Schwarz-type lemma contained in Theorem 6.15 and the following uniqueness result for harmonic maps in a given homotopy class. It is an improved version of the Schoen and Yau uniqueness theorem, [149], see also [148] Chapter XIII, which in turn extends to the non-compact setting a theorem by P. Hartman, [78].

Theorem 6.26. *Let $(M, \langle\,,\rangle)$ and $(N, (\,,\,))$ be complete Riemannian manifolds. Assume that M is parabolic and that N has strictly negative sectional curvatures. Let $f : M \to N$ be a non-constant, harmonic map of finite energy $|df| \in L^2(M)$. Then, there is no other harmonic map of finite energy in the (free) homotopy class of f, unless $f(M)$ is contained in a geodesic of N.*

Remark 6.27. The original statement required $\mathrm{vol}(M) < +\infty$; see Theorem 1 on page 321 in [148].This assumption was used only to prove that a subharmonic function φ with finite Dirichlet integral $|\nabla\varphi| \in L^2(M)$ is in fact constant; see Lemma 1, Corollary 1 and Lemma 2 on page 318 in [148]. Therefore, Theorem 6.25 follows from the Schoen–Yau proof using Corollary 7.28 in the next section.

We are now in a position to give the

Proof of Theorem 6.25. Note that, due to the dilation assumption, the rank of any $f \in \mathcal{H}_{k,T}$ is different from 1. Therefore, $f(M)$ is not a geodesic of N. By Theorem 6.26 it follows that different maps $f \neq g \in L^2\mathcal{H}_{k,T}$ represent distinct free homotopy classes $[f] \neq [g] \in [M; N]$. We have to show that they are finite in number.

Fix $x_0 \in M$ and $y_0 \in N$. As noted at the beginning of the proof of Theorem 6.21 in the previous section, the manifold N is aspherical and therefore, the set $[M; N]$ is in one-to-one correspondence with the conjugacy classes of homomorphisms $\pi_1(M, x_0) \to \pi_1(N, y_0)$. The finiteness result will be established once we prove that there are only a finite number of conjugacy classes of homomorphisms induced by maps in $L^2\mathcal{H}_{k,T}$. This will be done by interpreting the fundamental group as the deck transformation group of the universal covering.

So, let $(M', \langle\,,\rangle')$ and $(N', (\,,\,)')$ be the Riemannian universal coverings of M and N, respectively. Note that any smooth map $f : M \to N$ lifts to an $f_\#$-equivariant smooth map $f' : M' \to N'$. This means that, for every $\gamma \in \pi_1(M, x_0) = \mathrm{Deck}(M')$,

$$f'(\gamma(x')) = f_\#(\gamma)\, f'(x')$$

where (up to conjugation) $f_\#$ is the homomorphism induced by f. We choose a (relatively compact) fundamental region $\mathcal{F}$ of N' and we fix a point $x'_0 \in M'$ over $x_0 \in M$. Next, for every $f \in \mathcal{H}_{k,T}$, we consider the lifting f' of f with the property that $f'(x'_0) \in \mathcal{F}$. This can be done by composing any lifting f' with a deck transformation $\alpha \in \pi_1(N, y_0) = \mathrm{Deck}(N')$. Note that, in this way, f' becomes an $\alpha \circ f_\#$-equivariant map.

According to Theorem 6.15, there exists a constant $C > 0$ such that

$$\sup_{f \in \mathcal{H}_{k,T}} \sup_M |df|^2 \leq C.$$

Since Riemannian covering maps are local isometries, we deduce that the same L^∞-estimate holds true for the lifted maps f'. From this, the relative compactness of $\mathcal{F}$ and the fact that $f'(x_0') \in \mathcal{F}$, it easily follows that, for every $R > 0$ there is $\bar{R} > 0$ such that

$$f'\left(B_R^{M'}(x_0')\right) \subset B_{\bar{R}}^{N'},$$

for every $f \in \mathcal{H}_{k,T}$.

Let $S = \{\gamma_0 = 1, \ldots, \gamma_s\}$ be a (finite) set of generators for $\pi_1(M, x_0)$ and choose $R > 0$ large enough that

$$\{\gamma_0(x_0'), \ldots, \gamma_s(x_0')\} \subset B_R^{M'}(x_0').$$

Then, recalling the equivariant property, we have

$$\{f' \circ \gamma_0(x_0'), \ldots, f' \circ \gamma_s(x_0')\} \subset B_{\bar{R}}^{N'} \cap \{\alpha \circ f'(x_0') : \alpha \in \pi_1(N, y_0)\},$$

for every $f \in \mathcal{H}_{k,T}$. Let

$$G = \cup_{f \in \mathcal{H}_{k,T}} \left\{\alpha \in \pi_1(N, y_0) : \alpha \circ f'(x_0') \in B_{\bar{R}}^{N'}\right\}$$

which is a finite set because

$$G \subset \left\{\alpha \in \pi_1(N, y_0) : B_{\bar{R}}^{N'} \cap \alpha(\mathcal{F}) \neq \varnothing\right\}.$$

The conjugacy classes of the homomorphisms $f_\#$ associated to $f \in L^2\mathcal{H}_{k,T}$, are characterized by the restrictions

$$f_\#|_S : S \to G.$$

Since both S and G are finite, the number of these classes is also finite, as required to complete the proof. □

Chapter 7

Some topological applications

7.1 Ends and harmonic functions

Given a compact set K in M, an end $E(K)$ with respect to K is an unbounded connected component of $M \setminus K$. By a compactness argument, it is readily seen that the number of ends with respect to K is finite,

It is also clear that if $K \subset K'$, then every end $E(K')$ is contained in one end $E(K)$, so that the number of ends increases as the compact K enlarges.

We say that M has a finite number of ends if there exists a constant C such that for every compact K the number of ends with respect to K is bounded above by C. In this case there exists a compact K_o and a number N such that for every compact $K \supset K_o$ the number of ends with respect to K is exactly N, and we say that M has N ends.

In what follows we will fix a compact domain D with smooth boundary and by an end we will always mean an end with respect to D.

We recall that a Riemannian manifold $(M, \langle \, , \rangle)$ is parabolic if every subharmonic, bounded-above function on M is constant. See Section 7.2 for a description of some equivalent definitions of parabolicity. An end E will be said to be parabolic if the double of E (see, Section 7.3) is a parabolic manifold. Equivalently, if every positive superharmonic function u satisfying $\partial u/\partial \nu \geq 0$ on ∂E, ν being the unit outward normal to ∂E, is constant. Otherwise the end will be called non-parabolic. The non-parabolicity of E is also characterized by the existence of a (minimal) positive Green kernel satisfying Neumann boundary conditions on ∂E.

We are going to describe the connection between the number of ends and harmonic functions as developed by P. Li and L.-F. Tam and co-authors (see [99]–[102], [156], [104] and [105]). In particular, we are going to show that the number of non-parabolic ends of a manifold is controlled by the dimension of the space of bounded harmonic functions with finite Dirichlet integral. This, together with the vanishing, or finiteness theorems described above, allows us to obtain results on the number of non-parabolic ends.

We fix an exhaustion $\{D_i\}$ of M by relatively compact open domains with smooth boundary with $D \subset D_i \subset D_{i+1}$, and, given an end E with respect to D we write $E_i = E \cap D_i$, and $\partial E_i = \partial D_i \cap E$.

Our first task is to obtain a characterization of non-parabolic ends (see [100], Proposition 1.1), that will be instrumental in the constructions that follow. We begin by recalling the following equivalent characterization of parabolicity in terms of a maximum principle for unbounded domains (See [3], Theorem 6.C).

Lemma 7.1. *A manifold M is parabolic if and only if the following condition holds: Let ϕ be defined and subharmonic in a region $G \subset M$ with non-empty boundary ∂G. If ϕ is bounded above, and for every $\zeta \in \partial G$*

$$\liminf_{G \ni x \to \zeta} \phi(x) \leq m,$$

then $\phi \leq m$ in G

Proof. Let M be parabolic, and assume by contradiction that $\sup_G \phi > m + 2\epsilon$ for some $\epsilon > 0$. Define

$$v_\epsilon = \begin{cases} \max\{\phi, m+\epsilon\} & \text{in } G, \\ m+\epsilon & \text{in } M \setminus G. \end{cases}$$

Then v_ϵ is subharmonic and bounded above on M, and therefore it is constant on M. Since $\phi < m + \epsilon$ on ∂G, we conclude that $v_\epsilon \equiv m + \epsilon$, so that $\phi \leq m + \epsilon$ on G, a contradiction.

For the converse, assume that the condition holds, let v be a subharmonic bounded above function on M, and suppose by contradiction that v is not constant. Then there exist $\epsilon > 0$ and $x_o \in M$ such that $\phi(x_o) < \sup \phi - 2\epsilon$. Then, if G is a connected component of the non-empty set $\{x : \phi(x) > \sup \phi - 2\epsilon\}$, ϕ is subharmonic on G and $\phi \leq \sup \phi - \epsilon$ on ∂G. The condition implies that $\phi \leq \sup \phi - \epsilon$ on G contradicting the definition of G. $\square$

Applying the lemma to the double of an end we immediately obtain

Corollary 7.2. *Let E be an end of the manifold M. Then E is parabolic if and only if for every function $\phi : \bar{E} \to \mathbb{R}$ which is subharmonic and bounded above on E we have $\sup_E \phi = \max_{\partial E} \phi$.*

This allows us to deduce the characterization of non-parabolic ends mentioned above (see [100]).

Proposition 7.3. *Let E be an end of M. Then E is non-parabolic if and only if there exists a bounded harmonic function $h : \bar{E} \to \mathbb{R}$ such that $\min_{\partial E} h > \inf_E h = \liminf_{E \ni x \to \infty} h(x)$. In fact, if E is non-parabolic, it can be arranged that $h = 1$ on ∂E and $\inf_E h = 0$. Furthermore h has finite Dirichlet integral, and it is minimal in the sense that, if $\tilde{h} : \bar{E} \to \mathbb{R}$ is a positive harmonic satisfying $\tilde{h} \geq h$ on ∂E, then $\tilde{h} \geq h$ on E.*

Proof. Corollary 7.2 above implies that if there exists a bounded harmonic function h satisfying $\inf_E h < \min_{\partial E} h$, then E is non-parabolic.

Conversely, assume that E is non-parabolic. Then, again by Corollary 7.2, there exists a function $\psi : \bar{E} \to \mathbb{R}$ which is superharmonic and bounded below on E and such that $\inf_E \psi < \min_{\partial E} \psi$. By the maximum principle, this implies that the sequence $\min_{\partial E_i} \psi$ is eventually decreasing. By translating and scaling, we may assume that $\inf_E \psi = 0$ and $\psi \geq 1$ on ∂E.

Next, for every i, let h_i be the solution of the following boundary value problem:

$$\begin{cases} \Delta h = 0 \text{ in } E_i, \\ h = 1 \text{ on } \partial E, \quad h = 0 \text{ on } \partial E_i. \end{cases}$$

It follows from the maximum principle that $0 < h_i < 1$ on E_i and that the sequence $\{h_i\}$ is increasing, and therefore h_i converges to a positive harmonic function h on E satisfying $0 < h \leq 1$ and $h = 1$ on ∂E. On the other hand, again by the maximum principle, $h_i \leq \psi$ on $\bar{E}_i$, and passing to the limit, $h \leq \psi$. Thus $\inf_E h \leq \inf_E \psi = 0$.

We note in passing that since h is harmonic on E, it follows that the sequence $\min_{\partial E_i} h$ is strictly decreasing.

The minimality of h is proved in the same way. Indeed, let $\tilde{h}$ be as in the statement. Then, by the maximum principle $\tilde{h} \geq h_i$ for every i, and the conclusion follows by letting $i \to +\infty$.

Finally, since h_i is harmonic on E_i with boundary values $h_i = 1$ on ∂E, $h_i = 0$ on ∂E_i, integrating by parts and denoting by ν the outward unit normal to ∂E, it follows that, for every $i_o \leq i$,

$$\int_{E_{i_o}} |\nabla h_i|^2 \leq \int_{E_i} |\nabla h_i|^2 = -\int_{\partial E} h_i \frac{\partial h_i}{\partial \nu} = \int_{\partial E} \frac{\partial h_i}{\partial \nu}.$$

Now, by elliptic estimates up to the boundary, all derivatives of h_i are uniformly bounded in compact subsets of $\bar{E}_i$ (see, e.g., [15], Section 5.6), and therefore, by passing to a subsequence if necessary, h_i converges to h with all its derivatives uniformly on compact subsets to $\bar{E}$. Thus letting i tend to infinity in the above inequality we obtain

$$\int_{E_{i_o}} |\nabla h|^2 \leq \liminf \int_{E_i} |\nabla h_i|^2 = -\lim \int_{\partial E} h_i \frac{\partial h_i}{\partial \nu} = \int_{\partial E} h \frac{\partial h}{\partial \nu}.$$

An alternative proof, which does not use refined regularity estimates goes as follows: Note that h_i minimizes the Dirichlet integral $\int_{E_i} |\nabla u|^2$ subjected to the boundary conditions $u = 1$ on ∂E, $u = 0$ on ∂E_i. Thus, if $i < j$, and we extend h_i to E_j by setting it equal to zero in $E_j \setminus E_i$, we have

$$\int_{E_j} |\nabla h_j|^2 \leq \int_{E_j} |\nabla h_i|^2 = \int_{E_i} |\nabla h_i|^2$$

showing that the Dirichlet integral $\int_{E_j} |\nabla h_j|^2$ is a decreasing function of j, and therefore there exists a constant C such that

$$\int_{E_j} |\nabla h_j|^2 \leq C \quad \forall j.$$

Now, since h_j converges to h uniformly on compact subsets of E, $\nabla h_j \to \nabla h$ weakly in L^2_{loc}, and we deduce that for every compact K in E,

$$\int_K |\nabla h|^2 \leq \liminf_j \int_K |\nabla h_j|^2 \leq C.$$

The conclusion is obtained by exhausting E by a family of compacts. Observe that one could also use standard interior estimates to deduce that, by passing to a subsequence if needed, ∇h_i converges to ∇h locally uniformly in E. □

Definition 7.4. Let E be a non-parabolic end, and let h be the harmonic function constructed above. A sequence $\{x_n\} \subset E$ which tends to infinity is said to be regular if $\lim_n h(x_n) = \inf_E h = 0$.

Remark 7.5. Clearly, (a subsequence of) the sequence $g_i = 1 - h_i$ converges locally uniformly on $\overline{E}$ to a harmonic function g on E such that $0 < g < 1$ on E, $g = 0$ on ∂E and $\limsup_{E \ni x \to \infty} g(x) = 1$. For future use, we also note that given a relatively compact domain Ω with $D \subset\subset \Omega$, then $\inf_{E \setminus \Omega} g > 0$. Indeed, by construction $g_i \geq g$ on E_i, and $g > 0$ on $\partial\Omega \cap E$ so there exists a constant $c > 0$ such that $g_i \geq c$ on $\partial\Omega \cap E$. Moreover, by the maximum principle $g_i \geq \min_{\partial\Omega \cap E} g_i$ on $E_i \setminus \Omega$. Now, fix i_o such that $\bar{\Omega} \subset E_{i_o}$. Since g_i converges to g uniformly on E_{i_o}, $0 < g_i - g < c/2$ for every $i > i_o$ sufficiently large, and we conclude that

$$g = g_i - (g - g_i) \geq c/2 \quad \text{on } E_{i_o} \setminus \Omega.$$

Letting $i_o \to +\infty$ yields the required conclusion

Small modifications of the last part of the proof allow us to establish the validity of the following lemma ([100], Lemma 2.2).

Lemma 7.6. *Let E be an end of a non-compact manifold M, and let f be a harmonic function on E which is $C^{2,\alpha}$ up to the boundary for some $\alpha \in (0,1)$. Assume that there exist an exhaustion of M by relatively compact domains with smooth boundary D_i and a sequence of harmonic functions f_i on E_i, C^0 up to the boundary, satisfying the boundary conditions $f_i = f$ on ∂E and $f_i = 0$ on ∂E_i, and such that $\lim f_i = f$. Then f has finite Dirichlet integral and, if ν is the outward unit normal to ∂E, we have*

$$\int_E |\nabla f|^2 \leq -\int_{\partial E} f \frac{\partial f}{\partial \nu}. \tag{7.1}$$

The same conclusion holds if the end E is replaced by the complement of a relatively compact set D with smooth boundary.

Remark 7.7. The assumption that f is $C^{2,\alpha}$ up to the boundary ensures that so are the functions f_i (see [60], Theorem 6.14). This in turn yields, via boundary elliptic estimates (see [60], Theorem 6.6), that f_i together with its first and second derivatives are locally Hölder up to the boundary with constants independent of i, and therefore a subsequence of f_i converges locally uniformly in $C^2(\bar{E})$.

Using the function h constructed in Proposition 7.3 as barrier, we easily obtain the following lemma ([156], Lemma 2.2).

Lemma 7.8. *Let E be a non-parabolic end of M and let h be the harmonic function constructed in Proposition 7.3. Then:*

(i) *If f is harmonic on E, continuous up to ∂E, and satisfies $f \leq 0$ on ∂E and $f \leq Ch$ on E for some constant $C > 0$, then $f \leq 0$ on E. If $f = 0$ on ∂E and $|f| \leq Ch$ on E, then $f \equiv 0$ on E.*

(ii) *If f is bounded and harmonic on E, continuous up to ∂E and such that $|f| \leq Ch$ on E for some $C > 0$, then f has finite Dirichlet integral.*

Proof. We follow the proof in [156]. To prove (i) , consider the function $h - C^{-1}f$ which is positive on E and greater than or equal to h on ∂E. According to the minimality statement in the proposition, we conclude that $h - C^{-1}f \geq h$ and therefore $f \leq 0$ on E.

To prove (ii) , assume first that f is $C^{2,\alpha}$ up to the boundary. For every i let f_i be the harmonic function which solves the boundary problem $f_i = f$ on ∂E and $f_i = 0$ on ∂E_i. As in the proof of Proposition 7.3 a subsequence of f_i converges locally uniformly in $C^2(\bar{E})$ to a bounded harmonic function $\tilde{f}$ on E with finite Dirichlet integral. By the maximum principle, $|f_i| \leq C_1 h$ on E_i for some constant C_1 and passing to the limit $|\tilde{f}| \leq C_1 h$ on E. Thus, $|\tilde{f} - f| \leq (C + C_1)h$ on E and, by (i) , $f = \tilde{f}$ and f has finite Dirichlet integral. The conclusion in the case where f is only assumed to be continuous up to the boundary is obtained by applying the argument to $E \setminus U$ where U is a suitable neighborhood of ∂E. □

A variation of the argument above allows us to show that a bounded harmonic function defined on a parabolic end, has automatically finite Dirichlet integral (see, e.g., [156], Lemma 2.3).

Lemma 7.9. *Let E be a parabolic end of a manifold M. Assume that $f : \bar{E} \to \mathbb{R}$ is bounded harmonic on E and continuous up to the boundary. Then f has finite Dirichlet integral. In fact, if f is $C^{2,\alpha}$ up to the boundary, then f satisfies estimate* (7.1).

Proof. As above, we may assume that f is $C^{2,\alpha}$ up to the boundary of E. Further, by adding to f a suitable constant, we may suppose that f is positive on E. Arguing as in the proof of (ii) above yields a function $\tilde{f}$ which satisfies $\tilde{f} = f$ on ∂E, $0 \leq \tilde{f} \leq f$ on E and satisfies (7.1). Since $f - \tilde{f}$ is bounded on E and vanishes on ∂E, $f - \tilde{f} = 0$ on E by Corollary 7.2. □

We are now ready to present the proof that the number of non-parabolic ends of M is bounded above by the dimension of the space of bounded harmonic functions with finite Dirichlet integral (see [102] and [71]).

Proposition 7.10. *Let M be a complete non-compact Riemannian manifold, and assume M has two non-parabolic ends E_1 and E_2 (with respect to the compact* *domain D). Then there exists a bounded harmonic function $f : M \to \mathbb{R}$ with finite Dirichlet integral such that $0 < f < 1$, $\sup_{E_1} f = 1$, and $\inf_{E_2} f = 0$.*

Proof. Denote by e_a and E_A the parabolic and non-parabolic ends of M with respect to D, respectively, and, for every i, let f_i be the solution of the boundary value problem

$$\begin{cases} \Delta f_i = 0 \text{ in } D_i, \\ f_i = 1 \text{ on } \partial D_i \cap E_1, \quad f_i = 0 \text{ on } \partial D_i \cap \big(M \setminus E_1\big). \end{cases}$$

Then, $0 < f_i < 1$, and, by the Harnack principle, (a subsequence of) f_i converges, locally uniformly, to a harmonic function f on M, satisfying $0 \le f \le 1$.

For every A, let h_A be the harmonic function on E_A constructed in Proposition 7.3. By construction, and the maximum principle, for $A \neq 1$, $f_i < h_A$ on E_A. Thus $0 < f \le h_A$, so that $0 \le \inf_{E_A} f \le \inf_{E_A} h_A = 0$, and, by Lemma 7.8, f has finite Dirichlet integral on E_A.

Similarly, since $1 - f_i < h_1$ on E_1, we conclude that $\sup_{E_1} f = 1$, and that f has finite Dirichlet integral on E_1.

Finally, since f is bounded harmonic, by Lemma 7.9, it has finite Dirichlet integral on each parabolic end e_a. We may therefore conclude that f has finite Dirichlet integral on M, as required to complete the proof. □

Theorem 7.11. *Let $\mathcal{H}^\infty_\mathcal{D}(M)$ denote the space of bounded harmonic functions with finite Dirichlet integral on M, and by $N(D)$ the number of non-parabolic ends of M with respect to the relatively compact domain D. Then*

$$N(D) \le \dim \mathcal{H}^\infty_\mathcal{D}(M).$$

It follows that, if $\mathcal{H}^\infty_\mathcal{D}(M)$ is finite-dimensional, then the M has finitely many non-parabolic ends, whose number is bounded above by $\dim \mathcal{H}^\infty_\mathcal{D}(M)$.

Proof. Since the constants are in $\mathcal{H}^\infty_\mathcal{D}(M)$, we may assume that $N(D) \ge 2$. Let E_A, $A = 1, \dots, N(D)$ be the non-parabolic ends of M with respect to the relatively compact domain D. According to the previous proposition, for every A there exists a bounded harmonic function on M with finite Dirichlet integral f_A such that $\sup_{E_A} f_A = 1$ and $\inf_{E_B} f_A = 0$ if $B \neq A$.

To complete the proof it remains to be shown that the functions f_A are linearly independent. To this end note that for every A there exists a sequence of points $\{x_{A,n}\}$ going to infinity in the end E_A such that $\lim_n h_A(x_{A,n}) = 0$, and therefore, $\lim_n f_A(x_{A,n}) = 1$ and $\lim_n f_B(x_{A,n}) = 0$ if $B \neq A$.

Now if $\sum_{B=1}^{N(D)} c_B f_B = 0$, evaluating the sum along the sequence $x_{A,n}$ we obtain $c_A = 0$ for every A, proving that the functions f_A are linearly independent. □

An immediate application of Theorem 6.1 and Corollary 5.4 yields the following

Corollary 7.12. *Let $(M, \langle\,,\rangle)$ be a complete, non-compact, connected Riemannian manifold satisfying*

$$\mathrm{Ric} \geq -\rho(x) \tag{7.2}$$

for some non-negative continuous function ρ, and let $L = -\Delta - \rho$. Then

(i) *if $\lambda_1(L_M) \geq 0$, then M has at most one non-parabolic end;*

(ii) *if $\lambda_1(L_{M\setminus K}) \geq 0$ for some compact set K, then M has at most finitely many non-parabolic ends.*

We note that one may estimate the number of all ends, parabolic and non-parabolic, in terms of the dimension of suitable spaces of harmonic functions on M. More precisely, if we denote by $n(D)$ the number of parabolic ends with respect to D, and by $\mathcal{H}^+(M)$ the space spanned by the positive harmonic functions on M and by $\mathcal{H}^0(M, D)$ the space spanned by the harmonic functions which are bounded from one side on every end with respect to D, then Li and Tam prove that $n(D) + N(D) \leq \dim\mathcal{H}^0(M, D)$, $N(D) \leq \dim\mathcal{H}^\infty_{\mathcal{D}}(M)$ and if $N(D) \geq 1$, then $n(D) + N(D) \leq \dim\mathcal{H}^+(M)$ (see [102], Theorem 2.1).

It should be also noted that, more recently, C.-J. Sung L.-F. Tam and J. Wang improved on the above result (see Theorem 2.1 in [156]), by establishing isomorphisms between the spaces $\mathcal{H}^0(M, D)$, $\mathcal{H}^+(M, D)$ and $\mathcal{H}^\infty_{\mathcal{D}}(M)$ and direct sums of spaces of harmonic functions defined on each end, which yields:

(i) if $N(D) \geq 1$, then $\dim \mathcal{H}^\infty_{\mathcal{D}}(M) = \sum_{A=1}^{N(D)} \dim \mathcal{H}^\infty_{\mathcal{D}}(E_A)$,

(ii) if $N(D) \geq 1$, then $\dim \mathcal{H}^+(M) = \sum_{A=1}^{N(D)} \dim \mathcal{H}^\infty_{\mathcal{D}}(E_A) + \sum_{a=1}^{n(D)} \dim \mathcal{H}^+(e_a)$,

(iii) if $N(D) = 0$ and $n(D) \geq 2$, then $\dim \mathcal{H}^0(M, D) = \sum_{A=1}^{N(D)} \dim \mathcal{H}^+(e_a)$,

where the spaces $\mathcal{H}^\infty_{\mathcal{D}}(E_A)$, $\mathcal{H}^+(e_a)$ and $\mathcal{H}^+(e_a)$ are defined as in the case of the corresponding spaces of functions on M, but with the additional assumption that the functions vanish on the boundary of the end.

In order to deduce topological consequences from the results on the triviality respectively, finite-dimensionality, of the space of bounded harmonic functions with finite Dirichlet integral on M, we need to find conditions which ensure that all ends of M are non-parabolic.

We begin with the following lemma which states that if M supports an L^2-Sobolev inequality, then every end is either non-parabolic or has finite volume (see [25], Lemma 1 and [104], Theorem 3).

Lemma 7.13. *Let M be a complete Riemannian manifold, and assume that for some $0 \leq \alpha < 1$, there exists a constant $S(\alpha) > 0$ such that the L^2-Sobolev–Poincaré inequality*

$$S(\alpha)^{-1} \left(\int_M v^{\frac{2}{1-\alpha}} \right)^{1-\alpha} \leq \int_M |\nabla v|^2 \tag{7.3}$$

holds for every smooth function (H^1 is enough) compactly supported in the complement of a compact set K. Then every end E of M is either non-parabolic or it has finite volume.

Proof. Let E be an end with respect to the domain D. Without loss of generality, we may assume that $K \subset D$. Maintaining the notation introduced above, we proceed as in the proof of Proposition 7.3 and, for every i, we let h_i be the solution of the boundary value problem

$$\begin{cases} \Delta h_i = 0 \text{ in } E_i, \\ h_i = 1 \text{ on } \partial E, \quad h_i = 0 \text{ on } \partial E_i \end{cases}$$

so that $0 < h_i < 1$, and, by the gradient estimate, (a subsequence of) h_i converges, locally uniformly, to a harmonic function h on E, which satisfies $0 \leq h \leq 1$ and $h = 1$ on ∂E.

We claim that if $\operatorname{vol}(E) = +\infty$, then h is non-constant, and therefore E is non-parabolic by Corollary 7.2 (or directly, since h is non-constant, (super)harmonic and satisfies $\partial f / \partial \nu \geq 0$, ν being the outward unit normal to ∂E).

To this end, fix i, and let ϕ be a smooth cut-off function such that $\operatorname{supp} \phi \subset E$, $\phi = 1$ on $E \setminus E_i$, and $|\nabla \phi| \leq C$.

For every $j > i$ the function ϕh_j is compactly supported in E, and vanishes off E_j, so that, by the Sobolev inequality

$$\begin{aligned} \left(\int_{E_j} (\phi h_j)^{2/(1-\alpha)} \right)^{1-\alpha} &\leq S(\alpha) \int_{E_j} |\nabla (\phi h_j)|^2 \\ &= S(\alpha) \int_{E_j} \left(|\nabla \phi|^2 h_j^2 + \phi^2 |\nabla h_j|^2 + \frac{1}{2} \langle \nabla (\phi^2), \nabla (h_j^2) \rangle \right). \end{aligned}$$

On the other hand, since h_j is harmonic in E_j and vanishes on ∂E_j, while ϕ vanishes on ∂E, integrating by parts,

$$\int_{E_j} \phi^2 |\nabla h_j|^2 = - \int_{E_j} \phi^2 h_j \Delta h_j + h_j \langle \nabla (\phi)^2, \nabla h_j \rangle = -\frac{1}{2} \int_{E_j} \langle \nabla (\phi)^2, \nabla (h_j^2) \rangle$$

which, inserted into the above inequality, gives

$$\left(\int_{E_j} (\phi h_j)^{2/(1-\alpha)} \right)^{1-\alpha} \leq S(\alpha) \int_{E_j} |\nabla \phi|^2 h_j^2.$$

Since $|\nabla \phi| \leq C$ and $\nabla \phi = 0$ outside E_i, this in turn yields

$$\left(\int_{E_{jo} \setminus E_i} h_j^{2/(1-\alpha)} \right)^{1-\alpha} \leq S(\alpha) C \int_{E_i} h_j^2,$$

for every $j > j_o > i$.

To conclude, suppose that the limiting function h is constant, so that, by the boundary condition, $h \equiv 1$ on E, and $h_j \nearrow 1$ on E. Letting $j \to +\infty$ in the inequality above, and applying the monotone convergence theorem, we obtain that

$$\big(\mathrm{vol}\, E_{j_o} \setminus \mathrm{vol}\, E_i\big)^{1-\alpha} \leq S(\alpha) C \mathrm{vol}\,(E_i),$$

so that, letting $j_o \to +\infty$, we deduce that $\mathrm{vol}\,(E) < +\infty$. □

Remark 7.14. In the statement of Lemma 7.13 we assumed that the L^2-Sobolev inequality holds outside a compact set. It is worth noticing that G. Carron has shown that if an L^2-Sobolev inequality holds in the complement of a compact set, then it holds on M (see [26], Proposition 2.5).

Suppose now that M supports an L^1-Sobolev inequality outside a compact set K, namely for some $\alpha > 1$ and there exists a constant $S_1(\alpha) > 0$ such that

$$S_1(\alpha)^{-1}\Big(\int |u|^{\alpha}\Big)^{1/\alpha} \leq \int |\nabla u|, \tag{7.4}$$

for every smooth function u or, equivalently, for every $u \in W^{1,1}(M)$, compactly supported in $M \setminus K$.

As mentioned in Section 5.3, the validity of an L^1-Sobolev inequality like (7.4) is equivalent to isoperimetric inequality

$$S_1(\alpha)^{-1}\big(\mathrm{vol}(\Omega)\big)^{1/\alpha} \leq \mathrm{vol}\,(\partial\Omega) \tag{7.5}$$

for every relatively compact domain $\Omega \subset\subset M \setminus K$ with smooth boundary. In particular, if Ω is a ball of radius r, we obtain that in $\mathbb{R}^m$ the only α for which an L^1-Sobolev inequality may (and in fact does) hold is $\alpha = \frac{m}{m-1}$. On the other hand, recalling that the local geometry of a Riemannian manifold is almost Euclidean, we see that in a general setting α is restricted to satisfy the inequality $\alpha \leq \frac{m}{m-1}$.

Indeed, we have the following elementary lemma (whose first point is in fact implied by the more general (7.5)).

Lemma 7.15. *Suppose that the L^1-Sobolev inequality (7.4) holds for some $\alpha \in (1, \frac{m}{m-1}]$ and for every function $u \in W^{1,1}$ compactly supported in $M \setminus K$. Then the following holds:*

(i) *There exist a constant C depending only on α such that the volume growth estimate*

$$\mathrm{vol}\,(B_R(x)) \geq C R^{\alpha/(\alpha-1)}$$

holds for every geodesic ball $B_R(x) \subset M \setminus K$. In particular every end with respect to K has infinite volume.

(ii) *Assume that $\alpha < 2$ (this condition is automatically satisfied if $m > 2$). Then the L^2-Sobolev inequality*

$$S_2(\alpha)^{-1}\Big(\int |v|^{\frac{2\alpha}{2-\alpha}}\Big)^{\frac{2-\alpha}{\alpha}} \leq \int |\nabla v|^2$$

holds for every $v \in W^{1,2}(M)$ compactly supported in $M \setminus K$, and with

$$S_2(\alpha) = \Big(\frac{2S_1(\alpha)}{2-\alpha}\Big)^2.$$

Proof. Let ϕ_ϵ be a family of smooth cut-off functions such that $\phi_\epsilon = 1$ on $B_r(x)$, $= 0$ off $B_{r+\epsilon}(x)$ and $|\nabla \phi_\epsilon| \leq 2/\epsilon$, for every $\epsilon > 0$. Applying (7.4) we deduce that

$$S_1(\alpha)^{-1}\big(\operatorname{vol} B_r(x)\big)^{1/\alpha} \leq \frac{2}{\epsilon}\operatorname{vol}\big(B_{r+\epsilon}(x) \setminus B_r(x)\big).$$

Letting $\epsilon \to 0$ and applying the co-area formula yield, for a.e. r,

$$(2S_1(\alpha))^{-1}(\operatorname{vol} B_r(x))^{1/\alpha} \leq \operatorname{vol}(\partial B_r(x)) = \frac{d}{dr}(\operatorname{vol} B_r(x)),$$

whence (i) follows by integrating the differential inequality between 0 and R.

To prove (ii) one proceeds as in the proof of Theorem 5.11. We recall the argument: let v be a smooth function compactly supported in $M \setminus K$. Applying (7.4) to the function $u = |v|^{2/(2-\alpha)}$, and using Schwarz's inequality yield

$$\begin{aligned} S_1(\alpha)^{-1}\Big(\int |v|^{2\alpha/(2-\alpha)}\Big)^{1/\alpha} &\leq \int |\nabla |v|^{2/(2-\alpha)}| \\ &\leq \frac{2}{2-\alpha}\Big(\int |v|^{2\alpha/(2-\alpha)}\Big)^{1/2}\Big(\int |\nabla v|^2\Big)^{1/2}, \end{aligned}$$

and (ii) follows by rearranging and simplifying. □

As a consequence, it follows that if the Sobolev inequality (7.4) holds for some $\alpha \in \big(1, \frac{m}{m-1}\big]$ and for every function $u \in W^{1,1}$ compactly supported in $M \setminus K$, then every end of M with respect to K is non-parabolic and, therefore, if M has at least two ends, M supports a non-constant bounded harmonic function with finite Dirichlet integral. In particular, if M supports an L^1-Sobolev inequality off a compact set, and the version for harmonic functions of Theorem 6.1 is applicable, we may conclude that M has only one end. A situation where this occurs is described in the next section (see also Section 9.2 where a slightly better result is obtained by a different method).

As a matter of fact, using isoperimetric considerations (a la Faber–Krahn), Carron, [27], observed that even an L^2-Sobolev inequality is related to volume growth properties of M. More precisely, we have the following result that we state without proof.

Theorem 7.16. *Let $(M, \langle , \rangle)$ be a complete Riemannian manifold supporting the L^2-Sobolev inequality stated in* (ii) *of Lemma* 7.15. *Then, the volume growth estimate stated in* (i) *of Lemma* 7.15 *holds.*

As a corollary, the above considerations on the nature of the ends supporting an L^1-Sobolev inequality can be generalized in the following

Corollary 7.17. *Assume that the L^2-Sobolev inequality in* (ii) *of Lemma* 7.15 *holds for some $0 < \alpha < 1$ and every function $v \in W^{1,2}$ compactly supported in $M \backslash K$. Then, every end of M with respect to K is non-parabolic. In particular, if M has at least two ends, then M supports a non-constant harmonic function with finite Dirichlet integral.*

So far, we have been mainly concerned with non-parabolic situations. We now take a closer look at the parabolic setting. To begin with, we recall the following classical result by M. Nakai, [116], [144].

Theorem 7.18. *Let E be an end of the Riemannian manifold $(M, \langle\, , \rangle)$ with respect to a bounded domain $\overline{D}$ with smooth boundary. Then E is parabolic if and only if there exists a non-negative, continuous, proper function $f : \overline{E} \to \mathbb{R}_+$ which is harmonic on E and satisfies $f|_{\partial E} = 0$.*

The harmonic function f in the above statement is called an *Evans–Selberg potential* of E. Note that, by boundary elliptic regularity, $f \in C^\infty\left(\overline{E}\right)$; see, e.g., [9], Theorem 3.59.

Starting from Nakai's theorem, and using the gluing technique developed by L. Sario and collaborators, T. Napier and M. Ramachandran, observed the following (see the proof of Theorem 2.6 in [117]).

Theorem 7.19. *A Riemannian manifold $(M, \langle\, , \rangle)$, with at least two ends, is parabolic if and only if the following holds. Let $E_1, \ldots, E_l$, $l \geq 2$, be the ends of M with respect to a compact set $\overline{D}$ with smooth boundary. Then, given $b_1, \ldots, b_l \in \{\pm\infty\}$ (not all equal) there exist proper harmonic functions $f : M \to \mathbb{R}$ and $f_j : E_j \to \mathbb{R}$ satisfying the further properties:*

(i) *$f_j|_{\partial E_j} = 0$;*

(ii) *f_j has constant sign and therefore $f_j(x) \to b_j$ as x tends to ∞ in E_j;*

(iii) *$\sup_{E_j} |f - f_j| < +\infty$. In particular $f(x) \to b_j$ as x tends to ∞ in E_j.*

Imposing some further property on the underlying manifold $(M, \langle\, , \rangle)$, this result can be substantially improved by specifying the maximal growth rate of the energy of f. For instance, if $(M, \langle\, , \rangle)$ satisfies

$$\operatorname{inj}(M) > 0, \qquad \operatorname{Ric} \geq -K,$$

then T. Napier and M. Ramachandran are able to deduce that the energy of f must be finite. In the case of complete manifolds with almost non-negative Ricci curvature, we have the following remarkable result first observed by P. Li and M. Ramachandran; see the proof of the Main Theorem in [98].

Theorem 7.20. *For every $m \geq 2$, there exists a constant $\varepsilon = \varepsilon(m) > 0$ such that the following holds. Let $(M, \langle\ ,\rangle)$ be a complete, parabolic Riemannian manifold of dimension m. Assume that, for some reference origin $o \in M$, the Ricci curvature of M satisfies*

$$\mathrm{Ric} \geq -\frac{\varepsilon}{r(x)^2} \quad \text{for } r(x) >> 1 \tag{7.6}$$

where $r(x) = \mathrm{dist}(x, o)$. Let $f : M \to \mathbb{R}$ be one of the proper harmonic functions described in Theorem 7.19. Then

$$\int_{B_R(o)} |\nabla f|^2 = o\left(R^2\right) \quad \text{as } R \to +\infty. \tag{7.7}$$

Later on, in Section 8.2, we shall employ Theorem 7.20 to study the structure of complete Kähler manifolds. We are going to provide a somewhat detailed proof of Theorem 7.20. To this end, recall the volume growth estimate by Li–Schoen and Li–Ramachandran contained in Theorem 2.26 of Section 2.2, stating that for every $m \geq 2$, there exists a constant $\varepsilon = \varepsilon(m) > 0$ such that if M satisfies (7.6) with constant ε, then, for every $d > 1$,

$$\mathrm{vol}\left(B_{\frac{r(x)}{d}}(x)\right) \to +\infty, \text{ as } r(x) \to +\infty.$$

Theorem 7.20 is then an easy consequence of the next more general lemma which is modelled on some results by Li and Tam; see Lemma 3.2, Lemma 3.3, Corollary 3.2, Theorem 4.2 in [102].

Lemma 7.21. *Let $(M, \langle,\rangle)$ be a complete Riemannian manifold such that for some $o \in M$, $K > 0$, and $d > 2$,*

$$\mathrm{Ric}(x) \geq -\frac{K}{1 + r(x)^2}, \tag{7.8}$$

on M, and

$$\mathrm{vol}\left(B_{\frac{r(x)}{d}}(x)\right) \to +\infty, \text{ as } r(x) \to +\infty, \tag{7.9}$$

where $r(x) = \mathrm{dist}(x, o)$. Let E be a parabolic end of M with respect to a compact set $\overline{D}$ with smooth boundary, and let $f : \overline{E} \to \mathbb{R}_+$ be the Evans–Selberg potential of E given in Theorem 7.18. Then

$$\sup_{\partial B_R(o) \cap E} f = o\left(R^2\right) \quad \text{and} \quad \int_{B_R(o) \cap E} |\nabla f|^2 = o\left(R^2\right) \tag{7.10}$$

as $R \to +\infty$.

Let us show how to deduce Theorem 7.20 from Lemma 7.21.

Proof of Theorem 7.20. Having fixed a smooth compact domain $\overline{D} \subset M$, let $E_1, \dots, E_l$ denote the (parabolic) ends of M with respect to $\overline{D}$. Let f and $f_1, \dots, f_l$ be the smooth, proper harmonic functions constructed in Theorem 7.19. For any fixed j, since $f - f_j$ is a bounded harmonic function on the parabolic end E_j, by Lemma 7.9 we deduce that

$$\int_{E_j} |\nabla (f - f_j)|^2 < +\infty.$$

On the other hand, according to Theorem 2.26, the energy estimate of Lemma 7.21 applies to f_j and we have

$$\int_{B_R(o)\cap E_j} |\nabla f_j|^2 = o\left(R^2\right), \text{ as } R \to +\infty.$$

It follows that

$$\int_{B_R(o)\cap E_j} |\nabla f|^2 = o\left(R^2\right), \text{ as } R \to +\infty$$

which implies, for every $R >> 1$,

$$\begin{aligned}\int_{B_R(o)} |\nabla f|^2 &= \int_D |\nabla f|^2 + \int_{B_R(o)\cap\left(M\setminus\overline{D}\right)} |\nabla f|^2 \\ &= C + \sum_{j=1}^{l} \int_{B_R(o)\cap E_j} |\nabla f|^2 = o\left(R^2\right) \text{ as } R \to +\infty.\end{aligned}$$

□

It remains to prove Lemma 7.21. The argument makes an essential use of a version of the weak Harnack inequality for Schrödinger-type operators under curvature conditions. In the form we need, it follows applying to the function $g = u \exp\bigl(-At/(1 + R^2)\bigr)$ the parabolic Harnack inequality due to Li and Tam [101], Theorem 1.1.

Theorem 7.22. *Let $(M, \langle\, ,\, \rangle)$ be an m-dimensional Riemannian manifold with (possibly empty) boundary ∂M. Let $o \in M$ and $R > 0$ be such that the closed ball $\overline{B_{2R}(o)} \subset M \setminus \partial M$ is compact, and assume that*

$$\operatorname{Ric}(x) \geq -\frac{K}{1 + R^2} \quad \text{on } B_{2R}(o),$$

for some $K > 0$. Finally, let $A \in \mathbb{R}$ and $d > 1$. Then, there exists a constant $C > 0$ which depends on A, m, K, d but not on R such that, if $u \geq 0$ is a weak solution of

$$\Delta u + \frac{A}{1 + R^2} u \geq 0 \text{ on } B_{2R}(o),$$

then, for every $x \in B_R(o)$,

$$u(x) \leq \frac{C}{\operatorname{vol}\left(B_{\frac{R}{d}}(x)\right)} \int_{B_{\frac{R}{d}}(x)} u.$$

We will also need a localized version of the celebrated gradient estimate of Cheng Yau, which we recall for the convenience of the reader (see Schoen and Yau's book, [147], Theorem 3.1).

Theorem 7.23. *Let $(M, \langle\, , \rangle)$ be an m-dimensional Riemannian manifold with (possibly empty) boundary ∂M. Let $x_o \in M$ and $R > 0$ be such that the closed ball $\overline{B_{2R}(x_o)} \subset M \setminus \partial M$ is compact, and assume that*

$$\operatorname{Ric}(x) \geq -\frac{K}{1+R^2} \quad on\ B_{2R}(x_o),$$

for some $K > 0$. Then, there exists a constant C which depends only on m such that for every positive harmonic function f on $B_{2R}(x_o)$,

$$|\nabla f(x)| \leq C\Big(\frac{1+\sqrt{K}}{R}\Big) f(x),\ \forall x \in B_R(x_o).$$

We are now in a position to give the

Proof of Lemma 7.21. For ease of notation, in the sequel we will denote by B_R the geodesic ball of radius R centered at o. Assume that $D \subseteq B_{R_0}$. First, we show that, for any $R > R_0$,

$$\int_{B_R \cap E} |\nabla f|^2 \leq \sup_{\partial B_R \cap E} f \int_{\partial E} \left(-\frac{\partial f}{\partial \nu}\right), \tag{7.11}$$

where ν denotes the outward pointing unit normal to ∂E. Let

$$T = T(R) = \sup_{B_R \cap E} f = \sup_{\partial B_R \cap E} f,$$

so that

$$B_R \cap E \subset f^{-1}([0,T]).$$

For the sake of simplicity, we assume T is a regular value of f. The general case can be handled by choosing a sequence of regular values $\{T_n\} \searrow T$ by applying the reasonings below on $[0, T_n]$ and, finally, by taking the limit as $n \to +\infty$. We set

$$g = T - f.$$

Integrating by parts, and recalling that f is harmonic (enough superharmonic) and that $f^{-1}(\{0\}) = \partial E$, we obtain

$$\begin{aligned}\int_{B_R\cap E} |\nabla f|^2 &\le \int_{f^{-1}([0,T])} |\nabla f|^2 \\ &= \int_{f^{-1}([0,T])} |\nabla g|^2 \\ &= \int_{f^{-1}(T)} g\frac{\partial g}{\partial \nu} + \int_{f^{-1}(0)} g\frac{\partial g}{\partial \nu} \\ &= \int_{\partial E} T\left(-\frac{\partial f}{\partial \nu}\right) \\ &= \sup_{\partial B_R\cap E} f \int_{\partial E}\left(-\frac{\partial f}{\partial \nu}\right),\end{aligned}$$

proving (7.11).

Next, we show that there exists a constant $C > 0$ independent of R such that, whenever $R > R_0 d/(d-2)$,

$$\sup_{\partial B_R\cap E} |\nabla f|^2 \le \frac{C}{\inf\limits_{x\in\partial B_R} \operatorname{vol}\left(B_{\frac{R}{d}}(x)\right)}\left(\sup_{\partial B_R\cap E} f\right)\int_{\partial E}\left(-\frac{\partial f}{\partial \nu}\right). \tag{7.12}$$

Note that, for every $x \in \partial B_R$,

$$\operatorname{dist}(x, \partial B_{R_0}) = R - R_0 > \frac{2R}{d}$$

which implies

$$B_{\frac{2R}{d}}(x) \subset E.$$

Now, the Bochner formula and the curvature assumption imply that

$$\Delta |\nabla f|^2 + \frac{2K}{1+(d-2)^2R^2/d^2} |\nabla f|^2 \ge 0 \text{ on } B_{\frac{2R}{d}}(x),$$

where K is the constant in (7.8). Using the weak Harnack inequality of Theorem 7.22, we deduce that there exists a constant C_1 independent of R such that, for every $y \in B_{R/d}(x)$,

$$|\nabla f|^2(y) \le \frac{C_1}{\operatorname{vol}\left(B_{\frac{R}{d}}(y)\right)}\int_{B_{\frac{R}{d}}(y)} |\nabla f|^2. \tag{7.13}$$

Specializing (7.13) to $y = x \in \partial B_R(o)$ and using (7.11) we obtain

$$\begin{aligned}\sup_{\partial B_R \cap E} |\nabla f|^2 &\leq \frac{C_1}{\inf_{x \in \partial B_R \cap E} \operatorname{vol}(B_{\frac{R}{d}}(x))} \int_{B_{\frac{(d+1)R}{d}} \cap E} |\nabla f|^2 \\ &\leq \frac{C_1}{\inf_{x \in \partial B_R \cap E} \operatorname{vol}(B_{\frac{R}{d}}(x))} \Big(\sup_{\partial B_{\frac{(d+1)R}{d}} \cap E} f \Big) \int_{\partial E} \Big(-\frac{\partial f}{\partial \nu} \Big).\end{aligned}$$

Now, let x_o be the point where f attains its supremum on the set $\partial B_{(1+d)R/d} \cap E$. According to Theorem 7.23, there exists a constant C_2 which depends only on m and d and K such that

$$\frac{|\nabla f|}{f} \leq \frac{C_2}{R} \quad \text{on } B_{2R/d}(x_o). \tag{7.14}$$

Integrating this inequality along the unit speed minimizing geodesic γ joining o to x_o, for t between R and $(1+d)R/d$, yields

$$\sup_{\partial B_{\frac{(d+1)R}{d}} \cap E} f = f(x_o) \leq e^{C_2/d} f(\gamma(R)) \leq e^{C_2/d} \sup_{\partial B_R \cap E} f,$$

which, substituted into the above inequality, yields (7.12).

Similarly, integrating $|\nabla f|$ along the part contained in E of the unit speed geodesic joining o to $x \in \partial B_r \cap E$, recalling that $f|_{\partial E} = 0$, and using the fundamental theorem of calculus, we have

$$\sup_{\partial B_r \cap E} f = \sup_{B_r \cap E} f \leq r \sup_{B_r \cap E} |\nabla f| . \tag{7.15}$$

Therefore, inserting these inequalities into (7.12), we get

$$\sup_{\partial B_R \cap E} |\nabla f|^2 \leq \frac{CR}{\inf_{x \in \partial B_R(o) \cap E} \operatorname{vol}(B_{\frac{R}{d}}(x))} \Big(\sup_{B_R \cap E} |\nabla f| \Big) \int_{\partial E} \Big(-\frac{\partial f}{\partial \nu} \Big). \tag{7.16}$$

We claim that, in view of the volume growth assumption (7.9), this implies that

$$\sup_{B_R \cap E} |\nabla f| = o(R), \text{ as } r \to +\infty. \tag{7.17}$$

Indeed, suppose by contradiction that there exists a positive constant A such that, along some sequence $R_k \to +\infty$,

$$\sup_{B_{R_k} \cap E} |\nabla f| \geq A R_k .$$

Without loss of generality, we can assume

$$\sup_{B_{R_k} \cap E} |\nabla f| = \sup_{\partial B_{R_k} \cap E} |\nabla f| ,$$

and we deduce from (7.16) that

$$AR_k \leq \frac{CR_k}{\inf_{\partial B_{R_k}} \operatorname{vol}\left(B_{\frac{R_k}{d}}(x)\right)} \int_{\partial E} \left(-\frac{\partial f}{\partial \nu}\right).$$

Thus

$$\inf_{\partial B_{R_k}} \operatorname{vol}\left(B_{\frac{R_k}{d}}(x)\right) \leq CA^{-1} \int_{\partial E} \left(-\frac{\partial f}{\partial \nu}\right),$$

which contradicts (7.9). This proves (7.17). Hence, by (7.15),

$$\sup_{\partial B_R \cap E} f = o\left(R^2\right)$$

as $r \to +\infty$, which together with (7.11) yields

$$\int_{B_R \cap E} |\nabla f|^2 = o\left(R^2\right)$$

as $R \to +\infty$, as required to complete the proof. □

To complete the picture, we point out that, up to avoiding the properness assumption, the result of Theorem 7.20 still holds even in the non-parabolic setting. This observation is proved in the Li-Tam paper [102] in a quite implicit form and is further remarked in Li–Ramachandran, [98]. A direct, explicit construction using a gluing technique can be readily obtained following Napier–Ramachandran, [117].

Theorem 7.24. *For every $m \geq 2$, there exists a constant $\varepsilon = \varepsilon(m) > 0$ such that the following holds. Let $(M, \langle , \rangle)$ be a complete, non-parabolic Riemannian manifold of dimension m with at least two ends. Assume that, for some reference origin $p \in M$, the Ricci curvature of M satisfies*

$$\operatorname{Ric} \geq -\frac{\varepsilon}{r(x)^2} \text{ for } r(x) >> 1 \tag{7.18}$$

where $r(x) = \operatorname{dist}(x, p)$. Then, there exists a positive non-constant harmonic function f on M satisfying

$$\int_{B_R(p)} |\nabla f|^2 = o\left(R^2\right) \text{ as } R \to +\infty. \tag{7.19}$$

Proof. Having fixed a smooth compact domain D, we denote by $e_1, \ldots, e_{n(D)}$ and by $E_1, \ldots, E_{N(D)}$ respectively the parabolic and the non-parabolic ends of M with respect to D. According to Proposition 7.3, for every $A = 1, \ldots, N(D)$ and $a - A > 0$, we find a harmonic function $g_A = a_A(1 - h_A) : E_A \to \mathbb{R}$ satisfying

$$\text{(i) } 0 < g_A \leq 1, \text{ (ii) } g_A = 0 \text{ on } \partial E_A, \text{ (iii) } |\nabla g_A| \in L^2(E_A). \tag{7.20}$$

On the other hand, applying Lemma 7.21, for every $j = 1, \ldots, n(D)$, we get a proper harmonic function $f_j : e_j \to \mathbb{R}$ such that

$$\text{(i) } f_j > 0 \text{ on } e_j, \text{ (ii) } f_j = 0 \text{ on } \partial e_j, \text{ (iii) } \int_{B_R(p)\cap e_j} |\nabla f_j|^2 = o\left(R^2\right). \tag{7.21}$$

To get the desired function f we now glue the various harmonic functions together. Let Ω be a relatively compact domain with $D \subset\subset \Omega$, and note that according to Remark 7.5 there exists a constant $\varepsilon > 0$ such that $g_A \geq 3\varepsilon > 0$ on $E_A \setminus \Omega$. Next let $\chi : \mathbb{R} \to \mathbb{R}$ be a smooth function such that $\chi' \geq 0$, $\chi'' \geq 0$, $\chi(t) = 0$ for $t < \varepsilon$, $\chi(t) = t - 2\varepsilon$ for $t > 3\varepsilon$. We define a function $\alpha : M \to \mathbb{R}$ by setting

$$\alpha(x) = \begin{cases} \chi(f_j) & \text{on } e_j, \\ \chi(g_A) & \text{on } E_A, \\ 0 & \text{on } D. \end{cases} \tag{7.22}$$

Direct computations show that $\alpha(x)$ is a non-negative subharmonic function on M. Note that since f_j is proper, the level set $\{x \in e_j : 0 \leq f_j(x) \leq 3\epsilon\}$ is compact for every j. As noted above, this also holds for every g_A, and, since $\chi(t)$ is linear if $t \geq 3\varepsilon > 0$, we deduce that $\alpha(x)$ is harmonic off a large enough compact set, and therefore, $0 \leq \Delta\alpha \in C_c^\infty(M)$. Let $\beta : M \to \mathbb{R}$ be the solution of the Poisson equation

$$\Delta\beta = \Delta\alpha \tag{7.23}$$

given by

$$\beta(x) = -\int_M G(x,y)\,\Delta\alpha(y)\,dy,$$

where $G(x,y)$ is the positive Green function of M, which exists because of the non-parabolicity assumption. Recall that $\beta \leq 0$ is bounded, $|\nabla\beta| \in L^2(M)$ and $\beta(x_n) \to 0$ along any regular sequence. See Appendix 7.2. Finally, define $f : M \to \mathbb{R}$ as

$$f(x) = \alpha(x) - \beta(x).$$

Clearly, $f(x) \geq 0$ and, by (7.23), $f(x)$ is harmonic. Moreover, since $\beta(x)$ has finite Dirichlet integral, recalling definition (7.22) of $\alpha(x)$, and properties (7.20) (iii) and (7.21) (iii), we see that

$$\begin{aligned} \int_{B_R} |\nabla f|^2 &\leq C + \sum_{j=1}^{n(D)} \int_{B_R\cap e_j} |\nabla f_j - \nabla\beta|^2 + \sum_{A=1}^{N(D)} \int_{B_R\cap E_A} |\nabla g_A - \nabla\beta|^2 \\ &\leq C + 2\sum_{j=1}^{n(D)} \int_{B_R\cap e_j} |\nabla f_j|^2 + 2\sum_{A=1}^{N(D)} \int_{B_R\cap E_A} |\nabla g_A|^2 + 2\int_M |\nabla\beta|^2 \\ &= o\left(R^2\right), \text{ as } R \to +\infty. \end{aligned}$$

Finally, the non-constancy of f is obvious if M has at least one parabolic end. Otherwise, $\lim_n f(x_n) = \lim_n g(x_n) = a_A - 2\varepsilon$ along any regular sequence $\{x_n\} \subset E$, and it suffices to choose the a_A's not all equal to conclude. □

7.2 Appendix: Further characterizations of parabolicity

In Section 7.1 we gave a self-contained introduction to parabolic manifolds and parabolic ends using the point of view of function theory and maximum principles. As a matter of fact, parabolicity is related to a wide class of equivalent properties involving the Green kernel, Brownian motion, capacity of condensers, L^2 differential forms, and so on. We invite the interested reader to consult the very nice survey paper by A. Grigor'yan, [72]. Here, we collect those characterizations of a parabolic manifold that shall be employed at some isolated points in the book.

The Green kernel and the Poisson equation. Let $(M, \langle , \rangle)$ be a connected manifold of dimension $m = \dim M$. We do not assume that M is geodesically complete. Given a smooth, compact domain $\Omega \subset\subset M$, the Dirichlet Green kernel $G_\Omega(x,y)$ of Ω is defined as a positive, symmetric, fundamental solution of the Laplace–Beltrami operator satisfying Dirichlet boundary conditions. Namely $G_\Omega : \Omega \times \Omega \setminus \{(x,x) : x \in \Omega\} \to \mathbb{R}$ is a smooth function such that:

(P) (Positivity) $G_\Omega(x,y) > 0$.

(S) (Symmetry) $G_\Omega(x,y) = G_\Omega(y,x)$.

(H) (Harmonicity) $x \longmapsto G_\Omega(x,y)$ is harmonic, for every fixed $y \in \Omega$.

(A) (Asymptotic behavior)

$$G(x,y) \sim C(m) \begin{cases} -\log d(x,y) & m = 2 \\ d(x,y)^{2-m} & m \geq 3 \end{cases}, \text{ as } d(x,y) \to 0.$$

(F) (Fundamental solution) For every $f \in C_c^\infty(\Omega)$,

$$\Delta \int_\Omega G_\Omega(x,y) f(y)\, dy = \int_\Omega G_\Omega(x,y) \Delta f(y)\, dy = -f(x), \text{ on } \Omega.$$

(D) (Dirichlet condition) $G_\Omega(x,y) = 0$ on $x \in \partial\Omega$, for every fixed $y \in \Omega$.

By the maximum principle, G_Ω increases with Ω. Therefore, letting $\Omega \nearrow M$ (in the sense of an exhaustion), G_Ω converges pointwise to a well-defined limit $G : M \times M \to \mathbb{R}_{\geq 0} \cup \{+\infty\}$. In fact, by the Harnack principle, G_Ω converges uniformly on compact subsets of $M \times M \setminus \{(x,x) : x \in M\}$ and the limit function G is either identically equal to $+\infty$ or it is finite for every $x \neq y$. In this latter case, G is called the *positive Green function of* M, and satisfies conditions (P), (S), (H), (A), (F) listed above, with Ω replaced by M. Moreover, G enjoys the following further properties:

(M) (Minimality) If $\widetilde{G}(x,y)$ is a second function satisfying (P), (S), (H), (A), (F), then $\widetilde{G}(x,y) \geq G(x,y)$. In particular $\inf_M G(x,y) = 0$, for every $y \in M$.

Indeed, it is an easy consequence of the maximum principle that $\widetilde{G}(x,y) \geq G_{\Omega}(x,y)$ for every $x, y \in \Omega \subset\subset M$. As for the infimum property, note that, for every constant $C \in \mathbb{R}$, the function $G(x,y)+C$ is again a symmetric fundamental solution of the Laplacian, satisfying (A).

(LU) (Locally uniform integrability properties) Having fixed relatively compact domains $D_1 \subset\subset D_2 \subset\subset M$,

$$y \longmapsto G(x,y) \in L^1_{loc}(M) \cap L^\infty(M \setminus D_2),$$

uniformly for $x \in D_1$.

Indeed, the uniform, L^1_{loc}-integrability follows easily from (A). On the other hand, by the local Harnack inequality applied to the family of harmonic functions $\{G(\cdot,y)\}_{y\in M-D_2}$, there exists a constant C independent of $x \in D_1$ and such that

$$G(x,y) \leq CG(o,y), \ \forall y \in M - D_2.$$

Moreover

$$\sup_{M-D_2} G(o,y) < +\infty.$$

Indeed, let Ω_n be an exhaustion of M by open sets containing o and with smooth boundaries, and let G_n be the Dirichlet Green kernel of Ω_n, so that $G_n(x,y) \nearrow G(x,y)$ locally uniformly in $M - \{x\}$. Let $C > \sup_{\partial D_2} G(o,y)$. Then, for every sufficiently large n, $C > G(o,y) \geq G_n(o,y)$, for $y \in \partial D_2$ and, clearly, $C > G_n(o,y) = 0$ on $\partial\Omega_n$. Thus, by the comparison principle, $C > G_n(o,y)$ in $\Omega_n \setminus D_2$. Whence, letting $n \to +\infty$, $G(0,y) \leq C$ for every $y \in M \setminus D_2$. It follows that there exists a constant $C' > 0$ independent of $x \in D_1$ and $y \in M \setminus D_2$ such that

$$G(x,y) \leq C'.$$

This proves that $G \in L^\infty(M \setminus D_2)$ uniformly for $x \in D_1$.

(C) (Convergence along regular sequences) Let E be a non-parabolic end. Then $\{x_n\}$ is regular for E if and only if $G(x_n,y) \to 0$ as $n \to +\infty$ for every $y \in M$.

We state the following result whose proof can be found in [72].

Theorem 7.25. *The Riemannian manifold* $(M, \langle\,,\rangle)$ *is non-parabolic if and only if it supports a positive Green function.*

Property (F) above states that the positive Green function of a non-parabolic manifold inverts the Laplace–Beltrami operator and, for any $f \in C^\infty_c(M)$, enables us to solve the Poisson equation

$$\Delta u = f(x) \ \text{ on } M.$$

Using the integrability properties (LU) one can show that the same conclusion holds in the more general case $f(x) \in C^{1,\alpha}(M) \cap L^1(M)$; see Lemma 8.5. We explicitly note that, if f is compactly supported, then the solution u satisfies $|\nabla u| \in L^2(M)$. Moreover, if $\{x_n\}$ is a regular sequence relative to a non-parabolic end, then $u(x_n) \to 0$ as $n \to +\infty$. Indeed, let $f \in C_c^\infty(M)$, $f \geq 0$ and $f \not\equiv 0$. Then

$$u(x) = \int_M f(y)\, G(x,y)\, dy$$

is a smooth, positive, bounded solution of the Poisson equation

$$\Delta u = -f(x).$$

As remarked by T. Lyons–D. Sullivan, [110], in some limit sense and for the special solution u, one has

$$\int_M |\nabla u|^2 = \int_M u\Delta u = \int_M uf < +\infty.$$

To be precise, following Napier–Ramachandran, [117], consider a smooth exhaustion $\{\Omega_j\}$ of M such that $\operatorname{supp} f \subset \Omega_0$, let G_j denote the Dirichlet Green function corresponding to Ω_j. and recall that $G_j \nearrow G$. Next, define

$$u_j(x) = \int_M f(y)\, G_j(x,y)\, dy$$

and note that u_j is the unique positive solution of the boundary value problem

$$\begin{cases} \Delta u_j = -f & \text{on } \Omega_j, \\ u_j = 0 & \text{on } \partial\Omega_j. \end{cases}$$

In particular, u_j is a harmonic function on $\Omega_j \backslash \operatorname{supp} f$. The sequence $\{u_j\}$ is monotone increasing and converges to u, uniformly on compact subsets of $M\backslash\operatorname{supp} f$. Using the local gradient estimates by Cheng–Yau (or the Harnack principle), we deduce that $\nabla u_j \to \nabla u$ uniformly on compact subsets of $M\backslash\operatorname{supp} f$. Now, integrating by parts,

$$\int_{\Omega_j} u_j f = -\int_{\Omega_j} u_j \Delta u_j = \int_{\Omega_j} |\nabla u_j|^2.$$

Whence, letting $j \to +\infty$,

$$+\infty > \int_M uf = \lim_{j\to+\infty} \int_{\Omega_j} |\nabla u_j|^2 \geq \int_M \liminf_{j\to+\infty} \left\{ |\nabla u_j|^2 \chi_{\Omega_j\backslash\Omega_0} \right\} = \int_{M\backslash\Omega_0} |\nabla u|^2.$$

Finally, the last statement follows from (LU), (C) and dominated convergence.

Condensers and capacity. The next characterization of parabolicity that we need to recall involves the concept of (absolute) capacity of a compact set.

By a condenser we mean any couple $E = (K, \Omega)$ where K is compact and $\Omega \subseteq M$ is an open set with $K \subset \Omega$. The capacity of the condenser E is defined as

$$cap\,(K, \Omega) = \inf_{u \in L(K,\Omega)} \int_M |\nabla u|^2 ,$$

where $L\,(K, \Omega)$ is the space of Lipschitz functions $0 \leq u \leq 1$ on M which are compactly supported in $\bar{\Omega}$ and satisfy $u \equiv 1$ on K. In case $\Omega = M$ one can simply write $cap\,(K)$ and speak of the *(absolute) capacity of the compact set* K. We remark that if both K and Ω are smooth, relatively compact domains, then the infimum is realized by a function $u_0 \in C^\infty\,(\Omega \setminus K)$ solving the boundary value problem

$$\begin{cases} \Delta u_0 = 0 & \text{on } \Omega \setminus K, \\ u_0 = 0 & \text{on } \partial\Omega, \\ u_0 = 1 & \text{on } \partial K. \end{cases}$$

The extremal function u_0 is called the *equilibrium potential of* E. The following result can be found in [72].

Theorem 7.26. *The Riemannian manifold* $(M, \langle\, ,\rangle)$ *is non-parabolic if and only every compact set has positive capacity.*

L^2-vector fields. There is a further characterization of non-parabolicity that we would like to recall. Essentially, it represents a global form of the divergence theorem for L^2 vector fields. By comparison, recall that if $(M, \langle\, ,\rangle)$ is geodesically complete and X is a vector field on M satisfying $X, \mathrm{div} X \in L^1\,(M)$, then

$$\int_M \mathrm{div} X = 0.$$

Indeed, since M is geodesically complete there exists a family of smooth cut-off functions ϕ_R such that $\phi_R = 1$ on B_R, $\phi_R = 0$ off B_{2R} and $|\nabla \phi_R| \leq C/R$. Using the function ϕ_R as test function in the definition of weak divergence, we obtain

$$\int_M \phi_R \mathrm{div} X = -\int \langle \nabla \phi_R, X \rangle \leq \frac{C}{R} \int_{B_{2R} \setminus B_R} |X|$$

so that letting $R \to +\infty$ and using the dominated convergence theorem gives

$$\int_M \mathrm{div} X \leq 0.$$

Applying the same argument to $-X$ yields the required conclusion.

This is a celebrated result by P.M. Gaffney, [56]. In a sense, $X \in L^1\,(M)$ has vanishing "boundary values". In the L^2 setting, without the completeness assumption, we have the next result which is known in the literature as *the Kelvin–Nevanlinna–Royden criterion* for non-parabolicity; see [110], and [72] for historical remarks. We also advise the reader that there is a non-linear extension of the criterion due to Gold'stein and M. Troyanov, [62].

Theorem 7.27. *Let $(M, \langle , \rangle)$ be a complete manifold. Then M is non-parabolic if and only if there exists a (C^1) vector field X on M such that*

$$\begin{array}{rl} \text{(i)} & |X| \in L^2(M), \\ \text{(ii)} & \operatorname{div} X \in L^1_{loc}(M) \text{ and } (\operatorname{div} X)_- \in L^1(M), \\ \text{(iii)} & 0 < \int_M \operatorname{div} X \leq +\infty. \end{array}$$

Proof. Assume first that M is non-parabolic. Let f be a smooth, compactly supported, non-negative and non-identically zero function, and let $G(x,y)$ be the Green kernel of M. Let

$$u = -\int_M G(x,y) f(y) dy.$$

According to the discussion after the statement of Theorem 7.25 the vector field $X = \nabla u$ is in L^2 and $\operatorname{div} X = \Delta u = f$ is compactly supported, smooth, non-negative and not identically zero, as required.

To prove the reverse implication we show that if M is parabolic, and X is a vector field satisfying conditions (i) and (ii) , then $\int_M \operatorname{div} X \leq 0$. To this end, let $D = D_0 \subset\subset D_1 \cdots \subset\subset D_n$ be an exhaustion of M by relatively compact domains with smooth boundary, and, for every n, let ϕ_n be the solution of the Dirichlet problem

$$\begin{cases} \Delta \phi_n = 0 \text{ on } D_n \setminus D, \\ \phi_n = 1 \text{ on } \partial D, \quad \phi_n = 0 \text{ on } \partial D_n, \end{cases}$$

and extend it to a Lip function on M by setting it constant on D and on $M \setminus D_n$. The sequence ϕ_n converges monotonically to a harmonic function ϕ on $M \setminus \bar{D}$, and, since $\phi = 1$ on ∂D, then $\phi \equiv 1$, by the assumed parabolicity of M. Further, according to Remark 7.7, the convergence is locally uniform in $C^2(M \setminus D)$, so that, in particular $\nabla \phi_n \to 0$ uniformly on ∂D, and we have

$$\int_{D_n \setminus D} |\nabla \phi_n|^2 = -\int_{\partial D} \phi_n \frac{\partial \phi_n}{\partial \nu} \to 0, \qquad \text{as } n \to +\infty.$$

Using the Lip function ϕ_n as test function, and arguing as above, we write

$$\int \phi_n \operatorname{div} X = -\int \langle \nabla \phi_n, X \rangle \leq -||\nabla \phi_n||_{L^2} ||X||_{L^2}.$$

The required conclusion follows by noting that as $n \to +\infty$ the right-hand side tends to zero, while

$$\int \phi_n \operatorname{div} X \to \int \operatorname{div} X$$

by an application of the monotone convergence theorem to the positive and negative part of $\phi_n \operatorname{div} X$. □

As noted above, if we assume that M is parabolic and $\mathrm{div}X \in L^1$, applying the theorem to X and $-X$ we conclude that

$$\int_M \mathrm{div}X = 0.$$

As an immediate consequence, if $(M, \langle,\rangle)$ is parabolic, every subharmonic function $u : M \to \mathbb{R}$ satisfying $|\nabla u| \in L^2(M)$ must be harmonic. Indeed, u turns out to be constant

Corollary 7.28. *Let $(M, \langle\, ,\rangle)$ be a parabolic manifold. If $\varphi \in C^2(M, \mathbb{R})$ is a subharmonic function satisfying $|\nabla\varphi| \in L^2(M)$, then φ is constant.*

Proof. Let φ be subharmonic with gradient in L^2. According to the previous remark, φ is in fact harmonic. Consider now the smooth function $f = \left(1 + \varphi^2\right)^{\frac{1}{2}}$. Direct computations show that f is subharmonic. Moreover

$$|\nabla f| \leq |\nabla\varphi|$$

so that $|\nabla f| \in L^2(M)$. Applying the first part of the proof yields that f is harmonic. Since $f \geq 0$ and M is parabolic we conclude that f is constant. As a consequence, φ must be constant. □

We also observe that in the Kelvin–Nevanlinna–Royden criterion it suffices to consider gradient vector fields only. We refer to Appendix B for notation and basic facts concerning differential forms.

Proposition 7.29. *Let $(M, \langle\, ,\rangle)$ be a complete manifold of dimension $m \geq 3$. Then, for every vector field $X \in L^2(M)$ there is a function $u \in W^{1,2}_{loc}(M)$ satisfying $\nabla u \in L^2(M)$ and*

$$\mathrm{div}\, X = \Delta u \text{ weakly on } M.$$

Proof. By the Hodge–Kodaira decomposition (see Appendix B), given the L^2-differential form $\omega = X^\flat \in L^2\Lambda^1(M)$, there are sequences $\{u_k\} \subset C^\infty_c(M)$, $\{v_k\} \subset C^\infty_c\Lambda^2(M)$ and a harmonic 1-form $\gamma \in L^2\mathcal{H}^1(M)$ such that

$$\begin{aligned} &\text{(a) } du_k \overset{L^2}{\to} \alpha, \\ &\text{(b) } \delta v_k \overset{L^2}{\to} \beta, \end{aligned} \tag{7.24}$$

and

$$\omega = \alpha + \beta + \gamma.$$

Fix any $\Omega \subset\subset M$ with smooth boundary. Since $\{|du_k|\}$ is bounded in $L^2(\Omega)$, it follows from the local Poincaré inequality that $\{u_k\}$ is bounded in $W^{1,2}(\Omega)$. By the Rellich-Kondrakov compactness theorem,

$$\{u_{k_h}\} \overset{L^2}{\to} u$$

for some $u \in L^2(\Omega)$. As a matter of fact, according to (7.24) (a), $\{u_{k_h}\}$ is a convergent sequence in $W^{1,2}_{loc}$ and therefore $u \in W^{1,2}(\Omega)$ with $du = \alpha$. Repeating the argument on a smooth, compact exhaustion of M gives a function $u \in W^{1,2}_{loc}(M)$ satisfying $du = \alpha$ on M.

To conclude, note that since d is a closed operator on L^2, then so is δ, and $\delta\beta = 0$ by (7.24) (b). Obviously, $\delta\gamma = 0$. Thus, $\operatorname{div} X = \delta\omega = \delta\alpha = \delta du = -\Delta u$. □

7.3 Appendix: The double of a Riemannian manifold

Following J. Munkres, [115], we recall the construction of the Riemannian doubling of a manifold with boundary.

Let M be a differentiable, m-dimensional manifold with smooth compact boundary ∂M. The interior of M is denoted by $\overset{\circ}{M} = M \setminus \partial M$. Let us consider two copies of M, say $M \times 0$, $M \times 1$. The (topological) double of M is defined as the topological manifold (without boundary) $\mathcal{D}(M)$ obtained from $M \times 0 \cup M \times 1$, identifying points $x \times 0$ and $x \times 1$ for every $x \in \partial M$. The homeomorphic images of $\overset{\circ}{M} \times 0$ and $\overset{\circ}{M} \times 1$ in $\mathcal{D}(M)$ are denoted by $\mathcal{M}_0$ and $\mathcal{M}_1$. In order to introduce a smooth structure on $\mathcal{D}(M)$, recall that ∂M has a product neighborhood, i.e., we can find an open set $W \subset M$ and a smooth diffeomorphism $p : W \to \partial M \times [0,1)$ whose restriction $p|_{\partial M} : \partial M \to \partial M \times 0$ is the identity map $p(x) = x \times 0$. Without loss of generality, we can always assume that W is relatively compact. Applying this construction to each copy $M \times 0$ and $M \times 1$ of M we get product neighborhoods $p_0 : W_0 \to \partial M \times [0,1)$ and $p_1 : W_1 \to \partial M \times (-1, 0]$. Let $\mathcal{W}$ be the union of W_0 and W_1 in $\mathcal{D}(M)$ and note that the maps p_0 and p_1 induces a homeomorphism $p : \mathcal{W} \to \partial M \times (-1,1)$. A differentiable structure on $\mathcal{D}(M)$ is defined by imposing the following conditions:

(1) The homeomorphism p is a smooth diffeomorphism.

(2) The inclusions $i_0 : \partial M \times 0 \hookrightarrow \mathcal{D}(M)$ and $i_1 : \partial M \times 1 \hookrightarrow \mathcal{D}(M)$ are smooth imbeddings.

Although the differentiable structure thus obtained depends on the choice of the product neighborhoods, it can be shown that the resulting smooth manifolds are diffeomorphic to each other.

Suppose now that M is endowed with a smooth Riemannian metric $\langle , \rangle$. Obviously, this induces metrics $\langle , \rangle_0$, $\langle , \rangle_1$ on $M \times 0$ and $M \times 1$ which, in turn, give rise to the metrics $\langle , \rangle_{\mathcal{M}_0}$ and $\langle , \rangle_{\mathcal{M}_1}$ on $\mathcal{M}_0$ and $\mathcal{M}_1$. Thus, each $\left(\mathcal{M}_j, \langle , \rangle_{\mathcal{M}_j}\right)$ is an isometric copy of $\left(\overset{\circ}{M}, \langle , \rangle\right)$ in $\mathcal{D}(M)$. Let $\langle , \rangle_{\mathcal{W}}$ be the Riemannian metric on $\mathcal{W} \subset \mathcal{D}(M)$ obtained by pulling back via p the product metric $\langle , \rangle|_{\partial M} + dt \otimes dt$ of $\partial M \times (-1,1)$. Next, choose a relatively compact, open set $\mathcal{V} = p^{-1}(\partial M \times (-\varepsilon, \varepsilon)) \subset\subset \mathcal{W}$ and consider a smooth partition of unity $\{\varphi_0, \varphi_1, \varphi_{\mathcal{V}}\}$ of $\mathcal{D}(M)$, subordinated to the open cover $\{\mathcal{M}_0, \mathcal{M}_1, \mathcal{V}\}$. A Riemannian metric on $\mathcal{D}(M)$ is then defined by the formula

$$(,) = \varphi_0 \langle,\rangle_{\mathcal{M}_0} + \varphi_1 \langle,\rangle_{\mathcal{M}_1} + \varphi_2 \langle,\rangle_{\mathcal{W}}.$$

We remark that the parabolicity of the double D is independent of how the metric is defined in a compact neighborhood of the gluing, which makes the definition of a parabolic end consistent.

7.4 Topology at infinity of submanifolds of C-H spaces

In order to put the next results into the appropriate perspective, we recall that the topology at infinity of a submanifold M^m of $\mathbb{R}^n$ is influenced and, in some cases determined, by the size of its second fundamental tensor II.

In the setting of complete, minimal hypersurfaces in Euclidean space, J. Tysk, [158], has shown that the L^m-integrability of $|\mathrm{II}|$ forces the submanifold to possess only a finite number of ends. See also the more recent paper by L. Ni, [120], for a different proof and related results. If we also add the stability assumption, by a result of Y.-B. Shen and X.-H. Zhu, [151], the immersion is totally geodesic. On the other hand, H. Cao, Y. Shen and S. Zhu have shown in [25] that stability alone implies that the hypersurface has simple topology at infinity, i.e., it has only one end.

We note that a suitable control of the L^m size of $|\mathrm{II}|$ implies stability; see [153] and Lemma 7.34 below. According to an isolation phenomenon pointed out by M. Anderson, [7], and quantified by P. Berard, [12], if we allow the codimension of the minimal immersion to be greater than 1 and the L^m size of $|\mathrm{II}|$ is sufficiently small, then the submanifold is again an affine space; see also Theorem 7.36 below.

Finally, we know from the aforementioned paper by Ni, [120], that if we relax the bound on $|\mathrm{II}|$ the minimal submanifold still has only one end.

Our main purpose is to extend the results both in [25] and in [120] by showing that small perturbations of the minimal immersion (so that minimality is lost) do not modify the topology at infinity of the submanifold. In fact, we are able to quantify the amount of such perturbation and to replace the Euclidean ambient space with a Cartan–Hadamard manifold, i.e., a complete, simply connected Riemannian manifold of non-positive sectional curvature; see Theorems 7.31 and 7.35 below. We should remark that G. Carron has obtained in [28] a similar result, with a different method and a less precise condition on the second fundamental form of the immersion.

We recall that, according to D. Hoffman and J. Spruck, [83], if $f : (M, \langle\, ,\rangle) \to (N, (\, ,))$ is an isometric immersion of a complete manifold M of dimension $m \geq 2$ into a Cartan–Hadamard manifold N, and H denotes the mean curvature vector field of f, then the following L^1-Sobolev inequality holds:

$$S_1(m)^{-1}\left(\int_M |u|^{\frac{m}{m-1}}\right)^{\frac{m-1}{m}} \leq \int_M (|\nabla u| + |H|\,|u|),\ \forall u \in W_0^{1,1}(M) \tag{7.25}$$

with

$$S_1(m) = \frac{\pi 2^{m-1}}{\omega_m^{\frac{1}{m}}} \frac{(m+1)^{1+\frac{1}{m}}}{m-1}, \tag{7.26}$$

ω_m being the volume of the unit ball of $\mathbb{R}^m$. In particular, if we assume that $H \in L^m(M)$, so that, for a suitable compact K,

$$\|H\|_{L^m(M\setminus K)} < S_1(m)^{-1},$$

then, applying Hölder inequality, the term involving the mean curvature can be absorbed in the left-hand side, showing that the standard L^1-Sobolev inequality

$$C\left(\int |u|^{\frac{m}{m-1}}\right)^{\frac{m-1}{m}} \leq \int |\nabla u| \tag{7.27}$$

holds for every $u \in W_0^{1,1}(M)$ supported in $M \setminus K$. According to Lemma 7.15, every end of $M \setminus K$ is non-parabolic, and, if $m \geq 3$, the L^2-Sobolev inequality

$$C\left(\int |v|^{\frac{2m}{m-2}}\right)^{\frac{m-2}{m}} \leq \left(\int |\nabla v|^2\right)^{\frac{1}{2}} \tag{7.28}$$

holds in $M \setminus K$. We are therefore in the situation considered at the end of Section 7.1, and we have the following lemma.

Lemma 7.30. *Let $f : (M, \langle\,,\,\rangle) \to (N, (\,,\,))$ be an isometric immersion of a complete manifold M of dimension $m \geq 3$ into a Cartan–Hadamard manifold N. Denote by H the mean curvature vector field of f and assume that $H \in L^m(M)$. Then, each end of M is non-parabolic, and therefore the cardinality of the ends of M is bounded above by the dimension of the space of bounded harmonic functions on M with finite energy. In particular, if M has at least two ends, then M supports a non-constant, bounded, harmonic function with finite energy.*

We are now in a position to prove our first result on the topology at infinity of an immersed submanifold of a Cartan–Hadamard space with controlled extrinsic geometry, which generalizes [25], Theorem 1. We note that assumption (7.32) below is the counterpart of the stability condition assumed there.

Theorem 7.31. *Let $f : (M, \langle\,,\,\rangle) \to (N, (\,,\,))$ be an isometric immersion of a complete manifold M of dimension $m \geq 3$ into a Cartan–Hadamard manifold N whose sectional curvature (along f) satisfies*

$$(0 \geq)\ {}^N\mathrm{Sect}_{f(x)} \geq -\ {}^N R(x) \tag{7.29}$$

for some non-negative function ${}^N R \in C^0(M)$. Denote by H and II the mean curvature vector field and the second fundamental tensor of f respectively, and let $a(x) \in C^0(M)$ be the function defined by

$$a(x) = (m-1)\ {}^N R(x) + |\mathrm{II}|\,(|\mathrm{II}| + m\,|H|)(x). \tag{7.30}$$

Assume that $H \in L^m(M)$, and let $L = -\Delta - a(x)$.
If

$$\lambda_1 (L_{M \setminus K}) \geq 0, \tag{7.31}$$

for some compact set K, then M has only finitely many ends.
If

$$\lambda_1 (L_M) \geq 0, \tag{7.32}$$

then M has only one end.

Proof. Indeed the Gauss equations imply that the Ricci tensor of M satisfies

$$^M\mathrm{Ric}\,(x) \geq -a\,(x)\,.$$

According to Lemma 7.30, all ends of M are non-parabolic. The conclusion now follows from Corollary 7.12. □

At the end of [7], M. Anderson observed that a complete minimal submanifold $f : M^m \to \mathbb{R}^N$ of finite total scalar curvature $|\mathrm{II}| \in L^m(M)$ is necessarily totally geodesic provided it has only one end; see Theorem 5.2 in [7]. Combining this result with Theorem 7.31 we obtain the following theorem which was first proved, with different arguments and in case $N = m+1$, by Y.B. Shen and X.H. Zhu, [151].

Theorem 7.32. *Let $f : M^m \to \mathbb{R}^N$ be a complete, minimal submanifold of finite total scalar curvature. Suppose that f is stable, in the sense that $\lambda_1 \left(-\Delta - |\mathrm{II}|^2\right) \geq 0$. Then $f\,(M)$ is an affine m-plane.*

Our next task is to quantify the heuristic idea according to which the bottom of the spectrum of $L = -\Delta - a(x)$ is non-negative provided the norm of the function $a(x)$ in (7.30) is small.

It is well known that if an L^2-Sobolev inequality holds on M, then $\lambda_1 (L_M) \geq 0$ provided a suitable L^p-norm of a is strictly less than the Sobolev constant (see, e.g., [153]).

In the next lemma, we obtain the same conclusion in terms of an L^2-Sobolev inequality with a potential like (7.33) below.

Lemma 7.33. *Suppose that the following Sobolev-type inequality*

$$S(\alpha)^{-1} \left(\int_M v^{\frac{2}{1-\alpha}} \right)^{1-\alpha} \leq \int_M \left(|\nabla v|^2 + h(x) v^2 \right), \ \forall v \in C_c^\infty(M) \tag{7.33}$$

holds on M, where $0 < \alpha < 1$, $S\,(\alpha) > 0$ is a constant, and $h\,(x) \in C^0\,(M)$ is a non-negative function. Consider the Schrödinger operator

$$L = -\Delta - a(x)$$

with $a(x) \in C^0(M)$, and set $a_+(x) = \max\{a(x), 0\}$. If

$$\|h(x) + a_+(x)\|_{L^{\frac{1}{\alpha}}(M)} \leq S(\alpha)^{-1}, \tag{7.34}$$

then

$$\lambda_1(L_M) \geq 0.$$

If

$$\|h(x) + a_+(x)\|_{L^{\frac{1}{\alpha}}(M)} < +\infty, \tag{7.35}$$

then there exists a sufficiently large compact K such that

$$\lambda_1(L_{M\setminus K}) \geq 0.$$

Proof. Let K be a, possibly empty, compact set in M and let $\Omega = M \setminus K$. For every $v \in C_c^\infty(\Omega) \setminus \{0\}$, using (7.33) and applying Hölder inequality, we have

$$\begin{aligned}\int_\Omega (|\nabla v|^2 - a(x)v^2) &\geq \int_\Omega (|\nabla v|^2 + h(x)v^2) - \int_\Omega (a_+(x) + h(x))v^2 \\ &\geq \int_\Omega (|\nabla v|^2 + h(x)v^2) - ||a_+ + h||_{L^{\frac{1}{\alpha}}(\Omega)} \Big(\int_\Omega v^{2/(1-\alpha)}\Big)^{1-\alpha} \\ &\geq (1 - S(\alpha)||a_+ + h||_{L^{\frac{1}{\alpha}}(\Omega)}) \int_\Omega (|\nabla v|^2 + h(x)v^2).\end{aligned}$$

If (7.34) holds, then, taking $K = \emptyset$, from the variational characterization of the bottom of the spectrum, we immediately obtain that $\lambda_1(L_M) \geq 0$. On the other hand, if (7.35) holds, then we may find a compact set K such that the term in brackets on the right-hand side of the above inequality is positive, and we conclude that $\lambda_1(L_{M\setminus K}) \geq 0$. □

There are a number of geometric situations where the Sobolev inequality (7.33) is satisfied for some choices of $\alpha, S(\alpha), h(x)$. The interested reader can consult, e.g., [80].

Assume now that (M, g) is a submanifold of a Cartan–Hadamard manifold, so that the L^1-type Sobolev inequality (7.25) holds. As above, we apply this inequality to the function $u = |v|^{\frac{2(m-1)}{m-2}}$ with $v \in C_c^\infty(M)$ to get

$$S_1(m)^{-2} \left(\int_M |v|^{\frac{2m}{m-2}}\right)^{\frac{m-2}{m}} \leq \left\{\frac{2(m-1)}{m-2}\left(\int_M |\nabla v|^2\right)^{\frac{1}{2}} + \left(\int_M |H|^2 |v|^2\right)^{\frac{1}{2}}\right\}^2.$$

Expanding the square on the right-hand side and applying the inequality $2ab \leq \varepsilon^2 a^2 + \varepsilon^{-2} b^2$ with $\varepsilon > 0$, we finally obtain the L^2-Sobolev inequality

$$S_2(m, \varepsilon)^{-1} \left(\int_M |v|^{\frac{2m}{m-2}}\right)^{\frac{m-2}{m}} \leq \int_M |\nabla v|^2 + \varepsilon^2 \frac{(m-2)^2}{4(m-1)^2} \int_M |H|^2 v^2, \tag{7.36}$$

where

$$S_2(m,\varepsilon) = \frac{4(m-1)^2}{(m-2)^2}\frac{1+\varepsilon^2}{\varepsilon^2} S_1(m)^2. \tag{7.37}$$

We observe that in case f is a minimal immersion, the best L^2-Sobolev constant in (7.37) is achieved by choosing $\varepsilon = +\infty$. In this situation, we set

$$S_2(m) = S_2(m, +\infty).$$

In particular, from Lemma 7.33 we immediately conclude

Lemma 7.34. *Let $f : (M, \langle\, ,\, \rangle) \to (N, (\, ,\,))$ be an isometric immersion of a complete manifold M of dimension $m \geq 3$ into a Cartan–Hadamard manifold N. Denote by H the mean curvature vector field of f. Consider the Schrödinger operator*

$$L = -\Delta - a(x) \tag{7.38}$$

with $a \in C^0(M)$. If, for some $\varepsilon > 0$,

$$\left\| \frac{(m-2)^2\varepsilon^2}{4(m-1)^2} |H|^2 + a_+(x) \right\|_{L^{\frac{m}{2}}} \leq S_2(m,\varepsilon)^{-1}, \tag{7.39}$$

then

$$\lambda_1(L_M) \geq 0.$$

If

$$\left\| \frac{(m-2)^2\varepsilon^2}{4(m-1)^2} |H|^2 + a_+(x) \right\|_{L^{\frac{m}{2}}} < +\infty, \tag{7.40}$$

then, for a sufficiently large compact K,

$$\lambda_1(L_{M\setminus K}) \geq 0.$$

Lemma 7.34 allows us to obtain a version of Theorem 7.31 above, where the assumption on the bottom of the spectrum of $-\Delta - a(x)$ is replaced by a suitable control on the second fundamental tensor of the immersion. We note that our result extends the above mentioned result of Ni, [120], valid for a minimal submanifold of $\mathbb{R}^n$, and yields a quantitative improvement on the result of Carron [28] where the constant in the integral bound is not explicit. We also note that since

$$S_2(m,\varepsilon)^{-1} < \frac{(m-2)^2\varepsilon^2}{4(m-1)^2} S_1(m)^{-2},$$

the norm estimate in the statement of Theorem 7.35 below implies that $\|H\|_{L^m} < S_1(m)^{-1}$. Thus in this case, the standard L^1-Sobolev inequality (7.27) holds on M and we may conclude that L^2-Sobolev inequality (7.28) is valid on M, without having to appeal to Proposition 2.5 in [26].

Theorem 7.35. *Maintaining the notation of Theorem 7.31, assume that the sectional curvature (along f) of N satisfies (7.29) and that*

$$\left\| \frac{(m-2)^2\varepsilon^2}{4(m-1)^2} |H|^2 + (m-1)\ {}^N R(x) + |\mathrm{II}| (|\mathrm{II}| + m|H|)(x) \right\|_{L^{\frac{m}{2}}} \leq S_2(m,\varepsilon)^{-1}$$

for some $\varepsilon > 0$, where $S_2(m,\varepsilon)$ is the L^2-Sobolev constant defined in (7.37). Then M has only one end.

If the norm on the left-hand side is finite, then M has finitely many ends.

Finally, combining Lemma 7.34 with some careful computations of Berard, [12], we obtain the following isolation phenomena for minimal submanifolds of the Euclidean space. Note that our constant improves on that of Proposition II.1 in [12].

Theorem 7.36. *Let $f : (M, \langle\, ,\, \rangle) \to \mathbb{R}^n$ be a complete, minimal, immersed submanifold of dimension $m \geq 3$ whose second fundamental tensor* II *satisfies*

$$\left(\int_M |\mathrm{II}|^m\right)^{\frac{2}{m}} \leq \frac{2}{m\left(2 - \frac{1}{n-m}\right)} S_2(m)^{-1}. \tag{7.41}$$

Then f is totally geodesic.

Proof. From Proposition I.2 of [12] we know that the function $\psi = |\mathrm{II}|$ is a (weak) solution of

$$\psi\Delta\psi + \left(2 - \frac{1}{n-m}\right)|\mathrm{II}|^2\psi^2 \geq \frac{2}{(m+2)(n-m)-2}|\nabla\psi|^2.$$

Moreover, by (7.41), $\psi \in L^m(M)$ so that

$$\frac{1}{\int_{\partial B_r}\psi^m} \notin L^1(+\infty).$$

We define the differential operator

$$L = -\Delta - \frac{m}{2}\left(2 - \frac{1}{n-m}\right)|\mathrm{II}|^2,$$

and we note that, according to (7.41), we can apply Lemma 7.34 to obtain that $\lambda_1(L_M) \geq 0$. This means that there exists a positive solution $\varphi \in C^\infty(M)$ of the equation

$$\Delta\varphi + \frac{m}{2}\left(2 - \frac{1}{n-m}\right)|\mathrm{II}|^2\varphi = 0.$$

Applying Theorem 4.5, case 1, with the choices $a(x) = \left(2 - \frac{1}{n-m}\right)|\mathrm{II}|^2$, $H = \frac{m}{2}$, $\beta = \frac{m}{2} - 1$, $A = -\frac{2}{(m+2)(n-m)-2}$, we therefore conclude that ψ is constant and $a(x) \equiv 0$, i.e., $|\mathrm{II}| \equiv 0$. □

7.5 Line bundles over Kähler manifolds

Suppose $(M, \langle , \rangle , J)$ is a complete Kähler manifold of complex dimension m and Ricci form $R_{i\bar{j}}$. Let E be a holomorphic line bundle over M endowed with a Hermitian metric $(,)$ with curvature form $\Omega_{i\bar{j}}$. The complex vector space of L^p holomorphic $(k,0)$-forms with values in E is denoted by $L^p\Lambda^{(k,0)}_{holo}(E)$. We also set $L^p\Gamma_{holo}\left(\otimes^k E\right)$ for the space of L^p holomorphic sections of tensor powers of E.

In [121], L. Ni, Y. Shi and L.F. Tam investigate geometric conditions forcing $\dim L^{2p}\Lambda^{(k,0)}_{holo}(E) = 0$. In the L^2 setting, vanishing results and corresponding quantitative finiteness theorems for $\dim L^2\Gamma_{holo}\left(\otimes^k E\right)$ are established, e.g., in works by N. Mok, [112], and by L. Ni, [119]. In this section, as a direct application of Theorem 5.1, we prove qualitative L^p finite-dimensionality results in both these situations.

To begin with, we consider the space $L^p\Lambda^{(k,0)}_{holo}(E)$. To simplify the writings, let

$$a(x) = 4\left(s(x) - \min_{1\leq i_1<\cdots<i_k\leq m}(\gamma_{i_1} + \cdots + \gamma_{i_k})\right),$$

where $\gamma_1, \ldots, \gamma_m$ are the eigenvalues of the Ricci form $R_{i\bar{j}}$ of M and $s(x)$ is the trace of the curvature form $\Omega_{i\bar{j}}$ of E with respect to $(,)$.

We have the following

Theorem 7.37. *Suppose that $H > 0$ and let ${}^H L = -\Delta - Ha(x)$. Then, for every $0 < p \leq H$, if*

$$\operatorname{Ind}\left({}^H L\right) = 0, \quad \textit{then} \quad \dim L^{4p}\Lambda^{(k,0)}_{holo}(E) = 0.$$

If

$$\operatorname{Ind}\left({}^H L\right) < +\infty, \quad \textit{then} \quad \dim L^{4p}\Lambda^{(k,0)}_{holo}(E) < \infty.$$

Proof. Assume that $\operatorname{Ind}\left({}^H L\right) < +\infty$. As noted at the beginning of Section 5, (see Lemma 3.16, Corollary 3.18 and Lemma 3.10) the index assumption guarantees the existence of a positive solution φ of

$$\Delta\varphi + Ha(x)\varphi = 0 \quad \text{on } M \setminus K$$

for some compact set $K \subset M$. Moreover, the Kodaira–Bochner formula states that, for every $\xi \in \Lambda^{(k,0)}_{holo}(E)$, the smooth function $u = |\xi|^2$ satisfies

$$u\Delta u + a(x)u^2 - |\nabla u|^2 \geq 0 \quad \text{on } M;$$

see [113] Chapter 3, Section 6. Therefore, the result follows directly from (a slightly modified version of) Theorem 5.1. The case where $\operatorname{Ind}\left({}^H L\right) = 0$ is similar. □

We now come to the case of $L^p\Gamma_{holo}\left(\otimes^k E\right)$.

Theorem 7.38. *Let $k \in \mathbb{N}$, let $H > 0$, and define ${}^{H}L = -\Delta - 2Hks(x)$. Then, for every $0 < p \leq H$, if*

$$\operatorname{Ind}\left({}^{H}L\right) < +\infty, \quad \textit{then} \quad \dim L^{2p}\Gamma_{holo}\left(\otimes^{k}E\right) < +\infty.$$

If

$$\operatorname{Ind}\left({}^{H}L\right) = 0, \quad \textit{then} \quad \dim L^{2p}\Gamma_{holo}\left(\otimes^{k}E\right) = 0.$$

Proof. The proof goes as above. The only difference is that now we use the Bochner formula

$$|\xi|\left(\Delta|\xi| + 2ks(x)|\xi|\right) - |\nabla|\xi||^{2} \geq 0$$

which is valid, in the sense of distributions, for every $\xi \in \Gamma_{holo}\left(\otimes^{k}E\right)$; see [119]. □

7.6 Reduction of codimension of harmonic immersions

From a somewhat different perspective, Theorem 5.1 applies to codimensional problems for (non-isometric) harmonic immersions into Euclidean spaces.

R. Greene and H.H. Wu, [67], [68], proved that any m-dimensional Riemannian manifold $(M, \langle\,,\rangle)$ can be imbedded into $\mathbb{R}^{2m+1}$ and immersed into $\mathbb{R}^{2m}$ via a proper, harmonic immersion $f : M \to \mathbb{R}^{d}$ ($d = 2m, 2m+1$ respectively). The properness condition insures that the induced metric $f^{*}\mathrm{can}_{\mathbb{R}^{d}}$ is complete. Observe that, due to, e.g., volume growth restrictions, the immersion is in general non-homothetic, and, in particular, non-isometric; see Remark 7.42 below. Equivalently, in general $|df|^{2} \neq \mathrm{const}$.

Theorem 7.39. *Let $(M, \langle\,,\rangle)$ be a complete Riemannian manifold of dimension $m \geq 3$, satisfying*

$$\operatorname{Ric} \geq -a(x) \quad \textit{on } M \tag{7.42}$$

and assume that

$$\operatorname{Ind}\left(-\Delta - Ha(x)\right) < +\infty \tag{7.43}$$

for some $H \geq \frac{m-2}{m-1}$. Then, there exists a compact set $K \subset M$ and an integer $n \geq m$ depending on H and on the geometry of $(M, \langle\,,\rangle)$ in a neighborhood of K such that the following holds.

Let $f : M \to \mathbb{R}^{d}$, $d > n$ be a harmonic immersion whose energy density satisfies the growth condition

$$\int_{B_R} |df|^{2p} = o\left(R^{2}\right), \ \textit{as } R \to +\infty, \tag{7.44}$$

for some $\frac{m-2}{m-1} \leq p \leq H$. Then, there is an n-dimensional affine subspace $\mathbb{A}^{n}$ of $\mathbb{R}^{d}$ such that $f(M) \subset \mathbb{A}^{n}$.

Proof. Let $\mathcal{H}_p$ be the real vector space of harmonic functions $u : M \to \mathbb{R}$ satisfying

$$\int_{B_R} |du|^{2p} = o\left(R^2\right), \text{ as } R \to +\infty,$$

with p as in the statement. Define $\mathcal{V}_p = \operatorname{Im} d|_{\mathcal{H}_p}$, and observe that, by the Bochner formula and the refined Kato inequality, we have

$$|du|\left(\Delta|du| + a(x)|du|\right) - \frac{1}{m-1}|\nabla|du||^2 \geq 0$$

for every $du \in \mathcal{V}_p$. On the other hand, according to (7.43), there exists a solution $\varphi > 0$ of

$$\Delta\varphi + a(x)\varphi = 0 \quad \text{ on } M \setminus K,$$

for some compact set $K \subset M$. Thus, we can apply Theorem 5.1 to deduce that there exists $n \in \mathbb{N}$ depending on p and on the geometry of $(M, \langle , \rangle)$ in a neighborhood of K such that,

$$\dim \mathcal{V}_p \leq n. \tag{7.45}$$

Let $f = \left(f^i\right) : M \to \mathbb{R}^d$, $d > n$ be a harmonic immersion satisfying (7.44). Note that, for each i,

$$df^i \in \mathcal{V}_p,$$

and from the estimate (7.45) we deduce that

$$\operatorname{span}\{df^1, \ldots, df^d\} = \operatorname{span}\{df^{i_1}, \ldots, df^{i_n}\}$$

for some $i_1, \ldots, i_n$. Note that since f is assumed to be an immersion, $n \geq m$. Without loss of generality, we can assume $i_1 = 1, \ldots, i_n = n$, so that

$$df^\alpha = \sum_{i=1}^n \lambda_i^\alpha df^i, \quad \alpha = n+1, \ldots, d,$$

for some appropriate real coefficients $\{\lambda_i^\alpha\}$. This clearly implies that there exist suitable constants $\{c^\alpha\}$ such that

$$f^\alpha = \sum_{i=1}^n \lambda_i^\alpha f^i + c^\alpha.$$

It follows that $f(M)$ is contained in the affine subspace $\mathbb{A}^n$ of $\mathbb{R}^n$ passing through $\left(0, \ldots, 0, c^{n+1}, \ldots, c^d\right)^t$ and spanned by

$$\operatorname{span}\left\{e_i + \left(0, \ldots, 0, \lambda_i^{n+1}, \ldots, \lambda_i^d\right)^t : i = 1, \ldots, n\right\},$$

where $\{e_i\}$ is the standard basis of $\mathbb{R}^d$. □

Remark 7.40. The result of Theorem 7.39 is qualitative. It would be very interesting to get a quantitative version, where the dimension n of the affine ambient subspace is estimated in terms of the geometric data. In particular, forcing $n = \dim M$ would yield a Bernstein-type result.

Remark 7.41. The above arguments can be applied to obtain holomorphic immersions of Kähler manifolds into $\mathbb{C}^d$.

Remark 7.42. Note that according to Theorem 2.28 in Section 2.3, if $f : M \to \mathbb{R}^n$ is an isometric harmonic (=minimal) immersion, then

$$\operatorname{vol} B_R(x_o) \geq c_m R^m, \tag{7.46}$$

where c_m is the volume of the unit sphere in $\mathbb{R}^m$. It follows that the harmonic immersion f in the statement of Theorem 7.39 cannot be isometric, for otherwise (7.44) and $|df| = 1$ would imply

$$\operatorname{vol}(B_R) = o\left(R^2\right) \text{ as } R \to +\infty$$

contradicting (7.46). In fact, if $m \geq 3$, the immersion f cannot be conformal, because conformal harmonic maps are homothetic; see, e.g., [47].

Some restrictions apply even if f is only assumed to be bi-Lipschitz. Indeed, if there exist positive constants A and B such that

$$A\langle X, X\rangle \leq (df(X), df(X))_{\mathbb{R}^n} \leq B\langle X, X\rangle$$

for every $X \in TM$, then, again by Theorem 2.28,

$$\operatorname{vol} B_R(x_o) \geq CR^{\sqrt{m}AR/B} \quad \text{for} \quad R \geq 1,$$

which is compatible with the energy growth condition (7.44) in the statement of the theorem only if $\sqrt{m}AB^{-1} < 2$.

Here are some special situations where Theorem 7.39 applies.

Corollary 7.43. *Let $(M, \langle,\rangle)$ be a complete, m-dimensional Riemannian manifold satisfying both*

$$\operatorname{vol}(B_R) = o\left(R^2\right) \textit{ as } R \to +\infty,$$

and

$${}^M\mathrm{Ric} \geq 0 \textit{ on } M \setminus K$$

for some compact set $K \subset M$. Then, there exists $n \in \mathbb{N}$ depending on the geometry of $(M, \langle,\rangle)$ in a neighborhood of K such that, any (non-isometric) harmonic, Lipschitz immersion $f : M \to \mathbb{R}^d$, $d > n$, must satisfy $f(M) \subset \mathbb{A}^n$ for some n-dimensional affine subspace $\mathbb{A}^n \subset \mathbb{R}^d$.

We say that a Riemannian metric $(\, ,)$ on the smooth manifold M is dominated by the metric $\langle \, , \rangle$ if there exists $C > 0$ such that $(,) \leq C^2 \langle,\rangle$, in the sense of quadratic forms.

Corollary 7.44. *Let $(M, \langle\, ,\rangle)$ be a complete, m-dimensional Riemannian manifold satisfying*

$$^{M}\mathrm{Ric} \geq 0 \text{ on } M \setminus K$$

for some compact set $K \subset M$. Then, there exists $n \in \mathbb{N}$ depending on the geometry of $(M, \langle\, ,\rangle)$ in a neighborhood of K such that the following holds.

Let $f : M \to \mathbb{R}^d$, $n > d$, be a harmonic immersion whose induced metric $f^*\mathrm{can}$ is dominated by a metric $\widetilde{\langle\, ,\rangle}$ in the conformal class of $\langle\, ,\rangle$ satisfying

$$\widetilde{\mathrm{vol}}\,(M) < +\infty.$$

Then, f is in fact a harmonic immersion into some n-dimensional affine subspace $\mathbb{A}^n \subset \mathbb{R}^d$.

Proof. We set $\widetilde{\langle,\rangle} = u^2(x) \langle\, ,\rangle$ and we note that

$$\int_M |df|^m \,\mathrm{dvol} \leq C \int_M u^m \mathrm{dvol} = C\widetilde{\mathrm{vol}}\,(M)$$

for some constant $C > 0$. □

Suppose $f : (M, \langle\, ,\rangle_M) \to (N, (\, ,))$ is a smooth map between Riemannian manifolds of dimensions m and n respectively. As in Section 6.2, we denote by

$$\lambda_1(x) \geq \lambda_2(x) \geq \cdots \geq \lambda_m(x) \geq 0$$

the eigenvalues of the quadratic form $f_x^*(\, ,)$ and we say that f has bounded k^{th} dilation if

$$\lambda_1(x) \leq C_k \lambda_k(x) \text{ on } M$$

for some constant $C_k \geq 1$. When $k = m$ we say (perhaps improperly) that f is of bounded distortion (or, equivalently, quasi-regular).

Corollary 7.45. *Let $(M, \langle\, ,\rangle)$ be a complete, m-dimensional Riemannian manifold satisfying*

$$^{M}\mathrm{Ric} \geq 0 \text{ on } M \setminus K$$

for some compact set $K \subset M$. Then, there exists $n \in \mathbb{N}$ depending on the geometry of $(M, \langle\, ,\rangle)$ in a neighborhood of K such that, any (non-isometric) harmonic immersion $f : M \to \mathbb{R}^d$, $d > n$, of bounded distortion and satisfying

$$\mathrm{vol}_{f^*\mathrm{can}}(M) < +\infty,$$

is in fact a harmonic immersion into some n-dimensional affine subspace $\mathbb{A}^n \subset \mathbb{R}^d$.

Proof. We set $\widetilde{\langle,\rangle} = f^*\mathrm{can}$, where can denotes the canonical metric of $\mathbb{R}^d$, and we observe that

$$\frac{|df|^m}{\widetilde{\mathrm{dvol}}} = \frac{\{\mathrm{tr}\,(f^*\mathrm{can})\}^{\frac{m}{2}}}{\{\det\,(f^*\mathrm{can})\}^{\frac{1}{2}}} = \frac{(\sum \lambda_i)^{\frac{m}{2}}}{(\Pi\lambda_i)^{\frac{1}{2}}} \leq C\left(\frac{\lambda_1}{\lambda_m}\right)^{\frac{m}{2}} \leq CC_m.$$

Thus the energy density of f satisfies the integrability condition (7.44) and Theorem 7.39 applies. □

Chapter 8

Constancy of holomorphic maps and the structure of complete Kähler manifolds

8.1 Three versions of a result of Li and Yau

The aim of this section is to prove three versions of a vanishing result by Li and Yau for holomorphic maps. The first theorem is a strengthening of the original formulation and its proof follows the lines of [107].

Theorem 8.1. *Let $(M, \langle\,,\rangle, J_M)$ be a complete Kähler manifold of real dimension $m = 2s$ such that*

$$^{M}\mathrm{Ric} \geq -\rho(x) \tag{8.1}$$

for some $\rho \in C^0(M) \cap L^1(M)$. Assume that $\rho_+ \in L^1(M)$, satisfies

$$\int_M \rho(x) \leq 0, \tag{8.2}$$

and

$$\int_{B_r} \rho_+(x)^p = o\left(r^{\beta(p-1)}\right), \text{ as } r \to +\infty, \tag{8.3}$$

for some constants, $\beta > 0$ and $p > \frac{m}{2}$. Let $(N, (\,,), J_N)$ be a Hermitian manifold with holomorphic bisectional curvature bounded above by $k(z)$, $z \in N$, $k \in C^0(N)$. Let $f : M \to N$ be a holomorphic map such that

$$k(f(x)) < 0 \text{ on } M; \quad k(f(x)) \leq -\frac{B}{r(x)^{\mu}} \text{ on } M \setminus B_{R_0} \tag{8.4}$$

for some $R_0, B > 0$ and $\mu \geq 0$. If

$$\mu\frac{m}{2} + \beta(\frac{m}{2} - 1) < 2, \tag{8.5}$$

then f is constant.

Remark 8.2. As will be clear from the proof of Theorem 8.1, the same constancy result holds if f is a pluriharmonic function which is bounded on one side on a Kähler manifold M satisfying the curvature assumptions listed in the statement of the theorem with $\mu = 0$. Indeed, suppose for instance that $\inf_M f > -C$ for some constant $C > 0$. Then, according to Proposition 1.26, the function $u = |\nabla \log(f + C)|^2$ satisfies inequality (8.15) below, with $k(z) \equiv -1$. The constancy of $\log(f + C)$, and hence of f, now follows, applying without changes the arguments in the proof of Theorem 8.1. For later purposes, we also note that $u \leq |\nabla f|^2 / (f + C)^2$ and, therefore, u inherits the same integrability properties of $|\nabla f|^2$.

The proof of Theorem 8.1 is based on the important estimate contained in the next

Lemma 8.3 ([107]). *Let $(M, \langle , \rangle)$ be a complete Riemannian manifold. Let also $a(x), b(x) \in C^0(M)$ with $b(x) \geq 0$ on M and $\sigma \geq 1$. Assume that $u \geq 0$ is a C^2-solution of the differential inequality*

$$u\Delta u + a(x) u^2 \geq b(x) u^{\sigma+1} + |\nabla u|^2 \quad \text{on } M. \tag{8.6}$$

Set $a_+(x) = \max\{a(x), 0\}$, $a_-(x) = \max\{-a(x), 0\}$ so that $a(x) = a_+(x) - a_-(x)$. Suppose that

$$a_+(x) \in L^1(M) \tag{8.7}$$

and that

$$\int_{B_r} u^p = o\left(r^2\right), \quad \text{as } r \to +\infty \tag{8.8}$$

for some constant $p > 0$. Then, $a_-(x), b(x) u^{\sigma-1} \in L^1(M)$ and

$$\int_M b(x) u^{\sigma-1} \leq \int_M a(x). \tag{8.9}$$

Proof. We fix $R > r$ and let $\phi : M \to [0,1]$ be a smooth cut-off function such that

$$\phi \equiv 1 \text{ on } B_r; \quad \phi \equiv 0 \text{ on } M \setminus B_R; \quad |\nabla \phi| \leq \frac{\sqrt{3}}{R - r} \text{ on } M. \tag{8.10}$$

We fix $\varepsilon, \delta > 0$ and consider the vector field

$$W = \phi^2 \frac{(u+\delta)^{p-2}}{(u+\delta)^p + \varepsilon} u \nabla u.$$

We compute its divergence and use (8.6) and Schwarz's inequality to estimate

$$\begin{aligned} \operatorname{div} W \geq{} & \frac{\phi^2 (u+\delta)^{p-2}}{(u+\delta)^p + \varepsilon} \left(-a(x) u^2 + b(x) u^{\sigma+1}\right) - 2 \frac{\phi (u+\delta)^{p-2}}{(u+\delta)^p + \varepsilon} |\nabla u| \, |\nabla \phi| \\ & + \frac{\phi^2 (u+\delta)^{p-3}}{(u+\delta)^p + \varepsilon} \left\{ 2(u+\delta) + (p-2)u - p \frac{u(u+\delta)^p}{(u+\delta)^p + \varepsilon} \right\} |\nabla u|^2. \end{aligned}$$

Since the term in braces on the right-hand side is equal to

$$2\delta + \frac{\varepsilon p u}{(u+\delta)^p + \varepsilon} \geq \frac{\varepsilon p u}{(u+\delta)^p + \varepsilon},$$

rearranging and applying the divergence theorem on B_R yield

$$\int_{B_R} \phi^2 \frac{(u+\delta)^{p-2} u^{\sigma+1}}{(u+\delta)^p+\varepsilon} b(x) - \int_{B_R} \phi^2 \frac{(u+\delta)^{p-2} u^2}{(u+\delta)^p+\varepsilon} a(x)$$
$$\leq - \int_{B_R} \varepsilon p u \phi^2 \frac{(u+\delta)^{p-3} |\nabla u|^2}{[(u+\delta)^p+\varepsilon]^2} + 2 \int_{B_R} \phi \frac{(u+\delta)^{p-2} u}{(u+\delta)^p+\varepsilon} |\nabla \phi| |\nabla u|,$$

whence, using the elementary inequality,

$$2\alpha\beta \leq \varepsilon p \alpha^2 + \frac{1}{\varepsilon p}\beta^2, \quad p, \varepsilon > 0$$

in the second integral of the RHS, we finally get

$$\int_{B_R} \phi^2 \frac{(u+\delta)^{p-2} u^{\sigma+1}}{(u+\delta)^p+\varepsilon} b(x) - \int_{B_R} \phi^2 \frac{(u+\delta)^{p-2} u^2}{(u+\delta)^p+\varepsilon} a(x)$$
$$\leq \frac{1}{\varepsilon p} \int_{B_R} u (u+\delta)^{p-1} |\nabla \phi|^2.$$

Next, we let $\delta \to 0^+$ and we apply the Lebesgue dominated convergence theorem to deduce

$$\int_{B_R} \phi^2 \frac{u^{p+\sigma-1}}{u^p+\varepsilon} b(x) - \int_{B_R} \phi^2 \frac{u^p}{u^p+\varepsilon} a(x) \leq \frac{1}{\varepsilon p} \int_{B_R} u^p |\nabla \phi|^2.$$

We choose $R = 2r$ and we use the properties of ϕ to infer

$$\int_{B_r} \frac{u^{p+\sigma-1}}{u^p+\varepsilon} b(x) - \int_{B_{2r}} \frac{u^p}{u^p+\varepsilon} a_+(x) + \int_{B_r} a_-(x) \frac{u^p}{u^p+\varepsilon} \leq \frac{3}{\varepsilon p r^2} \int_{B_{2r}} u^p$$

and letting $r \to +\infty$, because of (8.8), we obtain

$$\int_M b(x) \frac{u^{p+\sigma-1}}{u^p+\varepsilon} + \int_M \frac{u^p}{u^p+\varepsilon} a_-(x) - \int_M a_+(x) \leq 0.$$

Now, as $\varepsilon \searrow 0$,

$$\frac{u^{p+\sigma-1}}{u^p+\varepsilon} \nearrow u^{\sigma-1} \text{ and } \frac{u^p}{u^p+\varepsilon} \nearrow 1$$

monotonically, and since $a_+(x) \in L^1(M)$ and $b(x) \geq 0$ by assumption, it follows from the monotone convergence theorem that $a_-(x)$ and $b(x)\, u^{\sigma-1} \in L^1(M)$ and that

$$\int_M b(x)\, u^{\sigma-1} + \int_M a_-(x) - \int_M a_+(x) \leq 0,$$

which is (8.9). □

In order to verify that condition (8.8) holds we will appeal to the following

Lemma 8.4. *Let $(M, \langle\, ,\, \rangle)$ be a complete Riemannian manifold, and let $a(x), b(x) \in C^0(M)$ with $b(x) > 0$ on M. Assume that $u \geq 0$ is a C^2 solution of the differential inequality*

$$u\Delta u + a(x)u^2 - b(x)u^{\sigma+1} \geq -A|\nabla u|^2, \tag{8.11}$$

for $A \leq 1$ and $\sigma > 1$. Then for every $q \geq 1$, $q > A+2$ there exist constants C_1, $C_2 > 0$ which depend only on q, σ and $R_0 > 0$ such that, for every $R \geq R_0$,

$$\int_{B_R} b(x)u^{q+\sigma-2} \leq C_1 R^{-2\frac{q+\sigma-2}{\sigma-1}} \int_{B_{2R}} b(x)^{-\frac{q-1}{\sigma-1}} + C_2 \int_{B_{2R}} \left(\frac{a_+(x)}{b(x)}\right)^{\frac{q-1}{\sigma-1}} a_+(x). \tag{8.12}$$

Proof. Observe first that we may assume that $u \not\equiv 0$, for otherwise there is nothing to prove. Thus, there exists $R_0 > 0$ such that $u \not\equiv 0$ on B_R for every $R \geq R_0$.

Next, for every $R \geq R_0$, let $\phi = \phi_R : M \to [0,1]$ be a smooth cut-off function such that

$$\phi \equiv 1 \quad \text{on } B_R, \quad \phi \equiv 0 \quad \text{on } M \setminus B_{2R}, \text{ and } |\nabla\phi| \leq \frac{C}{R}\phi^{\frac{q-1}{q+\sigma-2}} \quad \text{on } B_{2R}, \tag{8.13}$$

for some C which depends only on q and σ. Note that this is possible since the exponent $(q-1)/(q+\sigma-2)$ is strictly less than 1. Having fixed $\epsilon > 0$, we let W be the vector field defined by

$$W = \phi^2(u+\epsilon)^{q-3}u\nabla u.$$

A computation that uses (8.11), $q - A - 2 > 0$ and $u \geq 0$ yields

$$\begin{aligned} \mathit{div}\, W \geq{}& \phi^2(u+\epsilon)^{q-3}\{-a(x)u^2 + b(x)u^{\sigma+1} + (q-A-2)\frac{u}{u+\epsilon})|\nabla u|^2\} \\ &+ 2\phi(u+\epsilon)^{q-3}u\langle\nabla u, \nabla\phi\rangle. \end{aligned}$$

We use the Cauchy–Schwarz inequality and Young's inequality $2\alpha\beta \leq \lambda\alpha^2 + \lambda^{-1}\beta^2$ with $\lambda = q - 2 - A > 0$ to estimate the last term on the right-hand side from below and obtain

$$\mathit{div}\, W \geq \phi^2(u+\epsilon)^{q-3}\{-a_+(x)u^2 + b(x)u^{\sigma+1}\} - \frac{1}{q-2-A}u(u+\epsilon)^{q-2}|\nabla\phi|^2.$$

We integrate the above inequality, apply the divergence theorem, rearrange, let $\epsilon \to 0+$ and use the monotone and dominated convergence theorem, in that order, to deduce that

$$\int_{B_{2R}} b(x)\phi^2 u^{q+\sigma-2} \leq \frac{1}{q-2-A}\int_{B_{2R}} u^{q-1}|\nabla\phi|^2 + \int_{B_{2R}} \phi^2 a_+(x)u^{q-1}. \tag{8.14}$$

If $q = 1$, the conclusion follows immediately using (8.14). If $q > 1$, we denote by I and II the two integrals on the right-hand side, and use Hölder inequality with conjugate exponents

$$\frac{q+\sigma-2}{q-1}(>1) \quad \text{and} \quad \frac{q+\sigma-2}{\sigma-1},$$

and the assumption that $b(x) > 0$, to estimate

$$I \leq \Big(\int_{B_{2R}} b(x)\phi^2 u^{q+\sigma-2}\Big)^{\frac{q-1}{q+\sigma-2}} \Big(\int_{B_{2R}} \phi^{-2\frac{q-1}{\sigma-1}} b(x)^{-\frac{q-1}{\sigma-1}} |\nabla\phi|^{2\frac{q+\sigma-2}{\sigma-1}}\Big)^{\frac{\sigma-1}{q+\sigma-2}}$$

and

$$II \leq \Big(\int_{B_{2R}} b(x)\phi^2 u^{q+\sigma-2}\Big)^{\frac{q-1}{q+\sigma-2}} \Big(\int_{B_{2R}} \phi^2 a_+(x)^{\frac{q+\sigma-2}{\sigma-1}} b(x)^{-\frac{q-1}{\sigma-1}}\Big)^{\frac{\sigma-1}{q+\sigma-2}}.$$

Inserting the above into (8.14), noting that the integral on the left-hand side is strictly positive by the choice of R, and simplifying, we obtain

$$\begin{aligned}\int_{B_{2R}} b(x)\phi^2 u^{q+\sigma-2} \leq & \Big\{\frac{1}{q-2-A}\Big(\int_{B_{2R}} \phi^{-2\frac{q-1}{\sigma-1}} b(x)^{-\frac{q-1}{\sigma-1}} |\nabla\phi|^{2\frac{q+\sigma-2}{\sigma-1}}\Big)^{\frac{\sigma-1}{q+\sigma-2}} \\ & + \Big(\int_{B_{2R}} \phi^2 a_+(x)^{\frac{q+\sigma-2}{\sigma-1}} b(x)^{-\frac{q-1}{\sigma-1}}\Big)^{\frac{\sigma-1}{q+\sigma-2}}\Big\}^{\frac{q+\sigma-2}{\sigma-1}}.\end{aligned}$$

The required conclusion follows using again (8.13) and the elementary inequality $(a+b)^\tau \leq 2^\tau(a^\tau + b^\tau)$ valid for $a, b, \tau \geq 0$. □

Proof of Theorem 8.1. We let $u = |df|^2$, and use (8.1) and Corollary 1.29 to deduce that

$$u\Delta u + 2\rho(x)\, u^2 \geq -2k(\psi(x))\, u^3 + |\nabla u|^2 \tag{8.15}$$

so that (8.6) is satisfied with $a(x) = 2\rho(x)$, $b(x) = -2k(\psi(x))$ and $\sigma = 2$. Since $\rho_+ \in L^1(M)$ by assumption, (8.2) and (8.4) imply, via (8.9), that $u \equiv 0$ as desired, provided we can apply Lemma 8.3. In order to do this, we only need to show the existence of $q > 0$ such that

$$\int_{B_r} u^q = o\left(r^2\right), \text{ as } r \to +\infty. \tag{8.16}$$

Towards this end, we first observe that, since $\rho_+ \in L^1(M)$, by Hölder inequality, for every $1 < \gamma \leq p$ we have

$$\int_{B_r} \rho_+(x)^\gamma \leq \Big\{\int_{B_r} \rho_+(x)\Big\}^{\frac{p-\gamma}{p-1}} \Big\{\int_{B_r} \rho_+(x)^p\Big\}^{\frac{\gamma-1}{p-1}} \leq C\Big\{\int_{B_r} \rho_+(x)^p\Big\}^{\frac{\gamma-1}{p-1}},$$

so that, using (8.3) and (8.5) we may choose γ satisfying

$$\frac{m}{2} < \gamma \leq p, \quad \mu\gamma + \beta(\gamma - 1) \leq 2 \tag{8.17}$$

and for which

$$\int_{B_r} \rho_+ (x)^\gamma = o\left(r^{\beta(\gamma-1)}\right), \text{ as } r \to +\infty. \tag{8.18}$$

Next, (8.18) and $\gamma > \frac{m}{2}$ allows us to apply Corollary 2.21 to obtain

$$\text{vol} B_r = o\left(r^{2\gamma+\beta(\gamma-1)}\right), \text{ as } r \to +\infty. \tag{8.19}$$

Now, since u satisfies (8.15), using (8.4) we see that it also satisfies

$$u\Delta u + 2\rho_+ (x) u^2 \geq \frac{\tilde{B}}{[1 + r(x)]^\mu} u^3 + |\nabla u|^2 \text{ on } M \tag{8.20}$$

for an appropriate constant $\tilde{B} > 0$. Conditions (8.18) and (8.19) enable us to apply Lemma 8.4 with the choices $\sigma = 2$, $A = -1$, $a(x) = 2\rho(x)$, $b(x) = \tilde{B}(1 + r(x))^{-\mu}$, $q = \gamma > 1$ to obtain

$$\int_{B_r} u^\gamma = o\left(r^{\mu\gamma+\beta(\gamma-1)}\right) = o\left(r^2\right), \text{ as } r \to +\infty,$$

as required. □

In order to give a second version of this result we need to solve the Poisson equation on a generic non-parabolic manifold.

Lemma 8.5. *Let $(M, \langle , \rangle)$ be a complete, non-parabolic manifold and let $a(x) \in C^{0,\alpha}(M) \cap L^1(M)$, $0 < \alpha < 1$, be a non-negative function. Then, there exists a solution $v \in C^2(M)$ of the Poisson equation*

$$\Delta v = a(x) \text{ on } M \tag{8.21}$$

satisfying $v \leq 0$.

Proof. Let $G(x, y)$ be the positive Green kernel of M, which exists by the non-parabolicity assumption (see Appendix 7.2). Recall that:

i) $G(x, y)$ is symmetric.

ii) For any fixed $R > 0$,

$$y \longmapsto G(x, y) \in L^1_{loc}(M) \cap L^\infty(M \setminus B_{2R}),$$

uniformly for $x \in B_R$.

iii) If $\psi \in C_c^\infty(M)$, then $u(x) = -\int_M G(x,y)\psi(y)\,dy$ is smooth and satisfies

$$\Delta u(x) = -\int_M G(x,y)\Delta\psi(y)\,dy = \psi(x).$$

We claim that the function

$$v(x) = -\int_M G(x,y)a(y)\,dy$$

is a well-defined, locally bounded distributional solution of the desired Poisson equation. Then, by standard elliptic regularity theory it follows that $v(x)$ is a classical C^2 solution; see Theorem 3.54 in [9]. Let $R>0$ be fixed. If we write

$$\int_M G(x,y)a(y)\,dy = \int_{B_{2R}} + \int_{M\setminus B_{2R}} G(x,y)a(y)\,dy,$$

using the integrability properties ii) above, we can estimate each of the summands, uniformly for $x \in B_R$, and deduce $v \in L^\infty_{loc}(M)$. Moreover, for every $\varphi \in C_c^\infty(M)$,

$$G(x,y)a(y)\varphi(x) \in L^1(M\times M),$$

as one realizes using again property ii). Therefore, applying Fubini's theorem, and recalling property iii), for each $\psi \in C_c^\infty(M)$ we get

$$\begin{aligned}\int_M v(x)\Delta\psi(x)\,dx &= -\int_M \Delta\psi(x)\int_M G(x,y)a(y)\,dydx\\ &= -\int_M a(y)\int_M G(x,y)\Delta\psi(x)\,dxdy\\ &= \int_M a(y)\psi(y),\end{aligned}$$

proving the claim. □

Theorem 8.6. *Let $(M, \langle\,,\rangle, J_M)$ be a non-parabolic, complete, Kähler manifold such that*

$$^M\mathrm{Ric} \geq -\rho(x) \tag{8.22}$$

for some $\rho \in C^0(M)$ satisfying

$$\rho_+(x) \in L^1(M). \tag{8.23}$$

Suppose also that

$$\liminf_{r\to+\infty} \frac{\log \mathrm{vol} B_r}{r^{2-\mu}} < +\infty \tag{8.24}$$

for some $0 \leq \mu < 2$. Let $(N, (\,,), J_N)$ be a Hermitian manifold with holomorphic bisectional curvature bounded above by $k(z)$, $z \in N$, $k \in C^0(N)$. Let $f : M \to N$ be a holomorphic map such that

$$k(f(x)) < 0 \text{ on } M; \qquad k(f(x)) \leq -\frac{B}{r(x)^\mu} \text{ on } M \setminus B_{R_0} \tag{8.25}$$

for some $R_0, B > 0$. Then f is constant.

Remark 8.7. Note that, for non-parabolic manifolds, Theorem 8.6 extends Theorem 8.1. Indeed, in Theorem 8.6 the assumptions on μ are relaxed; condition (8.2) implies (8.23) and condition (8.3) implies, via Corollary 2.21, that M has at most polynomial growth so that (8.24) is satisfied (see 8.19).

Proof of Theorem 8.6. As before, we let $u = |df|^2$ so that, using (8.22) and (8.25) we deduce

$$u\Delta u + 2\rho_+(x)\, u^2 \geq \frac{2B'}{[1 + r(x)]^\mu} u^3 + |\nabla u|^2 \text{ on } M, \tag{8.26}$$

for some constant $B' > 0$. Since $\rho_+(x) \in L^1(M)$ we can apply Lemma 8.5 to deduce the existence of a solution $v \leq 0$ of

$$\Delta v = 2\rho_+(x) \text{ on } M.$$

We set $\varphi = e^{-v}$ so that, from the above, we get

$$\Delta\varphi + 2\rho_+(x)\,\varphi = \frac{|\nabla\varphi|^2}{\varphi} \text{ on } M \tag{8.27}$$

with $\varphi \geq 1$. We define $w = \varphi^{-1}u$ and we use (8.26) and (8.27) to compute

$$\begin{aligned}\Delta w &\geq \frac{2B'}{[1 + r(x)]^\mu}\varphi w^2 \\ &\geq \frac{2B'}{[1 + r(x)]^\mu} w^2.\end{aligned} \tag{8.28}$$

Assumption (8.24) and $0 \leq \mu < 2$ enable us to apply Theorem 6.16 and deduce that w, and therefore u, vanishes on M. □

As explained in Remark 8.2, the result still holds in the case where M is as in the theorem, and $f : M \to \mathbb{R}$ is a pluriharmonic function which is bounded on one side. Since this result will be used in the next section, we state it as a corollary.

Corollary 8.8. *Let $(M, \langle\,,\rangle, J_M)$ be a non-parabolic, complete Kähler manifold such that*

$${}^M\mathrm{Ric} \geq -\rho(x) \tag{8.29}$$

for some $\rho \in C^0(M)$ satisfying

$$\rho_+(x) \in L^1(M). \tag{8.30}$$

Suppose also that

$$\liminf_{r\to+\infty} \frac{\log \mathrm{vol} B_r}{r^2} < +\infty. \tag{8.31}$$

Then every pluriharmonic function $f : M \to \mathbb{R}$ which is bounded from below on M is constant.

Indeed, as already observed, according to Proposition 1.26, the function $u = |\nabla \log(f + C)|^2$ satisfies the inequality

$$u\Delta u + 2\rho_{+} u^2 \geq 2u^3 + |\nabla u|^2$$

and therefore (8.28) holds with $\mu = 0$.

In our third, and final, version of the theorem of Li–Yau, we avoid the request $\rho_+(x) \in L^1(M)$; see Theorem 8.11 below. Towards this end, we first prove

Theorem 8.9. *Let $\sigma > 1$ and $a(x), b(x) \in C^0(M)$ satisfy*

$$\text{i) } 0 \not\equiv b(x) \geq 0 \quad \textit{and} \quad \text{ii) } a(x) \leq Cb(x) \ \textit{ on } M \tag{8.32}$$

for some constant $C > 0$. Given $\alpha > 0$, suppose that, for some

$$H \geq \frac{(1+\alpha)^2}{4\alpha}, \tag{8.33}$$

the operator $L_H = -\Delta - Ha(x)$ satisfies

$$\lambda_1\left({}^H L_M\right) \geq 0. \tag{8.34}$$

Then the differential inequality

$$\Delta u + a(x)u \geq b(x)u^\sigma \tag{8.35}$$

has no bounded, non-negative, non-identically zero, C^2-solutions u satisfying

$$\text{i) } a(x)u^{1+\alpha} \in L^1(M); \qquad \text{ii) } \left\{\int_{\partial B_r} u^{1+\alpha}\right\}^{-1} \notin L^1(+\infty). \tag{8.36}$$

Remark 8.10. Note that (8.36) i), ii) are satisfied if $a(x)$ is bounded and $u^{1+\alpha} \in L^1(M)$. Note also that if $b(x) > 0$ on M and $a(x) \leq Cb(x)$ for $r(x) \gg 1$ and some constant $C > 0$, then condition (8.32) (ii) holds, possibly with a bigger constant $\tilde{C} > 0$.

Proof. Assume by contradiction that $u \in C^2(M)$ is a bounded, non-negative, non-identically zero solution of (8.35) satisfying (8.36). We claim that, without loss of generality, we may assume that

$$a(x) \leq b(x) \text{ on } M \tag{8.37}$$

and

$$\sup_M u < 1. \tag{8.38}$$

Indeed, let

$$A = \max\left\{\sup_M u, \frac{1}{2}C^{\frac{1}{\sigma-1}}, 1\right\}$$

and

$$\bar{u}(x) = (2A)^{-1} u(x).$$

Then, $\sup_M \bar{u} \leq 1/2$ and $\bar{u}$ satisfies

$$\Delta\bar{u} + a(x)\bar{u} \geq \bar{b}(x)\bar{u}^{\sigma} \tag{8.39}$$

with

$$\bar{b}(x) = (2A)^{\sigma-1} b(x) \geq Cb(x) \geq a(x)$$

on M, proving the claim.

We now proceed by dividing the argument in four steps.

Step 1. There exists a sequence $\{r_k\} \nearrow +\infty$ such that

$$\lim_{k\to+\infty}\int_{\partial B_{r_k}} u^{\alpha}|\nabla u| = 0 \tag{8.40}$$

Let $\varepsilon > 0$, define $W_{\varepsilon} = (u+\varepsilon)^{\alpha}\nabla u$, compute its divergence using (8.39) and apply the divergence theorem to get

$$\int_{\partial B_r}(u+\varepsilon)^{\alpha}\langle\nabla u, \nabla r\rangle - \alpha\int_{B_r}(u+\varepsilon)^{\alpha-1}|\nabla u|^2 \tag{8.41}$$

$$\geq \int_{B_r}(b(x)u^{\sigma} - a(x)u)(u+\varepsilon)^{\alpha} \tag{8.42}$$

$$\geq \int_{B_r} a(x)(u^{\sigma} - u)(u+\varepsilon)^{\alpha}$$

where, in the last inequality we have used 8.37. Letting $\varepsilon \to 0^+$ the last term on the RHS of (8.42) tends to the finite quantity

$$\int_{B_r} a(x)(u^{\sigma} - u)u^{\alpha}$$

and the first integral on the LHS tends to

$$\int_{\partial B_r} u^{\alpha}\langle\nabla u, \nabla r\rangle.$$

Thus, by the monotone (or dominated, depending on whether $\alpha < 1$ or > 1) convergence theorem ,

$$\int_{B_r}(u+\varepsilon)^{\alpha-1}|\nabla u|^2 \to \int_{B_r} u^{\alpha-1}|\nabla u|^2 < +\infty$$

as $\varepsilon \searrow 0$. In particular $u^{\alpha-1}|\nabla u|^2 \in L^1_{loc}(M)$. Moreover

$$\int_{\partial B_r} u^{\alpha}\langle \nabla u, \nabla r\rangle - \alpha \int_{B_r} u^{\alpha-1}|\nabla u|^2 \geq \int_{B_r} a(x)(u^{\sigma}-u)u^{\alpha}.$$

Now, since $u<1$, and $\sigma>1$, and u is bounded, it follows from (8.36) ii) that $a(x)\left(u^{\sigma+\alpha}-u^{1+\alpha}\right) \in L^1(M)$ and, letting $r \to +\infty$, we obtain

$$\liminf_{r\to+\infty}\left\{\int_{\partial B_r} u^{\alpha}\langle \nabla u, \nabla r\rangle - \alpha \int_{B_r} u^{\alpha-1}|\nabla u|^2\right\} = B > -\infty. \tag{8.43}$$

We claim that

$$u^{\alpha-1}|\nabla u|^2 \in L^1(M).$$

To see this, we define

$$G(r) = \int_{B_r} u^{\alpha-1}|\nabla u|^2.$$

Since $u^{\alpha-1}|\nabla u|^2 \in L^1_{loc}(M)$, it follows from the co-area formula that $G(r)$ is absolutely continuous and that

$$G'(r) = \int_{\partial B_r} u^{\alpha-1}|\nabla u|^2$$

is defined a.e. and is locally L^1. Assume by contradiction that $G(r) \to +\infty$ as $r \to +\infty$. Then, for large enough $r > R$, $G(r) > 0$ and $B > -(\alpha/2)G(r)$ so that, by (8.43),

$$\int_{\partial B_r} u^{\alpha}\langle \nabla u, \nabla r\rangle \geq \frac{\alpha}{2}G(r)$$

and therefore, by the Hölder inequality,

$$\left\{\frac{\alpha}{2}G(r)\right\}^2 \leq \left\{\int_{\partial B_r} u^{\alpha}\langle \nabla u, \nabla r\rangle\right\}^2 \leq G'(r)\int_{\partial B_r} u^{\alpha+1}.$$

It follows from this that

$$\frac{G'(r)}{G(r)^2} \geq \frac{\alpha^2}{4}\left\{\int_{\partial B_r} u^{\alpha+1}\right\}^{-1}$$

and, integrating over $[R, r]$, we get

$$-\frac{1}{G(r)} + \frac{1}{G(R)} \geq \frac{\alpha^2}{4}\int_R^r \left\{\int_{\partial B_t} u^{\alpha+1}\right\}^{-1} dt.$$

Whence, letting $r \to +\infty$, we contradict assumption (8.36) ii), proving the claim. It follows that $u^{\alpha-1}|\nabla u|^2 \in L^1(M)$ and therefore

$$\int_{\partial B_r} u^{\alpha-1}|\nabla u|^2 \in L^1((0,+\infty)).$$

Because of (8.36) ii), this in turn implies that there exists a sequence $\{r_k\} \nearrow +\infty$ such that

$$\lim_{k\to+\infty}\left(\int_{\partial B_{r_k}} u^{\alpha+1}\right)\left(\int_{\partial B_{r_k}} u^{\alpha-1}\left|\nabla u\right|^2\right) = 0.$$

To conclude we apply the Cauchy–Schwarz inequality,

$$\int_{\partial B_{r_k}} u^{\alpha}|\nabla u| \le \left\{\int_{\partial B_{r_k}} u^{\alpha+1}\right\}^{\frac{1}{2}}\left\{\int_{\partial B_{r_k}} u^{\alpha-1}\left|\nabla u\right|^2\right\}^{\frac{1}{2}} \to 0$$

as $k \to +\infty$.

Step 2. The condition $\lambda_1\left({}^H L_M\right) \ge 0$ implies that there exists $\varphi \in C^2(M)$, $\varphi > 0$, which is a solution of

$$\Delta\varphi + Ha(x)\varphi = 0 \text{ on } M. \tag{8.44}$$

We claim that

$$u^{\alpha+1}\varphi^{-2}\left|\nabla\varphi\right|^2 \in L^1(M). \tag{8.45}$$

Indeed, let $V = u^{\alpha+1}\varphi^{-1}\nabla\varphi$. Using Schwarz's inequality, (8.44) and the elementary inequality $2ab \le a^2 + b^2$, we compute

$$\begin{aligned}\mathrm{div}V &= (\alpha+1)\,u^{\alpha}\varphi^{-1}\left\langle\nabla u,\nabla\varphi\right\rangle + u^{\alpha+1}\varphi^{-1}\Delta\varphi - u^{\alpha+1}\varphi^{-2}\left|\nabla\varphi\right|^2 \\ &\le \frac{(\alpha+1)^2}{2}u^{\alpha-1}\left|\nabla u\right|^2 - Ha(x)\,u^{\alpha+1} - \frac{1}{2}u^{\alpha+1}\varphi^{-2}\left|\nabla\varphi\right|^2,\end{aligned} \tag{8.46}$$

and apply the divergence theorem,

$$\begin{aligned}&\int_{\partial B_r} u^{1+\alpha}\varphi^{-1}\left\langle\nabla\varphi,\nabla r\right\rangle + \frac{1}{2}\int_{B_r} u^{1+\alpha}\varphi^{-2}\left|\nabla\varphi\right|^2 \\ &\le -H\int_{B_r} a(x)\,u^{\alpha+1} + \frac{(\alpha+1)^2}{2}\int_{B_r} u^{\alpha-1}\left|\nabla u\right|^2 \le B \in \mathbb{R}\end{aligned}$$

where, in the last inequality, we have used the fact that $a(x)\,u^{\alpha+1}$, $u^{\alpha-1}\left|\nabla u\right|^2 \in L^1(M)$.

Setting, for ease of notation,

$$K(r) = \frac{1}{2}\int_{B_r} u^{1+\alpha}\varphi^{-2}\left|\nabla\varphi\right|^2,$$

we can write the above inequality in the form

$$\int_{\partial B_r} u^{1+\alpha}\varphi^{-1}\left\langle\nabla\varphi,\nabla r\right\rangle \le -K(r) + B.$$

Now, assume by contradiction that $K(r) \to +\infty$ as $r \to +\infty$. Thus, for each $r \ge R$ sufficiently large, $K(r) > \frac{1}{2}B$, so that

$$\int_{\partial B_r} u^{1+\alpha}\varphi^{-1}\left\langle\nabla\varphi,\nabla r\right\rangle \le -\frac{1}{2}K(r).$$

On the other hand, by Schwarz's inequality,

$$\frac{1}{2}K(r) \le -\int_{\partial B_r} u^{1+\alpha}\varphi^{-1}\langle\nabla\varphi,\nabla r\rangle \le \left\{\int_{\partial B_r} u^{1+\alpha}\right\}^{\frac{1}{2}}\left\{\int_{\partial B_r} u^{1+\alpha}\varphi^{-2}|\nabla\varphi|^2\right\}^{\frac{1}{2}},$$

that is,

$$K(r) \le \left\{\int_{\partial B_r} u^{1+\alpha}\right\}^{\frac{1}{2}}\{8K'(r)\}^{\frac{1}{2}}.$$

Squaring and integrating over $[R,r]$ give

$$K(R)^{-1} - K(r)^{-1} \ge \frac{1}{8}\int_R^r \left\{\int_{\partial B_t} u^{1+\alpha}\right\}^{-1} dt$$

and, letting $r \to +\infty$, we contradict (8.36) ii).

Step 3. We note that

$$\lim_{r\to+\infty}\int_{\partial B_r} u^{1+\alpha}\varphi^{-1}\langle\nabla\varphi,\nabla r\rangle = 0.$$

Indeed, by (8.45) of Step 2 and assumption (8.36) ii), there exists a sequence $\{r_k\} \nearrow +\infty$ such that

$$\lim_{k\to+\infty}\left\{\int_{\partial B_{r_k}} u^{1+\alpha}\right\}\left\{\int_{\partial B_{r_k}} u^{1+\alpha}\varphi^{-2}|\nabla\varphi|^2\right\} = 0$$

and therefore

$$\left\{\int_{\partial B_{r_k}} u^{1+\alpha}\varphi^{-1}\langle\nabla\varphi,\nabla r\rangle\right\}^2 \le \left\{\int_{\partial B_{r_k}} u^{1+\alpha}\right\}\left\{\int_{\partial B_{r_k}} u^{1+\alpha}\varphi^{-2}|\nabla\varphi|^2\right\} \to 0$$

as $k \to +\infty$. It remains to show that the desired limit exists. According to (8.44), (8.46) and the divergence theorem, we have

$$\begin{aligned}\int_{\partial B_r} u^{1+\alpha}\varphi^{-1}\langle\nabla\varphi,\nabla r\rangle &= (\alpha+1)\int_{B_r} u^{\alpha}\varphi^{-1}\langle\nabla\varphi,\nabla r\rangle \\ &\quad - H\int_{B_r} u^{1+\alpha}a(x) - \int_{B_r} u^{1+\alpha}\varphi^{-2}|\nabla\varphi|^2.\end{aligned} \tag{8.47}$$

Thus, we are reduced to showing that each of the (algebraic) summands on the RHS of (8.47) has a finite limit, as $r \to +\infty$. This easily follows from assumption (8.36) i), and from the following fact from Step 1:

$$\left|u^{\alpha}\varphi^{-1}\langle\nabla\varphi,\nabla r\rangle\right| \le \frac{1}{2}u^{\alpha-1}|\nabla u|^2 + \frac{1}{2}u^{1+\alpha}\varphi^{-2}|\nabla\varphi|^2 \in L^1(M).$$

Step 4. Fix $\varepsilon > 0$ so small that $\sup_M u + \varepsilon < 1$ and define the vector field

$$Z_\varepsilon = H^{-1}\frac{u^{\alpha+1}}{\varphi}\nabla\varphi - \frac{(u+\varepsilon)^\alpha}{1-(u+\varepsilon)^{\sigma-1}}\nabla u.$$

Then, using (8.39), (8.44) and (8.37), we compute

$$\begin{aligned}
\operatorname{div} Z_\varepsilon &= H^{-1}(\alpha+1)\,u^\alpha\varphi^{-1}\langle\nabla\varphi,\nabla u\rangle + H^{-1}u^{\alpha+1}\varphi^{-1}\Delta\varphi \\
&\quad - H^{-1}u^{\alpha+1}\varphi^{-2}|\nabla\varphi|^2 - \alpha\frac{(u+\varepsilon)^{\alpha-1}}{1-(u+\varepsilon)^{\sigma-1}}|\nabla u|^2 \\
&\quad - (\sigma-1)\frac{(u+\varepsilon)^{\alpha+\sigma-2}}{\left[1-(u+\varepsilon)^{\sigma-1}\right]^2}|\nabla u|^2 - \frac{(u+\varepsilon)^\alpha}{1-(u+\varepsilon)^{\sigma-1}}\Delta u \\
&\le H^{-1}(\alpha+1)\,\varphi^{-1}u^\alpha\langle\nabla\varphi,\nabla u\rangle - H^{-1}u^{\alpha+1}\varphi^{-2}|\nabla\varphi|^2 \\
&\quad - \alpha\frac{(u+\varepsilon)^{\alpha-1}}{1-(u+\varepsilon)^{\sigma-1}}|\nabla u|^2 - (\sigma-1)\frac{(u+\varepsilon)^{\alpha+\sigma-2}}{\left[1-(u+\varepsilon)^{\sigma-1}\right]^2}|\nabla u|^2 \\
&\quad + \frac{(u+\varepsilon)^\alpha}{1-(u+\varepsilon)^{\sigma-1}}a(x)\left(1-u^{\sigma-1}\right)u - a(x)\,u^{1+\alpha}.
\end{aligned}$$

Integrating over B_r and applying the divergence theorem give

$$\begin{aligned}
&H^{-1}\int_{\partial B_r} u^{\alpha+1}\varphi^{-1}\langle\nabla\varphi,\nabla r\rangle - \int_{\partial B_r}\frac{(u+\varepsilon)^\alpha}{1-(u+\varepsilon)^{\sigma-1}}\langle\nabla u,\nabla r\rangle \\
&\le \mathrm{I}+\mathrm{II}+\mathrm{III}
\end{aligned}$$

where we have set

$$\begin{aligned}
\mathrm{I} &= \int_{B_r} H^{-1}\left\{(\alpha+1)\,\varphi^{-1}u^\alpha\langle\nabla\varphi,\nabla u\rangle - u^{\alpha+1}\varphi^{-2}|\nabla\varphi|^2\right\}, \\
\mathrm{II} &= -\int_{B_r}\left\{\alpha\frac{(u+\varepsilon)^{\alpha-1}}{1-(u+\varepsilon)^{\sigma-1}} + (\sigma-1)\frac{(u+\varepsilon)^{\alpha+\sigma-2}}{\left[1-(u+\varepsilon)^{\sigma-1}\right]^2}\right\}|\nabla u|^2, \\
\mathrm{III} &= -\int_{B_r} a(x)\left\{u^{1+\alpha} - \frac{(u+\varepsilon)^\alpha}{1-(u+\varepsilon)^{\sigma-1}}\left(1-u^{\sigma-1}\right)u\right\}.
\end{aligned}$$

Letting $\varepsilon \searrow 0^+$ and applying the dominated convergence theorem, the LHS of the above converges to

$$H^{-1}\int_{\partial B_r} u^{\alpha+1}\varphi^{-1}\langle\nabla\varphi,\nabla r\rangle - \int_{\partial B_r}\frac{u^\alpha}{1-u^{\sigma-1}}\langle\nabla u,\nabla r\rangle$$

while

$$\mathrm{II} \to -\int_{B_r} \left\{ \alpha \frac{u^{\alpha-1}}{1-u^{\sigma-1}} |\nabla u|^2 + (\sigma-1) \frac{u^{\alpha+\sigma-2}}{[1-u^{\sigma-1}]^2} |\nabla u|^2 \right\},$$

and finally

$$\mathrm{III} \to 0.$$

Thus, using assumption (8.33):

$$H^{-1} \frac{(\alpha+1)^2}{4} \leq \alpha,$$

and completing the square in I, we obtain

$$\begin{aligned} &H^{-1} \int_{\partial B_r} u^{\alpha+1} \varphi^{-1} \langle \nabla\varphi, \nabla r \rangle - \int_{\partial B_r} \frac{u^{\alpha}}{1-u^{\sigma-1}} \langle \nabla u, \nabla r \rangle \\ &\leq -H^{-1} \int_{B_r} \left\{ \varphi^{-2} u^{\alpha+1} |\nabla\varphi|^2 + \frac{(\alpha+1)^2}{4} u^{\alpha-1} |\nabla u|^2 - (\alpha+1) \langle \nabla\varphi, \nabla u \rangle \varphi^{-1} u^{\alpha} \right\} \\ &\quad - \int_{B_r} \left\{ \alpha \frac{u^{\alpha-1}}{1-u^{\sigma-1}} + (\sigma-1) \frac{u^{\alpha+\sigma-2}}{[1-u^{\sigma-1}]^2} - H^{-1} \frac{(\alpha+1)^2}{4} u^{\alpha-1} \right\} |\nabla u|^2 \\ &\leq -H^{-1} \int_{B_r} u^{\alpha-1} \left| \frac{\alpha+1}{2} \nabla u - \frac{u}{\varphi} \nabla\varphi \right|^2 - \alpha \int_{B_r} \frac{1 + \frac{\sigma-1}{\alpha} - u^{\sigma-1}}{(1-u^{\sigma-1})^2} |\nabla u|^2 u^{(\alpha-1)+(\sigma-1)}. \end{aligned}$$

Now, letting $r \to +\infty$ along the sequence $\{r_k\}$ constructed in Step 1 and using Step 3 yields

$$0 \leq -H^{-1} \int_{M} u^{\alpha-1} \left| \frac{\alpha+1}{2} \nabla u - \frac{u}{\varphi} \nabla\varphi \right|^2 - \alpha \int_{M} \frac{1 + \frac{\sigma-1}{\alpha} - u^{\sigma-1}}{(1-u^{\sigma-1})^2} |\nabla u|^2 u^{\alpha+\sigma-2} \leq 0.$$

Let now $A \neq \emptyset$ be a connected component of the set

$$\{x \in M : u(x) > 0\}.$$

The above inequality forces $\nabla u = 0$ on A and thus $u = c_1$ on A. So, either $c_1 = 0$ on A, contradicting the definition of A, or else $A = M$ and $u = c_1 > 0$ on M. But then $\varphi^{-1}|\nabla\varphi| = 0$ on M and therefore $\varphi = c_2 > 0$ on M, and since $\Delta\varphi + Ha(x)\varphi = 0$ this implies $a(x) \equiv 0$ on M. Finally, since $0 = \Delta u \geq -a(x)\,u + b(x)\,u^{\sigma} = b(x)\,c_1^{\sigma}$, we conclude that $b(x) \equiv 0$ on M, contradicting (8.32) (i) . □

We are now ready to prove our third version of the Li–Yau vanishing result.

Theorem 8.11. *Let $(M, \langle\,,\rangle, J_M)$ be a complete Kähler manifold satisfying*

$${}^{M}\mathrm{Ric} \geq -\rho(x)$$

with $\rho \in C^0(M)$ *such that*

$$0 \le \rho(x) \le \frac{A}{r(x)^\mu},\ r(x) >> 1, \tag{8.48}$$

for some constants $0 \le \mu < 2$ *and* $A > 0$. *Assume also that, for some* $H \ge 2$, *the operator* ${}^H L = -\Delta - H\rho(x)$ *satisfies*

$$\lambda_1\left({}^H L_M\right) \ge 0.$$

Let $(N, (\,,\,), J_N)$ *be a Hermitian manifold with holomorphic bisectional curvature bounded above by* $k(z)$, $z \in N$, $k \in C^0(N)$. *Let* $f : M \to N$ *be a holomorphic map such that* (8.25) *holds true, i.e.,*

$$k(\psi(x)) < 0 \text{ on } M; \qquad k(\psi(x)) \le -\frac{B}{r(x)^\mu} \text{ on } M - B_{R_0} \tag{8.49}$$

for some $R_0, B > 0$. *If*

$$|df|^2 \in L^q(M) \tag{8.50}$$

for some

$$0 < q \le H + \sqrt{H(H-2)},$$

then ψ *is constant.*

Remark 8.12. As already remarked, this is the only instance where one allows integrability exponents greater than H.

We also note that even in this case the result holds in the case of a pluriharmonic function which is bounded on one side of M. See Remark 8.2.

Proof. As before, set $u = |df|^2$. Then, u satisfies

$$\Delta u + 2\rho(x)u \ge -k(f(x))u^2 \text{ on } M.$$

Our assumptions on the Ricci curvature and the volume comparison theorem contained in Corollary 2.17 imply that

$$\liminf_{r\to+\infty} \frac{\log \mathrm{vol} B_r}{r^{2-\mu}} < +\infty.$$

Setting $b(x) = -k(f(x))$ and $a(x) = 2\rho_+(x)$, and writing ${}^H L = -\Delta - \frac{H}{2}a(x)$ we see that (8.49) and (8.48) guarantee that the remaining assumptions of Theorem 6.16 are satisfied with $E = A/B$. It follows that $u \in L^\infty(M)$. Combining this fact with (8.50), recalling the range of q, and interpolating, we deduce

$$u \in L^{q_0}(M) \tag{8.51}$$

with

$$q_0 = H + \sqrt{H(H-2)}.$$

We set $\alpha = q_0 - 1$ and we note that

$$\frac{H}{2} = \frac{(1+\alpha)^2}{4\alpha}.$$

Now, (8.51) and the fact that $\rho \in L^\infty(M)$ imply the validity of (8.36) i) and ii) while (8.49) and (8.48) show that (8.32) holds with some constant $C > 0$. An application of Theorem 8.9 gives $u = 0$ on M, that is f is a constant map. □

Remark 8.13. Let $f : M \to N$ be a harmonic map with bounded dilation T between Riemannian manifolds M and N of dimensions m and n, respectively. According to formula (6.26) and to the refined Kato inequality, if the Ricci curvature of M is bounded below by $-\rho(x)$ and the sectional curvature of N is bounded above by a function $k(z)$ satisfying $k(f(x)) \leq 0$, then the function $u = |df|^2$ satisfies the inequality

$$u\Delta u \geq \frac{m}{4(m-1)}|\nabla u|^2 - 2\rho(x)\, u^2 + \frac{4k(f(x))}{T\min\{m,n\}^2}u^3.$$

It follows that one may give a version of the above theorem in this setting. Note that the coefficient of the gradient term on the right-hand side is now less than 1. It may be interesting to investigate if even the first two versions of the Li and Yau result admit similar extensions.

8.2 Plurisubharmonic exhaustions and a structure theorem for Kähler manifolds

A smooth function $\rho : M \to \mathbb{R}$ on a Riemannian manifold $(M, \langle,\rangle)$ is said to be an *exhaustion function* if, for every $r \in \mathbb{R}$, the sublevel sets $M_\rho(r) = \{x \in M : \rho(x) < r\}$ are relatively compact. In particular, by the well-known Sard theorem, $\partial M_\rho(r)$ are compact, smooth hypersurfaces for almost every $r \in \mathbb{R}$. Clearly, every open manifold $(M, \langle,\rangle)$ supports an exhaustion function. For instance, using a celebrated theorem by H. Whitney, one can embed M into some Euclidean space $\mathbb{R}^N$ via a proper map $f : M \to \mathbb{R}^N$ and the function $\rho(x) = |f(x)|^2$ gives the desired smooth exhaustion. In fact, according to R. Greene and H.H. Wu, [67], [68], we can require f to be a harmonic map and, therefore, $\rho(x) = |f(x)|^2$ to be a strictly subharmonic exhaustion function. Despite this fact, the possibility of constructing a special exhaustion usually reflects topological and geometrical properties of the underlying manifold. By way of example, if a complete manifold $(M, \langle,\rangle)$ supports a smooth, non-negative, strictly convex exhaustion function, then, by standard Morse theory, M must be simply connected. Indeed, according to work by Greene and Wu, [69], in this situation M is diffeomorphic to Euclidean space $\mathbb{R}^m$ and such a strictly convex exhaustion function exists, for instance, when M has strictly positive sectional curvature, thus recovering the conclusion of the Gromoll–Meyer theorem, [74].

As an application of the above versions of the Li–Yau vanishing theorem, see especially Corollary 8.8, this section aims to prove the existence of a plurisubharmonic exhaustion on a complete Kähler manifold with almost non-negative Ricci curvature; Theorem 8.17 below. The result will then be used to deduce information on the structure of such manifolds; Theorem 8.19 below.

To begin with, we point out the following

Lemma 8.14. *For every $s \geq 2$, there exists a constant $\varepsilon = \varepsilon(s) > 0$ such that the following holds. Let $(M, \langle\,,\rangle, J)$ be a complete Kähler manifold of real dimension $m = 2s$ and Ricci curvature satisfying*

$$\operatorname{Ric} \geq -\frac{\varepsilon}{r(x)^2} \text{ for } r(x) >> 1. \tag{8.52}$$

If $(M, \langle,\rangle, J)$ is parabolic, then there exists a smooth, proper, pluriharmonic function $f : M \to \mathbb{R}$ satisfying

$$\int_{B_R(o)} |\nabla f|^2 = o\left(R^2\right) \text{ as } R \to +\infty. \tag{8.53}$$

Proof. Simply recall that, by Theorem 7.20, $(M, \langle,\rangle)$ supports a proper harmonic function f satisfying (8.53). On the other hand, according to Theorem 1.20, a harmonic function with sub-quadratic energy growth on any Kähler manifold is necessarily pluriharmonic. □

The next parabolicity result is due to Li and Tam; see Theorem 4.2 in [102].

Lemma 8.15. *For every $s \geq 2$, there exists a constant $\varepsilon = \varepsilon(s) > 0$ such that the following holds. A complete, Kähler manifold $(M, \langle,\rangle, J)$ of real dimension $m = 2s$ is parabolic provided M has at least two ends, and Ricci curvature satisfying*

$$\operatorname{Ric} \geq -\rho(x) \text{ on } M, \tag{8.52}$$

where $0 \leq \rho(x) \in C^0(M)$ is such that

$$\text{(i) } \rho(x) \in L^1(M); \qquad \text{(ii) } \rho(x) \leq \frac{\varepsilon}{r(x)^2} \text{ as } r(x) >> 1. \tag{8.54}$$

Proof. By contradiction, suppose M is non-parabolic, that is, M has a non-parabolic end. Recall from Theorem 7.24 that M supports a non-constant, positive harmonic function f with sub-quadratic energy growth. According to Theorem 1.20, f is pluriharmonic. Since, by volume comparison (see Theorem 2.14 and Proposition 2.11), condition (8.54) (ii) ensures that the volume of balls grows at most polynomially, we may apply Corollary 8.8 to conclude that f is constant, a contradiction. □

Remark 8.16. Conditions (8.54) are both satisfied, e.g., if $\rho(x) = \alpha(r(x))$ for some non-increasing function $\alpha(t)$ such that

$$t^{m-1}\alpha(t) \in L^1(+\infty). \tag{8.55}$$

Indeed, because of the monotonicity of α, the above implies

$$\alpha(t) = o\left(t^{-m}\right) \qquad \text{as } t \to +\infty, \tag{8.56}$$

so that, in particular, (8.54) (ii) is met. Furthermore, by volume comparison,

$$\operatorname{vol}(\partial B_r) = O\left(r^{m-1}\right) \text{ as } r \to +\infty,$$

and therefore, integrating in polar coordinates and using the co-area formula yields

$$\int_M \rho = \int_0^{+\infty} \int_{\partial B_t} \alpha(t) \leq C \int_0^{+\infty} t^{m-1}\alpha(t) < +\infty,$$

for some positive constant C, proving the validity of (8.54) (i). For the sake of completeness, let us verify (8.56). By contradiction, suppose that $\alpha(t_k) \geq At_k^{-m}$ along some divergent sequence $\{t_k\} \nearrow +\infty$ and for some constant $A > 0$. Without loss of generality, we can assume $t_{k+1} - t_k > t_k$. Using the monotonicity of $\alpha(t)$ we conclude

$$\begin{aligned}
\int^{+\infty} t^{m-1}\alpha(t)\,dt &= \sum_k \int_{t_k}^{t_{k+1}} t^{m-1}\alpha(t)\,dt \\
&\geq \sum_k \int_{t_k}^{t_{k+1}} t^{m-1}\alpha(t_{k+1})\,dt \\
&\geq \frac{A}{m}\sum_k t_{k+1}^{-m}\left(t_{k+1}^m - t_k^m\right) \\
&\geq \frac{A}{m}\sum_k \left\{1 - \left(\frac{1}{2}\right)^m\right\} = +\infty,
\end{aligned}$$

contradicting (8.55).

Combining Lemma 8.14 and Lemma 8.15 yields the following existence result alluded to above. This is one of the main achievements of the paper [98] by P. Li and M. Ramachandran.

Theorem 8.17. *For every $s \geq 2$, there exists a constant $\varepsilon = \varepsilon(s) > 0$ such that the following holds. A complete, Kähler manifold $(M, \langle,\rangle, J)$ of real dimension $m = 2s$, with at least two ends, and Ricci curvature satisfying (8.52) and (8.54) supports a smooth, non-negative, plurisubharmonic exhaustion function.*

Proof. We first apply Lemma 8.15 to deduce that M is parabolic. Using Lemma 8.14, we next construct a proper, pluriharmonic function f on M. Finally, we observe that

$$\begin{aligned}\left(Ddf^2\right)^{(1,1)} &= 2f\,(Ddf)^{(1,1)} + 2\,(df)^{(1,0)} \otimes (df)^{(0,1)} \\ &= 2\,(df)^{(1,0)} \otimes (df)^{(0,1)} \geq 0\end{aligned}$$

in the sense of quadratic forms. Therefore $f^2(x) \geq 0$ is the desired plurisubharmonic exhaustion. □

Motivated by applications to Lefschetz-type theorems and to a conjecture of I.R. Shafarevich in complex geometry, in their paper [117], T. Napier and M. Ramachandran obtain structure results for complete Kähler manifolds $(M, \langle , \rangle, J)$. Their main assumption is that M *has a complex atlas of locally uniformly bilipschitz charts*, i.e., there exists an absolute constant $C > 0$ such that, for every $x \in M$, there is a biholomorphic map ψ_x of the unit Euclidean ball $\mathbb{B}_1(0) \subset \mathbb{C}^m$ onto an open neighborhood U_x of x, satisfying $\psi_x(0) = x$ and

$$C^{-1}\mathrm{can}_{\mathbb{C}^m} \leq \psi_x^* \langle , \rangle \leq C\mathrm{can}_{\mathbb{C}^m}, \text{ on } \mathbb{B}_1(0).$$

This is a special kind of 0-order bounded geometry in the complex setting which is obviously satisfied if M is a co-compact covering manifold. See also Appendix B.

In the same paper, [117], Napier–Ramachandran also consider the situation of manifolds supporting a plurisubharmonic exhaustion. In fact, they are able to develop a reduction procedure to the locally uniformly bilipschitz Euclidean case. The deep structure theorem by Napier-Ramachandran looks as follows. Since its proof is quite long and requires tools from complex geometry that go beyond the scope of the present notes, we limit ourselves to stating it.

Theorem 8.18. *Let $(M, \langle , \rangle, J)$ be a complete Kähler manifold of real dimension $m = 2s$. Assume that either M has a complex atlas of locally uniformly bilipschitz charts or M supports a plurisubharmonic exhaustion function.*

(1) *If M has at least three ends, then there exists a proper holomorphic map $\phi : M \to S$ onto a Riemann surface S, whose fibres are compact and connected.*

(2) *If $H^1(M, \mathbb{Z}) = 0$, then M has only one end.*

Note that, in case M is parabolic, the base Riemann surface S must be parabolic as well. To see this, observe that, given a holomorphic map $f : (M, \langle , \rangle, J_M) \to (N, (,), J_N)$ between Kähler manifolds and any function $u : N \to \mathbb{R}$, one has

$$Dd(u \circ f)^{(1,1)} = Ddu^{(1,1)}\left(df, \overline{df}\right).$$

In particular,

$$\Delta(u \circ f) = \mathrm{tr}_M \left\{ Ddu^{(1,1)}\left(df, \overline{df}\right)\right\}$$

proving that $\Delta(u \circ f)$ has constant sign whenever the complex Hessian $Ddu^{(1,1)}$ is semi-definite. In the special case where N is (complex) one-dimensional, i.e., N is a Riemann surface, we have $Ddu^{(1,1)} = \Delta u(,)$ and the above reduces to

$$\Delta(u \circ f) = |df|^2 \Delta u.$$

Summarizing, the holomorphic map ϕ in the statement of the theorem pulls (sub)superharmonic functions on S back to (sub)superharmonic functions on M.

On the base of this observation, combining Theorem 8.17 with Theorem 8.18, we get the following result by P. Li and M. Ramachandran, [98], that completes the section.

Theorem 8.19. *For every $s \geq 2$, there exists a constant $\varepsilon = \varepsilon(s) > 0$ such that the following holds. Let $(M, \langle,\rangle, J)$ be a complete Kähler manifold of real dimension $m = 2s$. Assume that, having set $r(x) = \operatorname{dist}(x,p)$ for some reference origin $p \in M$, the Ricci curvature of M satisfies*

$$\operatorname{Ric} \geq -\rho(x) \quad \text{on } M$$

where $0 \leq \rho(x) \in C^0(M)$ is such that

$$\text{(i) } \rho(x) \in L^1(M); \qquad \text{(ii) } \rho(x) \leq \frac{\varepsilon}{r(x)^2} \quad \text{as } r(x) >> 1.$$

(1) *If M has at least three ends, then M is parabolic and there exists a proper holomorphic map $\phi : M \to S$ onto a parabolic Riemann surface S, whose fibres are compact and connected.*

(2) *If $H^1(M,\mathbb{Z}) = 0$, then M has only one end.*

Chapter 9

Splitting and gap theorems in the presence of a Poincaré–Sobolev inequality

9.1 Splitting theorems

Up to now, we have been using Theorem 4.5 to show that solutions of a differential problem of the type

$$\begin{cases} \psi\Delta\psi + a(x)\psi^2 \geq -A|\nabla\psi|^2, \\ \psi \geq 0 \end{cases}$$

have to be identically zero. The aim of this section is to present a geometrical problem in which the second alternative of Theorem 4.5 does actually occur, that is, ψ becomes a positive solution of the linear equation

$$\Delta\psi + a(x)\psi = 0.$$

We shall focus our attention on splitting-type theorems depending on spectral and Ricci curvature bounds. To say that the complete manifold $(M, \langle , \rangle)$ *splits* usually means that M is isometric to the Riemannian direct product

$$\left(N_1 \times N_2, (,)_{N_1} + (,)_{N_2}\right)$$

for suitable complete Riemannian manifolds $\left(N_i, (,)_{N_i}\right)$, $i = 1, 2$. Manifolds that do not split are said to be irreducible. By way of example, the fundamental structure theorem by G. de Rham asserts that any complete, simply connected manifold splits, according to its holonomy, into simply connected, geodesically complete factors $(\mathbb{R}^n, \text{can})$, $\left(N_2, (,)_{N_2}\right), \ldots, \left(N_k, (,)_{N_k}\right)$; furthermore, the decomposition is uniquely determined and the complete manifolds $\left(N_i, (,)_{N_i}\right)$ are irreducible. See, e.g., [142].

Actually, the results we present in this section involve with a more general notion of splitting which allows warped factors. Accordingly, we say that $(M, \langle , \rangle)$ splits even if it is isometric to the warped product

$$\left(N_1 \times N_2, (,)_{N_1} + f^2 (,)_{N_2}\right)$$

where $\left(N_i, (,)_{N_i}\right)$, $i = 1, 2$, are complete Riemannian manifolds and $f \in C^\infty(N_1)$ is a suitable positive function.

Our purpose is to give a somewhat simplified proof of the following result due to P. Li and J. Wang, [103]. It extends previous work by X. Wang, [161] on the structure of conformally compact Einstein manifolds.

Theorem 9.1. *Let $(M, \langle , \rangle)$ be a complete manifold of dimension $m \geq 3$ satisfying*

$$\lambda_1 = \lambda_1 (-\Delta_M) > 0.$$

Suppose also

$$^M\mathrm{Ric} \geq -\left(\frac{m-1}{m-2}\right)\lambda_1. \tag{9.1}$$

Then, either

(1) *$(M, \langle , \rangle)$ has only one non-parabolic end,*

or

(2) *$(M, \langle , \rangle)$ splits as the warped product $\mathbb{R} \times \Sigma$ with metric*

$$\langle , \rangle = dt^2 + \cosh^2\left(t\sqrt{\tfrac{\lambda_1}{m-2}}\right)(,) \tag{9.2}$$

where $(\Sigma, (,))$ is a compact, isometrically imbedded hypersurface of $(M, \langle , \rangle)$ satisfying

$$^\Sigma\mathrm{Ric} \geq -\lambda_1. \tag{9.3}$$

Remark 9.2. According to Lemma 7.13 above, the presence of a Sobolev–Poincaré inequality implies that non-parabolic ends are precisely the infinite volume ends.

The proof of Theorem 9.1 has its root in the classical splitting theorem by J. Cheeger and D. Gromoll, [32]. This latter states that if a complete manifold $(M, \langle , \rangle)$ of non-negative Ricci curvature contains a line, i.e., a minimizing geodesic $\gamma : \mathbb{R} \to M$, then M splits as the Riemannian product $\mathbb{R} \times N$ where $N \subset M$ is a totally geodesic hypersurface. Note that in the case where M has more than one end, N is necessarily compact. The (very) original argument supplied by Cheeger-Gromoll relies on the existence of a harmonic function u of distance-type, i.e., satisfying the condition $|\nabla u| = 1$. A substantial part of the proof is devoted to showing that, under the assumptions of the theorem, the Busemann function corresponding to (a half-line in) γ has the desired properties. Note that the integral curves of the gradient vector field ∇u are geodesic lines of M, so that, in particular, ∇u is complete. Moreover the level sets of u are smooth hypersurfaces with Gauss map ∇u. Let N be such a level set. Using the flow ϕ_t of ∇u through N, one establishes a smooth diffeomorphism between $\mathbb{R} \times N$ and M. Finally, one observes that ϕ_t is, in fact, a Riemannian isometry with respect to the product metric on $\mathbb{R} \times N$. Indeed, obviously, $(\phi_t)_* \left(\frac{d}{dt}\right) = \nabla u$ which is a unit vector field of M.

Moreover, using the harmonic function u in the Weitzenbock formula and recalling the refined Kato inequality yield

$$|\mathrm{Hess}\,(u)|^2 = \frac{m}{m-1}\,|\nabla\,|\nabla u||^2 = 0,$$

i.e., ∇u is a parallel (hence a Killing) vector field, and N is totally geodesic. Thus, by the very definition of Lie derivative, for every $V, W \in T_x M$, it holds that

$$\frac{d}{dt}\left\langle (\phi_t)_* V, (\phi_t)_* W \right\rangle = 2\mathrm{Hess}\,(u)\left((\phi_t)_* V, (\phi_t)_* W\right) = 0$$

showing that, obviously, ϕ_t is a one-parameter group of isometries of M. In particular, for any orthonormal basis $e_1, \ldots, e_{m-1}$ of $T_x N = \langle \nabla u\,(x)\rangle^{\perp} \subset T_x M$, the tangent vectors $(\phi_t)_* e_1, \ldots, (\phi_t)_* e_{m-1} \in \langle \nabla u\,(\phi_t\,(x))\rangle^{\perp}$ remain orthonormal in $T_{\phi_t(x)}M$. This completes the proof.

We note in passing that a differentiable splitting can be obtained by simply assuming that there exists a smooth function u without critical points such that $|\nabla u|$ is constant on the level sets of u. Indeed, one considers the flow ϕ_t of the unit vector field $\nabla u/|\nabla u|$, which, by the assumption that $|\nabla u| = \alpha(u)$, moves level sets of u onto level sets of u. Therefore, having chosen a level set Σ_o, the map $\phi : \mathbb{R} \times \Sigma_o \to M$ realizes the splitting.

A generalization of this kind of argument led, first M. Cai and G.J. Galloway, [23], and later X. Wang, [161], and P. Li and J. Wang, [103], to obtain the following

Theorem 9.3. *Let $(M, \langle\,,\rangle)$ be a complete manifold of dimension $m \geq 2$ and Ricci curvature satisfying*

$${}^M\mathrm{Ric} \geq -\rho$$

for some constant $\rho > 0$. Suppose that $u \in C^\infty(M)$ is a non-constant harmonic function such that, setting

$$\psi = |\nabla u| \in Lip_{loc}\,(M)\,,$$

it holds that

$$\psi\Delta\psi + \rho\psi^2 = \frac{1}{m-1}\,|\nabla\psi|^2 \;\; on\ M, \tag{9.4}$$

in the weak sense. Then, the level sets of u are smooth hypersurfaces, and $(M, \langle\,,\rangle)$ splits as the warped product $\mathbb{R} \times \Sigma_0$ with metric

$$\langle\,,\rangle = dt^2 + w(t)\,(\,,)$$

for a suitable level set Σ_0 of u endowed with the inherited metric, and where

$$w(t) = \left\{ \frac{C_1 \exp\left(t\sqrt{\frac{\rho}{m-1}}\right) + C_2 \exp\left(-t\sqrt{\frac{\rho}{m-1}}\right)}{C_1 + C_2} \right\}^2, \tag{9.5}$$

for some non-negative constants C_1 and C_2 which are not both zero. Moreover, if M has at least two ends, then Σ_0 is compact and M has exactly two ends. If both ends of M have infinite volume, we may chose Σ_0 in such a way that

$$w(t) = \cosh^2\left(t\sqrt{\tfrac{\rho}{m-1}}\right). \tag{9.6}$$

If one end of M has finite volume, then, up to replacing t with $-t$, we have

$$w(t) = \exp\left(2t\sqrt{\tfrac{\rho}{m-1}}\right). \tag{9.7}$$

Remark 9.4. Since here $A = -1/(m-1)$, according to Lemma 4.12, for every $p > (m-2)/2(m-1)$, we have $\psi^p \in W^{1,2}_{loc}(M)$ and

$$\nabla\psi^p = p\psi^{p-1}\nabla\psi \in L^2_{loc}. \tag{9.8}$$

This fact will be crucial in the arguments below.

The proof of Theorem 9.3 is reached by means of two main steps: first, we show a differential splitting of the original manifold and, next, the metric rigidity of the differential decomposition.

We need to recall some facts of both topological and analytical nature. Although they are quite simple, we provide a proof for the sake of completeness.

Lemma 9.5. *Let Σ be a non-compact, connected manifold. Then, every compact set $[a,b]\times K$ of $\mathbb{R}\times\Sigma$ has a connected complement in $\mathbb{R}\times\Sigma$. In particular, $\mathbb{R}\times\Sigma$ has only one end.*

Proof. Let $P_j = (t_j, x_j) \in \mathbb{R}\times\Sigma - [a,b]\times K$, $j = 1,2$. We show that there is a continuous path γ in $\mathbb{R}\times\Sigma - [a,b]\times K$ connecting P_1 to P_2. Roughly speaking, γ is obtained by circumnavigating around $[a,b]\times K$. Four possibilities can occur: (i) $t_1, t_2 \in [a,b]$ and $x_1, x_2 \notin K$; (ii) $x_1, x_2 \in K$ and $t_1, t_2 \notin [a,b]$; (iii) up to interchanging P_1 with P_2, $t_1 \in [a,b]$, $x_1 \notin K$ and $t_2 \notin [a,b]$, $x_2 \in K$; (iv) $t_1, t_2 \notin [a,b]$ and $x_1, x_2 \notin K$. We limit ourselves to consider case (i), the other cases being similar and left to the reader. Fix $\bar{t} \notin [a,b]$ and define the paths $\Gamma_j(s) = (\bar{t}s + t_j(1-s)) \times x_j$, $j = 1,2$. Clearly, $\Gamma_j(s)$ lies in the complement of $[a,b]\times K$ and satisfies $\Gamma_j(0) = P_j$, $\Gamma_j(1) = (\bar{t}, x_j)$. Next, observe that Σ is connected and locally path connected, hence a path connected space. Choose a path σ in Σ satisfying $\sigma(0) = x_1$, $\sigma(1) = x_2$ and define $\Gamma_3(s) = \bar{t}\times\sigma(s)$. Since $\Gamma_3(s) \in \mathbb{R}\times\Sigma - [a,b]\times K$ and $\Gamma_3(0) = \Gamma_1(1)$, $\Gamma_3(1) = \Gamma_2(1)$, we can form a new path in $\mathbb{R}\times\Sigma - [a,b]\times K$ from $\gamma(0) = P_1$ to $\gamma(1) = P_2$ by setting $\gamma = \Gamma_1 * \Gamma_3 * \overline{\Gamma_2}$, where $*$ means juxtaposition and $\overline{\Gamma_2}$ is nothing but Γ_2 taken with opposite orientation. □

Lemma 9.6. *Let Σ be a closed connected submanifold of a connected Riemannian manifold $(M, \langle\,,\rangle)$. The normal exponential map $\exp^\perp : T^\perp\Sigma \to M$, defined as the restriction of* exp *to the normal bundle of Σ, is onto M.*

Proof. Since M is complete, given $x \in M$, a standard compactness argument shows that there exists $x_o \in \Sigma$ such that $d(x, x_o) = \mathrm{dist}(x, \Sigma)$. Again by completeness of M there exists a minimizing geodesic γ with $\gamma(0) = x_o$, $\gamma(1) = x$ and length$(\gamma) = d(x, x_o)$. Since, $\exp^{\perp}$ is a local diffeomorphism of a neighborhood of the zero section of $T^{\perp}\Sigma$ onto its image, and by Gauss' lemma geodesics normal to Σ locally minimize the distance from Σ, then $\gamma'(0) \perp T_{x_o}\Sigma$ and by definition $\exp^{\perp}(\gamma'(0)) = \gamma(1) = x$. □

Finally we shall need the following simple ODE result.

Lemma 9.7. *Let $\rho_1 > 1$, $\rho_2 > 0$ and $C_1, C_2 \geq 0$. Then, the solution of the differential equation*

$$\beta\beta'' - \rho_1 (\beta')^2 + \rho_2\beta^2 = 0$$

is given by

$$\beta(t) = \left\{ C_1 \exp\left(t\sqrt{\rho_2(\rho_1 - 1)}\right) + C_2 \exp\left(-t\sqrt{\rho_2(\rho_1 - 1)}\right) \right\}^{-\frac{1}{\rho_1 - 1}}.$$

We are now in a position to prove Theorem 9.3.

Proof of Theorem 9.3. For the sake of clarity, we divide the argument into several steps.

Step 1. The function $\psi = |\nabla u|$ satisfies $\psi > 0$ on M. In particular, ψ is smooth and the level sets of u are smooth hypersurfaces of M with (Gauss map) unit normal $\nabla u / |\nabla u|$.

Indeed, let

$$g = \psi^{\frac{m-2}{m-1}}.$$

We insert the test function

$$(\psi + \epsilon)^{-\frac{m}{m-1}}\lambda, \quad \text{with } \lambda \in Lip_c(M)$$

in the weak formulation of (9.4), and perform a computation, similar to that carried out in the proof of Theorem 4.5, to obtain

$$\int \psi^{-\frac{1}{m-1}} \left(\frac{\psi}{\psi + \epsilon}\right)^{\frac{m}{m-1}} \langle \nabla\psi, \nabla\lambda \rangle = \int \rho\lambda\psi^2(\psi + \epsilon)^{-\frac{m}{m-1}}$$
$$- \frac{m}{m-1} \int \lambda \left(\psi^{-\frac{m}{2(m-1)}} |\nabla\psi|\right)^2 \left(\frac{\psi}{\psi + \epsilon}\right)^{\frac{m}{m-1}} \frac{\epsilon}{\psi + \epsilon}.$$

Using (9.8) with $p = 1 - \frac{m}{2(m-1)}$ we may let $\epsilon \to 0$ and apply the dominated convergence theorem to deduce that the function g satisfies

$$\Delta g + \frac{m-2}{m-1}\rho g = 0$$

weakly on M. Thus, the desired positivity of ψ follows from a local Harnack inequality (see, e.g., [60]).

Step 2. Let $\{e_i\}$ be a local orthonormal frame of M such that

$$e_1 = \frac{\nabla u}{|\nabla u|} = \frac{\nabla u}{\psi}.$$

Then, denoting by u_{ij} the coefficients of Hess (u) with respect to this frame, it holds that

$$\begin{aligned} u_{1j} &= -(m-1)\,\mu\,\delta_{1j}, \quad j = 1, \dots, m, \\ u_{ij} &= \mu\,\delta_{ij}, \quad i, j = 2, \dots, m, \end{aligned} \tag{9.9}$$

for some smooth function μ. Equivalently,

$$\begin{aligned} &\operatorname{Hess}(u)(e_1, \cdot) = -(m-1)\,\mu\,\langle e_1, \cdot\rangle && \text{on } TM, \\ &\operatorname{Hess}(u)(\cdot, \cdot) = \mu\,\langle \cdot, \cdot\rangle && \text{on } \langle e_1\rangle^{\perp}. \end{aligned} \tag{9.10}$$

In particular, ψ is constant on each path-component of the smooth, level hypersurfaces $\{u = \text{const.}\}$ of u.

Indeed, note that (9.4) comes from the Bochner formula for harmonic functions, once we replace the usual inequalities with the equality sign. In particular, it forces equality in the refined Kato inequality of Proposition 1.3. Thus, setting $M = (u_{ij}) \in M_m(\mathbb{R})$ and $y = (u_i) \in \mathbb{R}^m$, we deduce that

$$\|M\|^2 = \frac{m}{m-1}\frac{|My|^2}{|y|^2}.$$

Application of the algebraic Lemma 1.5 with $A = 1$ enables us to obtain (9.9). As for the second assertion, simply note that if σ is any curve in a path-component of $\{u = \text{const.}\}$, then $\dot{\sigma}$ is orthogonal to e_1 and, therefore, by (9.9),

$$\frac{d}{dt}|\nabla u| \circ \sigma = \operatorname{Hess}(u)(e_1, \dot{\sigma}) = 0.$$

As a matter of fact, using a similar argument together with Sard's theorem, one can show that each of the path components of a.e. level set of $|\nabla u|$ coincides with a path component of some level set of u.

Step 3. Let $\{e_i\}$ be as in Step 2. Then

$$D_{e_1} e_1 = 0,$$

so that the integral curves of the global unit vector field $e_1 = \nabla u / |\nabla u|$ are geodesics of M.

Indeed, recall that

$$\begin{aligned} \nabla \left|\nabla u\right| &= \operatorname{Hess}(u)(e_1,\cdot)^{\#} \\ &= -(m-1)\,\mu e_1. \end{aligned} \tag{9.11}$$

Therefore, by a direct computation we get

$$\begin{aligned} D_{e_1}e_1 &= \frac{1}{|\nabla u|} D_{\nabla u} \frac{\nabla u}{|\nabla u|} \\ &= \frac{1}{|\nabla u|}\left\{ -\frac{1}{|\nabla u|^2}\left\langle \nabla u, \nabla\left|\nabla u\right|\right\rangle \nabla u + \frac{1}{|\nabla u|}\operatorname{Hess}(u)(\nabla u,\cdot)^{\#}\right\} \\ &= \frac{1}{|\nabla u|}\left\{ -\operatorname{Hess}(u)(e_1,e_1)\,e_1 + \operatorname{Hess}(u)(e_1,\cdot)^{\#}\right\} \\ &= \frac{1}{|\nabla u|}\left\{ (m-1)\,\mu e_1 - (m-1)\,\mu e_1\right\} = 0. \end{aligned}$$

Step 4. Let Σ_o be a connected component of a level set $\{u = u_o\}$ of u, and let ϕ_t be the flow of the vector field e_1. Then for every $x \in \Sigma_o$, $\phi_t(x)$ is defined for all t's, and $\phi : \mathbb{R} \times \Sigma_o \to M$ is a smooth diffeomorphism which realizes a differentiable splitting $\mathbb{R}\times\Sigma_o \underset{\text{diff}}{\approx} M$. Moreover, if M has more than one end, then Σ_o is compact, and M has exactly two ends.

Indeed, according to the previous step, for every x in M the integral curve $t \to \phi_t(x)$ of e_1 is a geodesic. Since M is complete, the flow ϕ_t is defined for every t (this actually follows directly from the fact that e_1 has bounded length), and for every fixed t gives rise to a global diffeomorphism of M.

In particular, if $x \in \Sigma_o$, then $\phi_t(x)$ coincides with the normal exponential map, namely $\phi_t(x) = \exp^{\perp}(te_1(x))$, and according to Lemma 9.6 the map ϕ : $\mathbb{R}\times\Sigma_o \to M$ is onto. We claim that it is also 1-1 so that it realizes a differentiable splitting $\mathbb{R} \times \Sigma_o \underset{\text{diff}}{\approx} M$. Note that since for fixed t, $\phi_t : M \to M$ is a global diffeomorphism, if $x_1 \neq x_2$, then $\phi_t(x_1) \neq \phi_t(x_2)$. On the other hand, since ∇u never vanishes, u is strictly increasing along the integral curves of e_1. Therefore if $t_1 < t_2$, then $u(\phi_{t_1}(x)) < u(\phi_{t_2}(x))$, and $\phi_{t_1}(x) \neq \phi_{t_2}(x)$. Thus, if ϕ is not $1-1$ on $\mathbb{R} \times \Sigma_o$, there exist $x_1 \neq x_2 \in \Sigma_o$ and $t_1 \neq t_2$ such that $\phi_{t_1}(x_1) = \phi_{t_2}(x_2) = \bar{x}$. But then, assuming, e.g., that $t_1 < t_2$ we have $x_2 = \phi_{-t_2}(\bar{x}) = \phi_{-t_2+t_1}(\phi_{-t_1}(\bar{x})) = \phi_{-t_2+t_1}(x_1)$ and this is impossible since x_1 and x_2 belong to the same level set of u, and u increases along integral lines of ϕ_t.

Note that, since u increases along the integral curve $\phi_t(x)$, the image of a level set of u cannot intersect the same level set. Thus Σ_o is necessarily connected. Since $\Sigma_o = \{u = t_o\}$ was arbitrary, all level sets of u are in fact connected, and $|\nabla u|$ is constant on every such set.

The last assertion follows from Lemma 9.5 and the fact that the flow of ϕ determines the ends of M.

Step 5. The map $\phi : \mathbb{R} \times \Sigma_o \to M$ moves Σ_o onto any other level set of u.

Since $|\nabla u|$ is constant on level sets of u, there exists a function α defined on the interval $u(M)$ such that $|\nabla u| = \alpha(u)$. For fixed $x \in M$ and every $t \in \mathbb{R}$ we have $|\nabla u|(\phi_t(x)) = \alpha(u(\phi_t(x))$, and since the right-hand side is a continuous function of t, while $t \to u(\phi_t(x))$ is a continuous bijection of $\mathbb{R}$ onto its image, we deduce that the function α is continuous. Moreover

$$\begin{aligned}\frac{d}{ds}u \circ \phi_s(x) &= \left\langle \nabla u(\phi_s(x)), \dot{\phi}_s(x) \right\rangle \\ &= |\nabla u(\phi_s(x))| \\ &= \alpha(u \circ \phi_s(x)).\end{aligned}$$

Whence, integrating and using the change of variable formula, we get

$$\int_{u(x)}^{u\circ\phi_t(x)} \frac{dy}{\alpha(y)} = t, \text{ on } \mathbb{R}.$$

In particular, the value $u(\phi_t(x))$ is independent of the point x in any fixed level set of u, showing that the image under ϕ_t of a level set of u is contained in a level set of u. The conclusion now follows from the fact that ϕ is a diffeomorphism of $\mathbb{R} \times \Sigma_o$ onto M.

Step 6. The smooth, positive function

$$\beta(t) = |\nabla u| \circ \phi_t(x) = \psi \circ \phi_t(x) \tag{9.12}$$

is independent of x varying in a given level set of u and satisfies the ODE

$$\beta'' - \frac{m}{m-1}\frac{(\beta')^2}{\beta} + \rho\beta = 0, \text{ on } \mathbb{R}. \tag{9.13}$$

The first assertion follows directly from the fact that ϕ_t moves level sets of u onto level sets of u, and that $|\nabla u|$ is constant on such level sets. As for the second assertion, direct computations that use (9.11) show

$$\begin{aligned}\beta'(t) &= \left\langle (\nabla\psi) \circ \phi_t(x), \frac{d}{dt}\phi_t(x) \right\rangle = \left\langle \nabla|\nabla u| \circ \phi_t(x), e_1 \right\rangle \\ &= \big(\mathrm{Hess}(u)(e_1, e_1)\big) \circ \phi_t(x) = -(m-1)\mu \circ \phi_t(x),\end{aligned} \tag{9.14}$$

and (again by (9.11))

$$|\nabla\psi|^2 \circ \phi_t(x) = (m-1)^2 \mu^2 \circ \phi_t(x),$$

so that, in particular,

$$\beta'(t)^2 = |\nabla\psi|^2 \circ \phi_t(x). \tag{9.15}$$

Moreover, since $t \to \phi_t(x)$ is a geodesic in M with tangent vector e_1,

$$\beta''(t) = \frac{d}{dt}\big(\langle \nabla\psi, e_1 \rangle (\phi_t(x))\big) = \mathrm{Hess}(\psi)(e_1, e_1)(\phi_t(x)). \tag{9.16}$$

To treat the Hessian term on the right-hand side we write

$$\operatorname{Hess}(\psi)(e_1, e_1) = \Delta\psi - \sum_{j=2}^{m} \operatorname{Hess}(\psi)(e_j, e_j) \tag{9.17}$$

and claim that

$$\operatorname{Hess}(\psi)(e_j, e_j) = -\frac{(m-1)\,\mu^2}{\psi}. \tag{9.18}$$

Indeed, using (9.11) and the fact that $\{e_i\}$ is an orthonormal frame, we obtain, for every $j \geq 2$,

$$\begin{aligned} \operatorname{Hess}(\psi)(e_j, e_j) &= \operatorname{Hess}(|\nabla u|)(e_j, e_j) \\ &= \left\langle D_{e_j} \nabla |\nabla u|, e_j \right\rangle \\ &= e_j \left\langle \nabla |\nabla u|, e_j \right\rangle - \left\langle \nabla |\nabla u|, D_{e_j} e_j \right\rangle \\ &= e_j \left(-(m-1)\,\mu\delta_{1j}\right) + (m-1)\,\mu \left\langle e_1, D_{e_j} e_j \right\rangle \\ &= -(m-1)\,\mu \left\langle D_{e_j} e_1, e_j \right\rangle. \end{aligned} \tag{9.19}$$

On the other hand, using again (9.11), a direct computation yields

$$\begin{aligned} D_{e_j} e_1 &= |\nabla u|^{-1} D_{e_j} \nabla u - |\nabla u|^{-2} e_j(|\nabla u|)\nabla u \\ &= |\nabla u|^{-1} \operatorname{Hess}(u)(e_j, \cdot)^{\#} + |\nabla u|^{-2} \operatorname{Hess}(u)(e_j, e_1)\nabla u \\ &= \psi^{-1} \operatorname{Hess}(u)(e_j, \cdot)^{\#}. \end{aligned}$$

Inserting this latter into (9.19) and using (9.9) establishes equation (9.18).

Now, substituting (9.18) into (9.17), inserting the resulting equality into (9.16) and recalling (9.14) and (9.4), we conclude that

$$\begin{aligned} \beta''(t) &= \left(\Delta\psi + (m-1)^2 \frac{\mu^2}{\psi}\right)(\phi_t(x)) \\ &= \left(\Delta\psi + \frac{|\nabla\psi|^2}{\psi}\right)(\phi_t(x)) \\ &= \left(-\rho\psi + \frac{m}{m-1} \frac{|\nabla\psi|^2}{\psi}\right)(\phi_t(x)) \\ &= -\rho\beta + \frac{m}{m-1} \frac{(\beta')^2}{\beta}. \end{aligned} \tag{9.20}$$

Step 7. If Σ_0 is a level set of M and $\phi : \mathbb{R} \times \Sigma_0 \to M$ realizes the differentiable splitting, then

$$\phi^* \langle\, , \rangle = dt \otimes dt + \left(\frac{\beta(t)}{\beta(0)}\right)^{-2/(m-1)} (\, ,). \tag{9.21}$$

Obviously, $\phi_* \frac{\partial}{\partial t} = e_1 \circ \phi$ so that

$$\phi^* \langle\, ,\rangle \left(\frac{\partial}{\partial t}, \frac{\partial}{\partial t}\right) = 1. \tag{9.22}$$

Moreover, for every $V \in T\Sigma_0 = \langle e_1 \rangle^{\perp}$,

$$\left\langle \phi_* \frac{\partial}{\partial t}, \phi_* V \right\rangle = 0. \tag{9.23}$$

Indeed, if σ is any curve in Σ starting with velocity V, by definition of β and the fact that it is independent of the point $x \in \Sigma_0$, we have

$$\left.\frac{d}{ds}\right|_{s=0} \left(|\nabla u| \circ \phi_t\right) \circ \sigma(s) = \left.\frac{d}{ds}\right|_{s=0} \beta(t) = 0. \tag{9.24}$$

On the other hand, using (9.10) and (9.14), we have

$$\begin{aligned} \left.\frac{d}{ds}\right|_{s=0} |\nabla u| \circ (\phi_t \circ \sigma(s)) &= \operatorname{Hess}(u)\left(e_1, (\phi_t)_* V\right) \\ &= -(m-1)\,\mu \circ \phi_t \left\langle e_1, (\phi_t)_* V \right\rangle \\ &= \beta'(t) \left\langle e_1, (\phi_t)_* V \right\rangle. \end{aligned} \tag{9.25}$$

Since $\beta'(t) \neq 0$ by (9.14), combining (9.24) and (9.25) gives (9.23), and proves that $\phi^* \langle , \rangle$ takes the form

$$\phi^* \langle\, ,\rangle = dt \otimes dt + (\phi_t)^* \langle\, ,\rangle.$$

Thus, the desired conclusion will follow once we show that

$$(\phi_t)^* \langle\, ,\rangle = \left(\frac{\beta(t)}{\beta(0)}\right)^{-2/(m-1)} (\, ,).$$

By the definition of Lie derivative we have, for all vectors V, W tangent to Σ_0,

$$\langle D_V e_1, W \rangle + \langle D_W e_1, V \rangle = L_{e_1} \langle , \rangle (V, W) = \left.\frac{d}{ds}\right|_{s=0} \left(\phi_s^* \langle , \rangle\right)(V, W).$$

On the other hand, recalling that $|\nabla u|$ is locally constant on the level sets of u, so that

$$D_V e_1 = D_V \left(\frac{\nabla u}{|\nabla u|}\right) = \frac{1}{|\nabla u|} D_V \nabla u,$$

and using (9.10), we deduce that

$$\langle D_V e_1, W \rangle = \frac{1}{|\nabla u|} \operatorname{Hess}(u)(V, W) = \frac{\mu}{|\nabla u|} \langle V, W \rangle$$

so that

$$\left.\frac{d}{ds}\right|_{s=0}\left(\phi_s^*\langle\,,\rangle\right)=\frac{2\mu}{|\nabla u|}\langle\,,\rangle\,.$$

As a consequence, recalling the definition (9.12) of β, (9.14) and the group property of ϕ_t, we get

$$\begin{aligned}\frac{d}{dt}\left(\phi_t^*\langle\,,\rangle\right)&=\frac{2\mu\circ\phi_t}{|\nabla u|\circ\phi_t}\,\phi_t^*\langle\,,\rangle\\&=\frac{-2\beta'(t)}{m-1}\frac{1}{\beta(t)}\,\phi_t^*\langle\,,\rangle\,.\end{aligned}$$

Integrating over $[0,t]$ we finally obtain

$$\phi_t^*\langle\,,\rangle=\left(\frac{\beta(t)}{\beta(0)}\right)^{-\frac{2}{m-1}}\phi_0^*\langle\,,\rangle$$

as required.

Step 8. In view of (9.13) and by Lemma 9.7, the positive function $\beta(t)=\psi\circ\phi_t(x)$ has the expression

$$\beta(t)=\left\{C_1\exp\left(t\sqrt{\tfrac{\rho}{m-1}}\right)+C_2\exp\left(-t\sqrt{\tfrac{\rho}{m-1}}\right)\right\}^{-(m-1)},$$

for some non-negative constants C_1, C_2 which are independent of $x\in\Sigma_o$, since so is β, and cannot both vanish because $\psi=|\nabla u|>0$. Up to replacing t with $-t$, we may assume that $C_1>0$. Thus

$$w(t)=\left(\frac{\beta(t)}{\beta(0)}\right)^{-\frac{2}{m-1}}=\left\{\frac{C_1\exp\left(t\sqrt{\frac{\rho}{m-1}}\right)+C_2\exp\left(-t\sqrt{\frac{\rho}{m-1}}\right)}{C_1+C_2}\right\}^2,$$

as required to prove (9.5).

Step 9. Assume now that M has two ends. Note that if both the constants C_1 and C_2 are different from zero, then the positive function $\beta(t)$ tends to zero as $t\to\pm\infty$ and β attains its positive maximum B at some $t_0\in\mathbb{R}$. Without loss of generality, up to using the translated flow ϕ_{t+t_0}, we can assume $t_0=0$. Accordingly,

$$\beta(0)=B>0,\quad \beta'(0)=0$$

which implies that $\beta(t)$ has the expression

$$\beta(t)=B\cosh^{-(m-1)}\left(t\sqrt{\tfrac{\rho}{m-1}}\right)$$

and

$$w(t)=\left(\frac{\beta(t)}{\beta(0)}\right)^{-2/(m-1)}=\cosh^2\left(t\sqrt{\tfrac{\rho}{m-1}}\right).$$

On the other hand, if $C_2 = 0$, then

$$w(t) = \exp\left(2t\sqrt{\tfrac{\rho}{m-1}}\right).$$

Since Σ_0 is compact, in the former case both ends of M have infinite volume. In the latter case, $\operatorname{vol}\left((-\infty, 0] \times \Sigma_0\right) < +\infty$, and one of the ends of M has finite volume. □

In view of Theorem 9.3, the strategy of the proof of Theorem 9.1 is to use the assumptions on the geometry of M to produce a non-constant harmonic function satisfying the differential equation (9.4). This requires some preliminary results on the energy of a special class of harmonic functions considered in Section 7.1.

We recall briefly the construction. Let D be a relatively compact domain D with smooth boundary, and fix an exhaustion $\{D_n\}$ of M by relatively compact open domains with smooth boundary and $D \subset\subset D_n \subset\subset D_{n+1}$.

According to Proposition 7.10, if M has at least two non-parabolic ends E_1 and E_2, then the sequence of functions u_n which solve the boundary value problem

$$\begin{cases} \Delta u_n = 0 \text{ in } D_n, \\ u_n = 1 \text{ on } \partial D_n \cap E_1, \quad u_n = 0 \text{ on } \partial D_n \cap \left(M \setminus E_1\right) \end{cases}$$

has a subsequence which converges locally uniformly to a bounded harmonic function u with finite Dirichlet integral such that $0 < u < 1$, $\sup_{E_1} u = 1$, and $\inf_{E_2} u = 0$.

Lemma 9.8. *Maintaining the notation introduced above, let E be an end of M with respect to D, satisfying the Poincaré inequality*

$$0 < \lambda_1 = \lambda_1\left(-\Delta_E\right) = \inf \frac{\int_E |\nabla\varphi|^2}{\int_E \varphi^2} \tag{9.26}$$

where the infimum is taken over $C_c^\infty(E) \setminus \{0\}$. Let u be the harmonic function obtained with the approximation procedure described above, and let $0 < \delta < \sqrt{\lambda_1}$. Then, there exists a constant $C = C(u) > 0$ such that

$$\int_E e^{2\delta r}(u-\alpha)^2 \leq \frac{C}{\left(\sqrt{\lambda_1} - \delta\right)^2}, \quad \textit{where } \alpha = \begin{cases} 1 & \textit{if } E = E_1, \\ 0 & \textit{otherwise.} \end{cases} \tag{9.27}$$

Remark 9.9. We observe that a similar statement, with appropriate values of α, holds for linear combinations of the harmonic functions as considered in the statement, that is, for harmonic functions obtained by the approximation procedure assigning to each u_n a constant boundary value on $E \cap \partial D_n$ (necessarily equal to zero if E is parabolic).

Proof. We consider the case where $E \neq E_1$, so that the approximating sequence u_n satisfies the boundary condition $u_n = 0$ on $E \cap \partial D_n$. The other case is dealt with by replacing u with $1 - u$. Let $o \in D$ and R_0 such that $\bar{D} \subset B_{R_0}(o)$, fix $R \geq 2R_0$, and choose $n \in \mathbb{N}$ sufficiently large that $\bar{B}_R(o) \subset D_n$. We set $E_n = E \cap D_n$, $\partial E_n = E \cap \partial D_n$, $E(R) = E \cap B_R(o)$ and $\partial E(R) = \partial B_R(o) \cap E$. Let also $\varphi : M \to [0,1]$ be a smooth cut-off function such that

$$\varphi = 0 \text{ on } B_{R_0}, \quad \varphi = 1 \text{ off } B_R, \quad |\nabla\varphi| \leq \frac{2}{R - R_0} \text{ on } M.$$

Fix $0 < \delta < \sqrt{\lambda_1}$ and integrate over E_n to obtain

$$\begin{aligned}
\int_{E_n} \left|\nabla\left(\varphi e^{\delta r} u_n\right)\right|^2 &= \int_{E_n} u_n^2 \left|\nabla\left(\varphi e^{\delta r}\right)\right|^2 + 2\int_{E_n} u_n \varphi e^{\delta r} \left\langle \nabla u_n, \nabla\left(\varphi e^{\delta r}\right)\right\rangle \\
&\quad + \int_{E_n} \varphi^2 e^{2\delta r} |\nabla u_n|^2 \\
&= \int_{E_n} u_n^2 \left|\nabla\left(\varphi e^{\delta r}\right)\right|^2 + \int_{E_n} \varphi^2 e^{2\delta r} |\nabla u_n|^2 \\
&\quad + \frac{1}{2}\int_{E_n} \left\langle \nabla\left(u_n^2\right), \nabla\left(\varphi^2 e^{2\delta r}\right)\right\rangle .
\end{aligned}$$

Applying the divergence theorem, noting that u_n vanishes on ∂E_n, while $\phi^2 e^{2\delta r}$ is zero on ∂E, and using the harmonicity of u_n in E_n, we have

$$\begin{aligned}
\frac{1}{2}\int_{E_n} \left\langle \nabla\left(u_n^2\right), \nabla\left(\varphi^2 e^{2\delta r}\right)\right\rangle &= -\frac{1}{2}\int_{E_n} \varphi^2 e^{2\delta r} \Delta\left(u_n^2\right) \\
&= -\int_{E_n} \varphi^2 e^{2\delta r} u_n \Delta u_n - \int_{E_n} \varphi^2 e^{2\delta r} |\nabla u_n|^2 \\
&= -\int_{E_n} \varphi^2 e^{2\delta r} |\nabla u_n|^2 ,
\end{aligned}$$

which, inserted into the above identity, yields

$$\int_{E_n} \left|\nabla\left(\varphi e^{\delta r} u_n\right)\right|^2 = \int_{E_n} u_n^2 \left|\nabla\left(\varphi e^{\delta r}\right)\right|^2 .$$

By Young's inequality, for every $\varepsilon > 0$ the right-hand side is estimated from above by

$$(1+\varepsilon)\,\delta^2 \int_{E_n} u_n^2 \varphi^2 e^{2\delta r} + \left(1+\varepsilon^{-1}\right)\int_{E_n} u_n^2 e^{2\delta r} |\nabla\varphi|^2 ,$$

so, using the Poincaré inequality to estimate the left-hand side from below and rearranging, we get

$$\left(\lambda_1 - (1+\varepsilon)\,\delta^2\right)\int_{E_n} u_n^2 \varphi^2 e^{2\delta r} \leq \left(1+\varepsilon^{-1}\right)\frac{4}{(R-R_0)^2}\int_{E(R)\setminus E(R_0)} u_n^2 e^{2\delta r}.$$

Choosing

$$\varepsilon = \frac{\sqrt{\lambda_1} - \delta}{\delta}$$

we obtain

$$\left(\sqrt{\lambda_1} - \delta\right)^2 \int_{E_n} u_n^2 \varphi^2 e^{2\delta r} \leq \frac{4}{(R - R_0)^2} \int_{E(R)\backslash E(R_0)} u_n^2 e^{2\delta r},$$

whence, recalling that φ is non-negative and identically equal to 1 outside B_R,

$$\left(\sqrt{\lambda_1} - \delta\right)^2 \int_{E_n\backslash E(R)} u_n^2 e^{2\delta r} \leq \frac{4}{(R - R_0)^2} \int_{E(R)\backslash E(R_0)} u_n^2 e^{2\delta r}.$$

Now, we choose $R = 2R_0$ and we let $n \to +\infty$ to get

$$\left(\sqrt{\lambda_1} - \delta\right)^2 \int_{E\backslash E(2R_0)} u^2 e^{2\delta r} \leq \frac{4}{R_0^2} \int_{E(2R_0)\backslash E(R_0)} u^2 e^{2\delta r},$$

and (9.27) follows. □

Lemma 9.10. *Assume that M has at least two non-parabolic ends, and that E is an end of M with respect to D satisfying the Poincaré inequality* (9.26)*. Let u be a harmonic function as in the statement of Lemma* 9.8*. Then there exists a constant $C = C\left(u, \sqrt{\lambda_1}\right) > 0$, such that for every sufficiently large R,*

$$\int_{E_R\backslash E_{R-1}} e^{2\sqrt{\lambda_1} r} (u - \alpha)^2 \leq C, \tag{9.28}$$

where α is as in Lemma 9.8*, namely, $\alpha = 1$ if u is the limit of a sequence of harmonic functions u_n equal to* 1 *on $E \cap \partial D_n$, and $\alpha = 0$ otherwise.*

Proof. Again we consider the case where $\alpha = 0$. Let $\varphi \in Lip_c(E)$. A straightforward computation shows the validity of the equality

$$\begin{aligned}\int_E \left|\nabla\left(e^{\sqrt{\lambda_1} r}\varphi u\right)\right|^2 = \int_E \Big\{& e^{2\sqrt{\lambda_1} r} u^2 \left|\nabla\varphi\right|^2 + \lambda_1 e^{2\sqrt{\lambda_1} r}\varphi^2 u^2 \\ &+ 2\sqrt{\lambda_1} e^{2\sqrt{\lambda_1} r}\varphi u^2 \left\langle\nabla\varphi, \nabla r\right\rangle + e^{2\sqrt{\lambda_1} r}\varphi^2 \left|\nabla u\right|^2 \\ &+ 2e^{2\sqrt{\lambda_1} r}\varphi u \left\langle\nabla\varphi, \nabla u\right\rangle + 2\sqrt{\lambda_1} e^{2\sqrt{\lambda_1} r}\varphi^2 u \left\langle\nabla u, \nabla r\right\rangle\Big\}.\end{aligned}$$

The last three terms under the integral on the right-hand side can be written in the form

$$\left\langle\nabla u, \nabla\left(e^{2\sqrt{\lambda_1}\, r}\varphi^2 u\right)\right\rangle,$$

so that, integrating by parts and using the fact that u is harmonic, they cancel out and the above equality reduces to

$$\int_E \left|\nabla\left(\varphi e^{\sqrt{\lambda_1} r} u\right)\right|^2 = \int_E e^{2\sqrt{\lambda_1} r}\left(u^2 \left|\nabla\varphi\right|^2 + 2\sqrt{\lambda_1}\varphi u^2 \left\langle\nabla\varphi, \nabla r\right\rangle + \lambda_1\varphi^2 u^2\right).$$

We now apply the Poincaré inequality to the left-hand side and rearrange to obtain

$$-2\sqrt{\lambda_1}\int_E e^{2\sqrt{\lambda_1}r}u^2\varphi\left\langle\nabla\varphi,\nabla r\right\rangle\le\int_E e^{2\sqrt{\lambda_1}r}u^2\left|\nabla\varphi\right|^2. \tag{9.29}$$

We let $R_0<R_1<R$ and we choose φ to be given by the formula

$$\varphi\left(r\right)=\begin{cases}\frac{r(x)-R_0}{R_1-R_0} & \text{on } E(R_1)\setminus E(R_0),\\ \frac{R-r(x)}{R-R_1} & \text{on } E(R)\setminus E(R_1),\\ 0 & \text{elsewhere.}\end{cases}$$

Substituting into (9.29) we have

$$\begin{aligned}&\frac{2\sqrt{\lambda_1}}{\left(R-R_1\right)^2}\int_{E(R)\setminus E(R_1)}e^{2\sqrt{\lambda_1}r}\left(R-r\right)u^2 \qquad (9.30)\\ &\le\frac{1}{\left(R-R_1\right)^2}\int_{E(R)\setminus E(R_1)}e^{2\sqrt{\lambda_1}r}u^2+\frac{1}{\left(R_1-R_0\right)^2}\int_{E(R_1)\setminus E(R_0)}e^{2\sqrt{\lambda_1}r}u^2\\ &\quad+\frac{2\sqrt{\lambda_1}}{\left(R_1-R_0\right)^2}\int_{E(R_1)\setminus E(R_0)}e^{2\sqrt{\lambda_1}r}\left(r-R_0\right)u^2.\end{aligned}$$

We fix $0<t<R-R_1$ and we observe that

$$\frac{2t}{\left(R-R_1\right)^2}\int_{E(R-t)\setminus E(R_1)}e^{2\sqrt{\lambda_1}r}u^2\le\frac{2}{\left(R-R_1\right)^2}\int_{E(R-t)\setminus E(R_1)}\left(R-r\right)e^{2\sqrt{\lambda_1}r}u^2.$$

Thus, from (9.30) we get

$$\begin{aligned}\frac{2t\sqrt{\lambda_1}}{\left(R-R_1\right)^2}\int_{E(R-t)\setminus E(R_1)}e^{2\sqrt{\lambda_1}r}u^2&\le\frac{1}{\left(R-R_1\right)^2}\int_{E(R)\setminus E(R_1)}e^{2\sqrt{\lambda_1}r}u^2\\ &\quad+\left[\frac{1}{\left(R_1-R_0\right)^2}+\frac{2\sqrt{\lambda_1}}{R_1-R_0}\right]\int_{E(R_1)\setminus E(R_0)}e^{2\sqrt{\lambda_1}r}u^2. \qquad (9.31)\end{aligned}$$

Let $R>\max\left\{2R_0,R_1+1/\sqrt{\lambda_1}\right\}$. We choose $R_1=R_0+1$, $t=1/\sqrt{\lambda_1}$ and define

$$g\left(R\right)=\int_{E(R)\setminus E(R_0+1)}e^{2\sqrt{\lambda_1}r}u^2.$$

The above inequality gives

$$\frac{2}{\left(R-R_0-1\right)^2}g\left(R-\frac{1}{\sqrt{\lambda_1}}\right)\le\frac{1}{\left(R-R_0-1\right)^2}g\left(R\right)+2A$$

for some $A=A\left(R_0,\sqrt{\lambda_1},u\right)>0$, and therefore

$$g\left(R\right)\le\frac{1}{2}g\left(R+\frac{1}{\sqrt{\lambda_1}}\right)+A\left(R+\frac{1}{\sqrt{\lambda_1}}\right)^2. \tag{9.32}$$

Now we let k be a positive integer and we iterate (9.32) k-times to obtain

$$g(R) \leq \frac{1}{2^k} g\left(R + \frac{k}{\sqrt{\lambda_1}}\right) + AR^2 \sum_{i=1}^{k} \frac{\left(1 + \frac{i}{R\sqrt{\lambda_1}}\right)^2}{2^{i-1}}$$

and therefore

$$g(R) \leq \frac{1}{2^k} g\left(R + \frac{k}{\sqrt{\lambda_1}}\right) + CR^2$$

for some constant $C > 0$ which depends only on R_o, $\sqrt{\lambda_1}$ and u. We now use assumption (9.27) to show that

$$\lim_{k \to +\infty} \frac{1}{2^k} g\left(R + \frac{k}{\sqrt{\lambda_1}}\right) = 0.$$

Indeed, if $0 < \delta < \lambda_1$, applying (9.27) we have

$$\begin{aligned} g\left(R + \frac{k}{\sqrt{\lambda_1}}\right) &= \int_{B_{R+\frac{k}{\sqrt{\lambda_1}}} \setminus B_{R_0+1}} e^{2\sqrt{\lambda_1} r} u^2 \\ &= \int_{B_{R+\frac{k}{\sqrt{\lambda_1}}} \setminus B_{R_0+1}} e^{2\left(\sqrt{\lambda_1} - \delta\right) r} u^2 e^{2\delta r} \\ &\leq C\left(\sqrt{\lambda_1} - \delta\right)^{-2} e^{2\left(R + \frac{k}{\sqrt{\lambda_1}}\right)\left(\sqrt{\lambda_1} - \delta\right)}. \end{aligned}$$

Therefore

$$\frac{1}{2^k} g\left(R + \frac{k}{\sqrt{\lambda_1}}\right) \leq C\left(\sqrt{\lambda_1} - \delta\right)^{-2} e^{2R\left(\sqrt{\lambda_1} - \delta\right)} e^{2\frac{k}{\sqrt{\lambda_1}}\left(\sqrt{\lambda_1} - \delta\right) - k \log 2} \to 0$$

as $k \to +\infty$, provided δ is sufficiently near to $\sqrt{\lambda_1}$. Recalling the definition of $g(R)$ we have thus proved

$$\int_{E(R) \setminus E(R_0+1)} e^{2\sqrt{\lambda_1} r} u^2 \leq \hat{C} R^2, \; R >> 1,$$

for some constant $\hat{C} = \hat{C}\left(u, R_0, \sqrt{\lambda_1}\right) > 0$ and therefore

$$\int_{E(R)} e^{2\sqrt{\lambda_1} r} u^2 \leq CR^2, \;\; R >> 1, \tag{9.33}$$

for some $C > 0$. To improve (9.33) we use again inequality (9.31). For R large enough, we choose $R_1 = R_0 + 1$, $t = R/2$ in (9.31) to obtain

$$R\sqrt{\lambda_1} \int_{E\left(\frac{R}{2}\right) \setminus E(R_0+1)} e^{2\sqrt{\lambda_1} r} u^2 \leq \int_{E(R) \setminus E(R_0+1)} e^{2\sqrt{\lambda_1} r} u^2 + 2AR^2$$

and then applying (9.33)

$$\int_{E(\frac{R}{2})\setminus E(R_0+1)} e^{2\sqrt{\lambda_1}r}u^2 \leq CR$$

or, equivalently,

$$\int_{E(R)} e^{2\sqrt{\lambda_1}r}u^2 \leq CR,\ R >> 1. \tag{9.34}$$

To conclude we set $R_1 = R - 4/\sqrt{\lambda_1}$, $t = 2/\sqrt{\lambda_1}$ in (9.31), and deduce that for sufficiently large R,

$$\begin{aligned}\int_{E(R-\frac{2}{\sqrt{\lambda_1}})\setminus E(R-\frac{4}{\sqrt{\lambda_1}})} & e^{2\sqrt{\lambda_1}r}u^2 \\ & \leq \frac{1}{4}\int_{E(R)\setminus E(R-\frac{4}{\sqrt{\lambda_1}})} e^{2\sqrt{\lambda_1}r}u^2 + \frac{C}{R}\int_{E(R-\frac{2}{\sqrt{\lambda_1}})\setminus E(R_0)} e^{2\sqrt{\lambda_1}r}u^2\end{aligned}$$

for some constant $C = C\left(u, R_0, \sqrt{\lambda_1}\right) > 0$. Thus, using (9.34),

$$\int_{E(R-\frac{2}{\sqrt{\lambda_1}})\setminus E(R-\frac{4}{\sqrt{\lambda_1}})} e^{2\sqrt{\lambda_1}r}u^2 \leq \frac{1}{3}\int_{E(R)\setminus E(R-\frac{2}{\sqrt{\lambda_1}})} e^{2\sqrt{\lambda_1}r}u^2 + C.$$

We iterate this inequality k-times to obtain, with the aid of (9.34),

$$\begin{aligned}\int_{E(R+\frac{2}{\sqrt{\lambda_1}})\setminus E(R)} e^{2\sqrt{\lambda_1}r}u^2 & \leq C\sum_{i=1}^{k}\frac{1}{3^{i-1}} + \frac{1}{3^k}\int_{E(R+\frac{2k}{\sqrt{\lambda_1}})\setminus E(R+\frac{2(k-1)}{\sqrt{\lambda_1}})} e^{2\sqrt{\lambda_1}r}u^2 \\ & \leq C + C\frac{1}{3^k}\left(R + \frac{2k}{\sqrt{\lambda_1}}\right),\end{aligned}$$

whence, letting $k \to +\infty$ we deduce that the integral on the left-hand side is bounded above by C. The required conclusion now follows at once. □

Lemma 9.11. *Let $(M, \langle , \rangle)$, u and E be as in the previous lemmas. Then, there exists a constant $C > 0$ independent of R such that, for R sufficiently large,*

$$\int_{E(R)} e^{2\sqrt{\lambda_1}r}\left|\nabla u\right|^2 \leq CR. \tag{9.35}$$

In particular $\left|\nabla u\right|^2 \in L^1(E)$.

Proof. Let φ be the cut-off function

$$\varphi(x) = \begin{cases} r(x) - R + 1 & \text{on } E(R)\setminus E(R-1), \\ 1 & \text{on } E(R+1)\setminus E(R), \\ R - 2 - r(x) & \text{on } E(R+2)\setminus E(R+1), \\ 0 & \text{elsewhere.}\end{cases}$$

Consider the vector field

$$Z = \varphi^2 u \nabla u.$$

Using the divergence theorem, the fact that u is harmonic and the Cauchy–Schwarz inequality, we have

$$\begin{aligned}\int_E \varphi^2 |\nabla u|^2 &= -2\int_E \varphi u \langle \nabla\varphi, \nabla u\rangle - \int_E \varphi^2 u \Delta u \\ &\le \frac{1}{2}\int_E \varphi^2 |\nabla u|^2 + 2\int_E |\nabla\varphi|^2 u^2.\end{aligned}$$

Thus, using (9.28) and the definition of φ,

$$\int_{E(R+1)\setminus E(R)} |\nabla u|^2 \le 4\int_{E(R+2)\setminus E(R-1)} u^2 \le Ce^{-2\sqrt{\lambda_1}(R-1)}.$$

It follows that

$$\int_{E(R+1)\setminus E(R)} e^{2\sqrt{\lambda_1}r} |\nabla u|^2 \le e^{2\sqrt{\lambda_1}(R+1)} \int_{E(R+1)\setminus E(R)} |\nabla u|^2 \le C.$$

We set $R = R_0 + i$ and we sum over $1 \le i \le k$ to get

$$\int_{E(R_0+k)\setminus E(R_0+1)} e^{2\sqrt{\lambda_1}r} |\nabla u|^2 \le Ck \le C\left(R_0 + k\right).$$

We have thus proved (9.35). The second assertion follows writing

$$\begin{aligned}\int_{E\setminus E(R_0)} |\nabla u|^2 &\le \sum_k \int_{E(R_0+k+1)\setminus E(R_0+k)} e^{-2\sqrt{\lambda_1}(R_0+k)} \int_{\partial B_r} e^{2\sqrt{\lambda_1}r}|\nabla u|^2 \\ &\le C\sum_k (R_0+k+1)e^{-2\sqrt{\lambda_1}(R_0+k)} < +\infty.\end{aligned}$$

□

We are now ready to give the

Proof of Theorem 9.1. Assume that M has at least two non-parabolic ends, and that $\lambda_1 = \lambda_1(-\Delta_M) > 0$. Then, there exists a non-constant harmonic function u on M which is obtained as the limit of a sequence $\{u_n\}$ as in the assumptions of Lemma 9.8. Having set $\psi = |\nabla u|$, and applying Lemma 9.11 to each end E of M with respect to D (note that $\lambda_1(-\Delta_E) \ge \lambda_1 > 0$ by domain monotonicity), we have the energy estimate

$$\int_{B_R} e^{2\sqrt{\lambda_1}r}\psi^2 \le CR \tag{9.36}$$

with R sufficiently large. From Bochner's formula and assumption (9.1) on the Ricci curvature of M, we obtain

$$\psi\Delta\psi + \frac{m-1}{m-2}\lambda_1\psi^2 \ge \frac{1}{m-1}|\nabla\psi|^2. \tag{9.37}$$

Now, the operator

$$L = -\Delta - H\frac{m-1}{m-2}\lambda_1 \qquad \left(H = \frac{m-2}{m-1}\right)$$

clearly satisfies $\lambda_1(L_M) = 0$. Furthermore, using Hölder inequality and (9.36),

$$\begin{aligned}\int_{B_R} \psi^{2\frac{m-2}{m-1}} &= \int_{B_R} \left(e^{2\sqrt{\lambda_1}r}\psi^2\right)^{\frac{m-2}{m-1}} e^{-2\frac{m-2}{m-1}\sqrt{\lambda_1}r} \\ &\leq CR^{\frac{m-2}{m-1}} \left(\int_{B_R} e^{-2(m-2)\sqrt{\lambda_1}r}\right)^{\frac{1}{m-1}}.\end{aligned}$$

By the co-area formula, (9.1) and the Bishop comparison theorem we have

$$\int_{B_R} e^{-2(m-2)\sqrt{\lambda_1}r} = \int_0^R \int_{\partial B_t} e^{-2(m-2)\sqrt{\lambda_1}r} dt \leq C \int_0^R e^{2\sqrt{\lambda_1}t\left(\frac{m-1}{2\sqrt{m-2}}-m+2\right)} dt.$$

Thus

$$\int_{B_R} \psi^{2\frac{m-2}{m-1}} \leq C \begin{cases} R & \text{for } m=3, \\ R^{\frac{m-2}{m-1}} & \text{for } m \geq 4. \end{cases}$$

It both cases it follows that

$$\frac{r}{\int_{B_r} \psi^{2\frac{m-2}{m-1}}} \notin L^1(+\infty)$$

and therefore

$$\frac{1}{\int_{\partial B_r} \psi^{2\frac{m-2}{m-1}}} \notin L^1(+\infty).$$

Since $\psi \not\equiv 0$, applying Theorem 4.5 with

$$\beta = A = H - 1 = -\frac{1}{m-1},$$

we deduce that equality holds in (9.37). The result now follows from Theorem 9.3, noting that, since $\lambda_1 > 0$, the infinite volume ends are precisely the non-parabolic ends. □

9.2 Gap theorems

We have seen in Chapter 4 that vanishing theorems may be obtained by imposing spectral conditions, namely, letting ${}^H L$ be the Schrödinger operator $-\Delta - Ha(x)$, then $\lambda_1({}^H L_M) \geq 0$ implies that non-negative L^p solutions of the differential inequality

$$\psi\Delta\psi + a(x)\psi^2 + A|\nabla\psi|^2 \geq 0$$

vanish identically provided the coefficients A, a and H satisfy suitable conditions. On the other hand, it was shown in Section 7.1 that the validity of Sobolev-type inequalities can be used to obtain spectral information on the operator ${}^{H}L$. In particular, if an appropriate integral norm of (the positive part of) the potential $Ha(x)$ is less than or equal to the L^2-Sobolev constant, then $\lambda_1({}^{H}L_M) \geq 0$ (see Lemma 7.33).

In this section we show that a direct use of a Poincaré–Sobolev inequality can be used to obtain similar gap theorems, requiring that the integral norm of $Ha(x)$ be smaller than a suitable multiple(> 1) of the Poincaré–Sobolev constant.

We are going to apply this result to the investigation, already considered in Section 7.4, of the topology at infinity of immersed submanifold, as well as to other geometric situations like characterization of space forms and gap theorems for harmonic maps.

Theorem 9.12. *Let $(M, \langle\, , \rangle)$ be a complete manifold and assume that, for some $0 \leq \alpha < 1$ and some non-negative function h, the inhomogeneous Sobolev–Poincaré-type inequality*

$$\int_M \left(|\nabla\varphi|^2 + h\varphi^2\right) \geq S(\alpha)^{-1} \left\{\int_M |\varphi|^{\frac{2}{1-\alpha}}\right\}^{1-\alpha} \tag{9.38}$$

holds for every $\varphi \in C_c^\infty(M)$ with a positive constant $S(\alpha) > 0$. Suppose that $\psi \in \mathrm{Lip}_{loc}(M)$ is a positive solution of

$$\psi\Delta\psi + a(x)\,\psi^2 + A\,|\nabla\psi|^2 \geq 0 \qquad \textit{(weakly) on } M \tag{9.39}$$

satisfying

$$\int_{B_r} |\psi|^\sigma = o\left(r^2\right) \quad \textit{as } r \to +\infty \tag{9.40}$$

with $A \in \mathbb{R}$, $\sigma - A - 1 > 0$, $\sigma \neq 0$, and $a(x) \in C^0(M)$. Then

$$\left\| a_+(x) + \frac{4(\sigma - A - 1)}{\sigma^2} h \right\|_{L^{\frac{1}{\alpha}}(M)} \geq \frac{4(\sigma - A - 1)}{\sigma^2} S(\alpha)^{-1}. \tag{9.41}$$

Furthermore, if ψ is assumed to be non-negative and not identically zero, then (9.41) holds under the further assumption that $\sigma > 0$.

We remark for future use that in the case where $\alpha = 0$, then the constant $S(0)^{-1}$ coincides with the bottom of the L^2-spectrum $\lambda_1(L_M)$ of the Schrödinger operator $L = -\Delta + h$.

We also remark that in the case where $h = 0$, that is when a standard Sobolev inequality is assumed to hold, then the theorem states that there are no non-zero solutions of (9.39) satisfying (9.40) if the $L^{1/\alpha}$-norm of the coefficient a is smaller than a multiple, depending on the integrability exponent σ, of the Sobolev constant. Thus the result compares directly with Theorem 4.5, replacing the spectral assumption with the validity of a Sobolev–Poincaré inequality and a norm estimate on the potential.

Proof. Note that the conclusion certainly holds if either a_+ or h are not in $L^{1/\alpha}(M)$, so we assume that $a_+ + h \in L^{1/\alpha}$.

We consider the case where ψ is only assumed to be non-negative, and $0 < \sigma < 2$, the other cases being easier. Next, recall that, by (9.39), for every test function $0 \leq \rho \in H^1_c$ we have

$$\int a_+\psi^2\rho \geq \int \psi\langle\nabla\psi, \nabla\rho\rangle + (1-A)\rho|\nabla\psi|^2.$$

Let $\phi = \phi_r \in C^\infty_c(M)$ be a family of cut-off functions satisfying

$$\phi_r \equiv 1 \text{ on } B_r, \qquad \phi_r \equiv 0 \text{ off } B_{2r}, \qquad |\nabla\phi_r| \leq \frac{4}{r} \text{ on } M,$$

and apply the above inequality to the test function $(\psi^2+\eta)^{\frac{\sigma-2}{2}}\phi^2$ to obtain, after some manipulations,

$$\begin{aligned}\int a_+\phi^2(\psi^2+\eta)^{\frac{\sigma}{2}}\frac{\psi^2}{\psi^2+\eta} &\geq 2\int(\psi^2+\eta)^{\frac{\sigma-2}{2}}\psi\phi\langle\nabla\phi,\nabla\psi\rangle \\ &+ \int(\psi^2+\eta)^{\frac{\sigma-2}{2}}\phi^2|\nabla\psi|^2\Big\{1-A+(\sigma-2)\frac{\psi^2}{\psi^2+\eta}\Big\}. \end{aligned} \tag{9.42}$$

We use the fact that

$$0 \leq \frac{\psi^2}{\psi^2+\eta} \leq 1 \tag{9.43}$$

and that $\sigma - 2 < 0$, to estimate the LHS from above, and the second integral on the RHS from below. Also, by Young's inequality and (9.43), for every $\epsilon > 0$, the first integral on the RHS is estimated from below by

$$-\frac{1}{\epsilon}\int(\psi^2+\eta)^{\frac{\sigma}{2}}|\nabla\phi|^2 - \epsilon\int(\psi^2+\eta)^{\frac{\sigma-2}{2}}.$$

Inserting the resulting inequalities and rearranging, we conclude that

$$\begin{aligned}\int a_+\phi^2(\psi^2+\eta)^{\frac{\sigma}{2}} &+ \frac{1}{\epsilon}\int(\psi^2+\eta)^{\frac{\sigma}{2}}|\nabla\phi|^2 \\ &\geq (\sigma-A-1-\epsilon)\int\phi^2(\psi^2+\eta)^{\frac{\sigma-2}{2}}|\nabla\psi|^2. \end{aligned} \tag{9.44}$$

Fix $\epsilon > 0$ small enough that $\sigma - A - 1 - \epsilon > 0$. As $\eta \searrow 0$, by dominated convergence, the LHS converges to

$$\int a_+\phi^2\psi^\sigma + \frac{1}{\epsilon}\int\psi^\sigma|\nabla\phi|^2,$$

while, since $\sigma - 2 < 0$, by monotone convergence, the RHS converges to

$$(\sigma-A-1-\epsilon)\int\phi^2\psi^{\sigma-2}|\nabla\psi|^2.$$

We may therefore conclude that

$$\int a_+(x)\,\phi^2\psi^\sigma + \frac{1}{\varepsilon}\int \psi^\sigma\,|\nabla\phi|^2 \geq (\sigma - A - 1 - \varepsilon)\int \varphi^2\psi^{\sigma-2}\,|\nabla\psi|^2. \qquad (9.45)$$

According to Lemma 4.13, the function $\psi^{\sigma/2} \in W^{1,2}_{loc}$ and $\nabla\big(\psi^{\sigma/2}\big) = \frac{\sigma}{2}\psi^{(\frac{\sigma}{2}-1)}\nabla\psi$. By the Poincaré–Sobolev inequality, Young inequality and (9.45), we estimate

$$\begin{aligned} S(\alpha)^{-1}\Big\{\int \big(\psi^{\frac{\sigma}{2}}\phi\big)^{\frac{2}{1-\alpha}}\Big\}^{1-\alpha} &\leq \int \big(|\nabla\big(\psi^{\frac{\sigma}{2}}\phi\big)|^2 + h\phi^2\psi^\sigma\big) \\ &\leq (1+\delta)\frac{\sigma^2}{4}\int \psi^{\sigma-2}\phi^2|\nabla\psi|^2 + (1+\frac{1}{\delta})\int \psi^\sigma|\nabla\phi|^2 + \int h\phi^2\psi^\sigma \\ &\leq C^{-1}_{\delta,\epsilon}\int \big(a_+ + C_{\delta,\epsilon}h\big)\phi^2\psi^\sigma + \Big(\frac{1}{\epsilon C_{\delta,\epsilon}} + 1 + \frac{1}{\delta}\Big)\int \psi^\sigma|\nabla\phi|^2, \end{aligned}$$

where

$$C_{\delta,\epsilon} = \frac{4}{\sigma^2}\frac{\sigma - A - 1 - \epsilon}{1+\delta}.$$

Using Hölder's inequality in the first integral,

$$\int \big(a_+ + C_{\delta,\epsilon}h\big)\phi^2\psi^\sigma \leq ||a_+ + C_{\delta,\epsilon}h||_{L^{1/\alpha}}\Big\{\int \big(\psi^{\frac{\sigma}{2}}\phi\big)^{\frac{2}{1-\alpha}}\Big\}^{1-\alpha},$$

and rearranging, we finally obtain

$$\begin{aligned} \Big\{S(\alpha)^{-1} - C^{-1}_{\delta,\epsilon}||a_+ + C_{\delta,\epsilon}h||_{L^{1/\alpha}}\Big\}\Big(\int \big(\psi^{\frac{\sigma}{2}}\phi\big)^{\frac{2}{1-\alpha}}\Big\}^{1-\alpha} \\ \leq \Big\{\frac{1}{\epsilon C_{\delta,\epsilon}} + 1 + \frac{1}{\delta}\Big\}\frac{C}{R^2}\int_{B_{2r}} \psi^\sigma. \end{aligned}$$

To conclude, assume that (9.41) does not hold. Since, by dominated convergence,

$$C^{-1}_{\delta,\epsilon}||a_+ + C_{\delta,\epsilon}h||_{L^{1/\alpha}} \to \frac{\sigma^2}{4(\sigma - A - 1)}\left\|a_+(x) + \frac{4(\sigma - A - 1)}{\sigma^2}h\right\|_{L^{\frac{1}{\alpha}}(M)},$$

as $\epsilon, \delta \to 0+$, we may choose δ and ϵ small enough that the coefficient on the LHS is positive. Since ψ does not vanish identically, there exists R such that, for every $r > R$ the integral is strictly positive. The required contradiction follows by noting that, according to (9.40), the right-hand side tends to zero as $r \to +\infty$. □

Theorem 9.12 allows us to obtain a quantitative improvement on the results obtained in Section 7.4

Recall that if $(M, \langle\,,\,\rangle)$ is isometrically immersed into a Cartan–Hadamard manifold N, the following L^2-type Sobolev inequality holds:

$$S_2(m,\varepsilon)^{-1}\Big(\int_M |v|^{\frac{2m}{m-2}}\Big)^{\frac{m-2}{m}} \leq \int_M |\nabla v|^2 + \frac{\varepsilon^2(m-2)^2}{4(m-1)^2}\int_M |H|^2 v^2, \qquad (9.46)$$

where

$$S(m,\varepsilon)=\left(\frac{2\sqrt{1+\varepsilon^{-2}}\,(m-1)}{(m-2)\,S_1(m)^{-1}}\right)^2, \tag{9.47}$$

and H is the mean curvature vector field of the immersion. We also recall that if the immersion is minimal, then the best constant in (9.47) is achieved by choosing $\varepsilon=+\infty$, and in this situation, we write $S_2(m)=S_2(m,+\infty)$.

Applying Theorem 9.12 instead of Theorem 4.5, we obtain the following improvements of Theorem 7.36 and Theorem 7.35.

Theorem 9.13. *Let $f:(M^m,\langle\,,\rangle)\to\mathbb{R}^n$ be a complete, minimal, immersed submanifold of dimension $m\geq 3$ whose second fundamental tensor* II *satisfies*

$$\left(\int_M |\mathrm{II}|^m\right)^{\frac{2}{m}} < 4S_2(m)^{-1}\,\frac{\frac{2}{(m+2)(n-m)-2}+m-1}{\left(2-\frac{1}{n-m}\right)m^2}. \tag{9.48}$$

Then f is totally geodesic.

Theorem 9.14. *Let $f:(M^m,\langle\,,\rangle)\to(N^n,(\,,))$ be an isometric immersion of the complete manifold M of dimension $m\geq 3$ into the Cartan–Hadamard manifold N whose sectional curvature (along f) satisfies*

$$(0\geq)\ {}^N\mathrm{Sect}_{f(x)}\geq -\ {}^N R(x) \tag{9.49}$$

for some function ${}^N R\in C^0(M)$. Denote by H and II respectively the mean curvature vector field and the second fundamental tensor of f. Assume that, for some $\varepsilon>0$,

$$\left\|\frac{m(m-2)^2\varepsilon^2}{4(m-1)^3}H^2+(m-1)\ {}^N R(x)+|\mathrm{II}|\left(|\mathrm{II}|+m\,|H|\right)(x)\right\|_{L^{\frac{m}{2}}} < \frac{m}{m-1}S_2(m,\varepsilon)^{-1}$$

Then M has only one end.

Remark 9.15. For the sake of comparison, recall that Theorem 7.36 yields the same conclusion under the stronger assumption that

$$\left\|\frac{(m-2)^2\varepsilon^2}{(m-1)^2}|H|^2+(m-1)\ {}^N R(x)+|\mathrm{II}|\left(|\mathrm{II}|+m\,|H|\right)(x)\right\|_{L^{\frac{m}{2}}} \leq S_2(m,\varepsilon)^{-1}. \tag{9.50}$$

Observe that the proof of Theorem 7.36 relies on a vanishing result, Theorem 4.5, which depends on the assumption that the bottom of the spectrum of a suitable Schrödinger operator is non-negative. This in turn is obtained by combining the integral bound (9.50) and the Sobolev inequality. By contrast, the argument in Theorem 9.12 uses the Sobolev inequality in a more direct way, and allows us to improve the constant.

Our next result is a further application of Theorem 9.12, and it implies a vanishing result for harmonic forms in L^2.

Theorem 9.16. *Let $(M, \langle\, ,\, \rangle)$ be a complete Riemannian manifold supporting the Poincaré–Sobolev inequality* (9.38) *for some $0 \leq \alpha < 1$, and assume that*

$$\mathrm{Ric} \geq -\rho(x)$$

for some continuous function ρ satisfying

$$||\rho_+ + \frac{4}{p^2}\frac{(m-1)(p-1)+1}{m-1}h||_{L^{1/\alpha}} < \frac{4}{p^2}\frac{(m-1)(p-1)+1}{m-1}S(\alpha)^{-1}, \tag{9.51}$$

for some $p > (m-2)/(m-1)$. If ω is a closed and co-closed 1-form satisfying

$$\int_{B_r} |\omega|^p = o\big(r^2\big),$$

then $\omega \equiv 0$.

The proof follows by noting that the norm of a closed and co-closed 1-form satisfies the differential inequality

$$\psi\Delta\psi + \rho\,(x)\,\psi^2 \geq \frac{1}{(m-1)}\,|\nabla\psi|^2$$

and applying Thoerem 9.12. Noting that a harmonic L^2 form is automatically closed and co-closed, we recover the vanishing results of P.Li and J. Wang, [103], Theorem 4.2, for manifolds with a positive spectral gap. In the case of L^p harmonic 1-forms, which are not necessarily closed and co-closed, and therefore do not satisfy a refined Kato inequality, one has a vanishing result provided the right-hand side of (9.51) is replaced by $4(p-1)S(\alpha)^{-1}/p^2$.

Similar results can be given for L^p harmonic forms of any degree, or for harmonic maps with L^p energy density, provided one uses the appropriate Weitzenböck formula. For instance, we have the following

Theorem 9.17. *Let $(M, \langle\, ,\, \rangle)$ be an m-dimensional complete Riemannian manifold, supporting the Poincaré–Sobolev inequality* (9.38) *for some $0 \leq \alpha < 1$, and assume that*

$$\mathit{Ricci} \geq -\rho(x).$$

Let $f : M \to N$ be a harmonic map into an n-dimensional non-positively curved Riemannian manifold N. If $|df| \in L^p(M)$, for some $p > \frac{m-2}{m-1}$ and

$$||\rho_+ + \frac{4}{p^2}\frac{(m-1)(p-1)+1}{(m-1)}h||_{L^{1/\alpha}} < \frac{4}{p^2}\frac{(m-1)(p-1)+1}{m-1}S(\alpha)^{-1},$$

then f is constant.

9.3 Gap Theorems, continued

In this section we will employ Theorem 9.12 to obtain isolation phenomena for the Ricci tensor of a scalar flat, conformally flat manifold.

A Riemannian manifold $(M, \langle\,,\,\rangle)$ of dimension m is said to be locally conformally flat if a neighborhood of each point of M can be conformally immersed into the standard sphere $\mathbb{S}^m$. When $m \geq 4$ this is equivalent to the fact that the Weyl tensor vanishes identically. The category of locally conformally flat spaces contains the manifolds of constant sectional curvature, hence, in particular, the space-forms $\mathbb{R}^m$, $\mathbb{H}^m_{-k^2}$, $\mathbb{S}^m_{k^2}$, where the subscripts refer to the constant sectional curvature of the space. Note that, according to the orthogonal decomposition of the Riemann tensor into its irreducible components, a conformally flat manifold of dimension $m \geq 3$ has constant sectional curvature if and only if it is Einstein, i.e., the traceless part of its Ricci tensor is identically equal to zero. As a consequence, by the H. Hopf classification theorem, the space forms are (up to isometries) the only complete, simply connected, locally conformally flat, Einstein manifolds. In this section we investigate other possible characterizations of space forms from the conformally-flat viewpoint.

Let $(M, \langle\,,\rangle)$ be a conformally flat manifold of dimension $m \geq 3$ with constant scalar curvature S. We carry out the computations that follow assuming that $m \geq 4$, but note that the conclusions we obtain also hold when $m = 3$. Conformal flatness and the decomposition of the Riemann tensor into its irreducible components yield

$$R_{ijkl} = \frac{1}{m-2}\left(R_{ik}\delta_{jl} - R_{il}\delta_{jk} + R_{jl}\delta_{ik} - R_{jk}\delta_{il}\right) - \frac{S}{(m-1)(m-2)}\left(\delta_{ik}\delta_{jl} - \delta_{il}\delta_{jk}\right), \tag{9.52}$$

where we have denoted with R_{ij} the components of the Ricci tensor. Taking covariant derivatives, tracing, and using the fact that S is constant give

$$\sum_i R_{ijkl,i} = \frac{1}{m-2}\Big(\sum_i R_{ik,i}\delta_{jl} - \sum_i R_{il,i}\delta_{jk} + R_{jl,k} - R_{jk,l}\Big). \tag{9.53}$$

Now, note that tracing with respect to the indices i, m in the second Bianchi identities

$$R_{ijkl,m} + R_{ijmk,l} + R_{ijlm,k} = 0$$

yields

$$\sum_i R_{ijkl,i} = -\sum_i R_{ijik,l} - \sum_i R_{ijli,k} = -R_{jk,l} + R_{jl,k}, \tag{9.54}$$

and tracing again with respect to j, l and using the fact that S is constant we deduce that the Ricci tensor satisfies

$$R_{ik,k} = 0.$$

Inserting this and (9.54) in (9.53) we conclude that

$$\frac{m-3}{m-2}(R_{jl,k} - R_{jk,l}) = 0,$$

showing that Ric is a Codazzi tensor, namely, the following commutation relations for the coefficients of the covariant derivative of Ric hold,

$$R_{ij,k} = R_{ik,j}. \tag{9.55}$$

As a consequence, the traceless part

$$T = \mathrm{Ric} - \frac{S}{m}\langle\, , \rangle$$

is again Codazzi so that, as first observed by J.P. Bourguignon, [18], we have the validity of the refined Kato-type inequality

$$|DT|^2 \geq \frac{m+2}{m} \, |\nabla |T||^2 . \tag{9.56}$$

Further covariant differentiation of Ric yields the commutation formulas

$$R_{ij,kl} - R_{ij,lk} = R_{it}R_{tjkl} + R_{tj}R_{tikl}. \tag{9.57}$$

Using (9.52), (9.55), (9.57), and the fact that S is constant, we compute

$$\begin{aligned} \frac{1}{2}\Delta |Ric|^2 = |DRic|^2 &+ \frac{m}{m-2} tr\left(\mathrm{ric}^{(3)}\right) \\ &- \frac{(2m-1)\,S}{(m-2)\,(m-1)}\,|\mathrm{Ric}|^2 + \frac{S^3}{(m-1)\,(m-2)} \end{aligned} \tag{9.58}$$

where $\mathrm{ric}^{(3)}$ is the third composition power of the Ricci endomorphism. Using the identity

$$|T|^2 = |\mathrm{Ric}|^2 - \frac{S^2}{m},$$

and expressing $tr\left(\mathrm{ric}^{(3)}\right)$ in terms of T we obtain, with the obvious meaning of the symbols,

$$\frac{1}{2}\Delta |T|^2 = |DT|^2 + \frac{m}{m-2} tr\left(T^{(3)}\right) + \frac{S}{m-1}\,|T|^2 .$$

A simple algebraic lemma due to M. Okumura, [122], shows that

$$tr\left(T^{(3)}\right) \geq -\frac{m-2}{\sqrt{m\,(m-1)}}\,|T|^3 . \tag{9.59}$$

Inserting (9.59) into (9.58) gives

$$\frac{1}{2}\Delta |T|^2 \geq |DT|^2 - \frac{m}{\sqrt{m(m-1)}} |T|^3 + \frac{S}{m-1} |T|^2 .$$

Setting $u = |T|$ and using (9.56), we rewrite the above inequality in the form

$$u\Delta u + \left(\sqrt{\frac{m}{m-1}} u - \frac{S}{m-1}\right) u^2 \geq \frac{2}{m} |\nabla u|^2, \tag{9.60}$$

pointwise on $\{x \in M : |T|(x) \neq 0\}$ and weakly on all of M.

In the late 1960s M. Tani, [157], showed that the universal cover of a compact, orientable, m-dimensional, locally conformally flat Riemannian manifold $(M, \langle\, , \rangle)$ with positive Ricci curvature and constant scalar curvature S is isometrically a sphere. This result has been generalized by S.I. Goldberg, [61], in the complete (non-necessarily orientable) case under the additional assumption that, for some $\varepsilon > 0$,

$$\frac{S^2}{m-1} - |\mathrm{Ric}|^2 \geq \varepsilon > 0, \qquad \text{on } M \tag{9.61}$$

(see also [79]). In fact, combining a classification theorem by S. Zhu, [170], with a PDEs global symmetry result due to L. Caffarelli, B. Gidas and J. Spruck, [22], we prove that the above characterization holds by merely assuming that the left-hand side of (9.61) is strictly positive at one point.

Theorem 9.18. *Let $(M, \langle\, , \rangle)$ be a complete, locally conformally flat Riemannian manifold of dimension $m \geq 3$ and with constant scalar curvature $S > 0$. If*

$$\frac{S^2}{m-1} - |\mathrm{Ric}|^2 \geq 0, \qquad \textit{on } M \tag{9.62}$$

and the strict inequality holds at some point, then the universal cover of $(M, \langle\, , \rangle)$ is isometrically a sphere.

Proof. We note that, by a lemma of Okumura, [123], inequality (9.62) implies that $\mathrm{Ric} \geq 0$ on M. Therefore, according to [170], the universal cover $\tilde{M}$ of M is either isometric to $\mathbb{R} \times \mathbb{S}^{m-1}_{S/(m-1)(m-2)}$ or conformally equivalent to $\mathbb{R}^m$ or $\mathbb{S}^m_1$. Since, by assumption, inequality (9.62) is strict somewhere, the first case is excluded. On the other hand, $\tilde{M}$ cannot be conformally diffeomorphic to $\mathbb{R}^m$. In fact, from Theorem 8.1 in [22] we know that a Riemannian metric on $\mathbb{R}^m$ which is of constant, positive scalar curvature and conformally related to the canonical metric must be a spherical metric, hence incomplete. It follows that $\tilde{M}$ is conformally a sphere, hence M is compact. The conclusion now follows from (the easy case of) Goldberg's argument. Namely, keeping the notation introduced above, we have only to show that $u \equiv 0$, i.e., that M is Einstein. From (9.60) we obtain

$$\Delta u^2 \geq b(x) u^2 \qquad \text{on } M \tag{9.63}$$

with

$$b(x) = \frac{2}{m-1}\left(S - \sqrt{m(m-1)}u(x)\right).$$

On the other hand, according to (9.62) and the fact that the strict inequality holds somewhere, we have

$$0 \not\equiv b(x) \geq C\left\{S^2 - (m-1)\left|\mathrm{Ric}\right|^2(x)\right\} \geq 0,$$

for some absolute constant $C > 0$. Since M is compact, applying the usual maximum principle we conclude that $b(x)$ is a positive constant and $u \equiv 0$, as desired. □

Analogous characterizations hold in the case where the scalar curvature is non-positive.

Our first result in this direction deals with Euclidean space and can be thought of as an extension of Theorem 9.18 to the scalar flat case.

Theorem 9.19. *Let $(M, \langle\, ,\rangle)$ be a complete, simply connected, locally conformally flat Riemannian manifold of dimension $m \geq 3$ and zero scalar curvature. Assume that*

$$\left\|\mathrm{Ric}\right\|_{L^{\frac{m}{2}}(M)} < \frac{2\omega_m^{\frac{2}{m}}(m-2)^3\sqrt{m-1}}{m^2\sqrt{m}} \tag{9.64}$$

where ω_m denotes the volume of the standard sphere $\mathbb{S}_1^m$. Then $(M, \langle\, ,\rangle)$ is isometric to Euclidean space.

Proof. Maintaining the notation introduced above, we observe that in the present setting $u = |\mathrm{Ric}|$ while (9.60) becomes

$$u\Delta u + a(x)u^2 \geq \frac{2}{m}\left|\nabla u\right|^2 \tag{9.65}$$

with

$$a(x) = \sqrt{\frac{m}{m-1}}u(x).$$

Again, we have to show that $u \equiv 0$. To this end we reason by contradiction. Since $(M, \langle\, ,\rangle)$ is simply connected and locally conformally flat, by a result of N. Kuiper, [90], there is a (global) conformal immersion (in fact an embedding) of M into the standard sphere $\mathbb{S}_1^m$. It follows by a result of R. Schoen and S.T. Yau, [147], that the Yamabe invariant $Q(M)$ of M defined by

$$Q(M) = \inf_{\phi \in C_c^\infty(M)\setminus\{0\}} \frac{\int\left(|\nabla\varphi|^2 + \frac{m-2}{4(m-1)}S\phi^2\right)}{\left(\int|\phi|^{\frac{2m}{m-2}}\right)^{\frac{m-2}{m}}}$$

satisfies

$$Q(M) = Q\left(\mathbb{S}_1^m\right) = \frac{m(m-2)\omega_m^{\frac{2}{m}}}{4}. \tag{9.66}$$

Thus, since $S=0$, we have

$$\int |\nabla\varphi|^2 \geq Q\left(\mathbb{S}_1^m\right)\left(\int \varphi^{\frac{2m}{m-2}}\right)^{\frac{m-2}{m}}, \text{ for each } \varphi \in C_c^\infty(M)\setminus\{0\}$$

which is the Sobolev inequality (9.38) with $h\equiv 0$, $\alpha=\frac{2}{m}$ and $S(\alpha)=Q\left(\mathbb{S}_1^m\right)^{-1}$. According to (9.65), conditions (9.39) and (9.40) are satisfied with $A=-\frac{2}{m}$ and $\sigma=\frac{m}{2}$. By Theorem 9.12 and (9.66) we conclude that

$$\|\mathrm{Ric}\|_{L^{\frac{m}{2}}(M)} \geq \frac{2\omega_m^{\frac{2}{m}}(m-2)^3\sqrt{m-1}}{m^2\sqrt{m}},$$

contradicting (9.64). □

In the case of negative scalar curvature, we establish the following result:

Theorem 9.20. *Let $(M,\langle\,,\rangle)$ be a complete, locally conformally flat Riemannian manifold of dimension $m\geq 4$ and constant scalar curvature $S<0$. Assume that, for some fixed $\varepsilon\geq 0$ and p, satisfying $m-1<mp<(m-1)(m-2)$, the following conditions hold:*

$$\text{i) } \left|\mathrm{Ric}-\frac{S}{m}\langle\,,\rangle\right| \leq -\varepsilon S, \qquad \text{ii) } \left|\mathrm{Ric}-\frac{S}{m}\langle\,,\rangle\right| \in L^p(M) \tag{9.67}$$

and furthermore

$$\lambda_1(-\Delta) > \frac{p^2}{4}\frac{m}{m(p-1)+2}(-S)\left\{\frac{\varepsilon\sqrt{(m-1)\,m}+1}{(m-1)}\right\} \tag{9.68}$$

where $\lambda_1(-\Delta)$ denotes the bottom of the spectrum of the (positive definite) Laplace operator $-\Delta$. Then, the universal cover of $(M,\langle\,,\rangle)$ is isometric to m-dimensional Hyperbolic space.

We remark that the restrictions on m and p follow from substituting into (9.68) the values of the scalar curvature and the bottom of the spectrum of m-dimensional hyperbolic space, for which (9.67) i) and ii) hold with $\varepsilon=0$ and every $p>0$.

Proof. As in the previous arguments the key point is to show that u, i.e., the length of the traceless Ricci tensor, vanishes identically.

Using (9.67) i) in (9.60) we see that u satisfies

$$u\Delta u - S\left(\varepsilon\sqrt{\frac{m}{m-1}}+\frac{1}{m-1}\right)u^2 \geq \frac{2}{m}|\nabla u|^2,$$

and $u\in L^p(M)$ by assumption (9.67) ii). If u were not identically zero, an application of Theorem 9.12, with the choices $A=-\frac{2}{m}$ and $a(x)=-S\left(\varepsilon\sqrt{\frac{m}{m-1}}+\frac{1}{m-1}\right)$, would contradict (9.68). Thus, $u\equiv 0$ as required. □

Appendix A

Unique continuation

It is well known that harmonic functions possess the unique continuation property, namely, if they vanish of infinite order at a point, then they vanish identically in the connected component containing the point. In fact unique continuation is shared by solutions to a wide class of differential elliptic differential systems satisfying the structural assumptions of Aronszajn–Cordes, and proves to be a vital tool in a variety of localization techniques. The vanishing and finite-dimensionality results which have been presented in previous chapters are no exceptions.

In this Appendix we describe, with only minor changes, the approach to unique continuation due to J. Kazdan, [87]. We begin by setting some preliminary definitions.

Definition A.1. Suppose we are given the elliptic system

$$\Delta u(x) = F(x, u, \nabla u) \quad u = (u_1, \dots, u_l) : M^m \to \mathbb{R}^l. \tag{A.1}$$

We say that (A.1) satisfies the *structural conditions of Aronszajn–Cordes* if, for every $p \in M$, we can find a small geodesic ball $B_R(p)$ and a smooth function $f : [0, R] \to \mathbb{R}$ satisfying

$$\text{(i) } f'(t) \geq 0, \qquad \text{(ii) } f(0) = 0, \qquad \text{(iii) } \frac{f(t)}{t} \in L^1(0^+), \tag{A.2}$$

such that

$$|F(x, u, \nabla u)| \leq \begin{cases} \text{(i) } a\frac{f(r(x))}{r(x)^2}|u(x)| + b\frac{f(r(x))}{r(x)}|\nabla u(x)| & m \geq 3, \\ \text{(ii) } a\frac{f(r(x))}{\left[r(x)\log\left(\frac{2R}{r(x)}\right)\right]^2}|u(x)| + b\frac{f(r(x))}{r(x)\log\left(\frac{2R}{r(x)}\right)}|\nabla u(x)| & m = 2 \end{cases} \tag{A.3}$$

for every $x \in B_R(p)$ and some constants $a, b > 0$. Here $r(x) = \operatorname{dist}(x, p)$ denotes the geodesic distance from p (which is Lipschitz on M and smooth in $B_R(p) \setminus \{p\}$, for $R > 0$ small enough).

Definition A.2. A vector-valued function $u \in W^{1,2}_{loc}(M; \mathbb{R}^l)$ is said to have a zero of infinite order at $p \in M$ if, for every $k \in \mathbb{N}$, we have

$$\liminf_{r \to 0^+} \frac{1}{r^k} \int_{\partial B_r(p)} |u|^2 = 0. \tag{A.4}$$

Remark A.3. It is easy to check, using Taylor expansion, that if derivatives of all orders of $u \in C^{\infty}\left(M;\mathbb{R}^l\right)$ vanish at p, then u has a zero of infinite order. In particular, if u vanishes identically on a domain $\Omega \subseteq M$, then u has a zero of infinite order at any point of $\overline{\Omega}$.

Definition A.4. Let $u = (u_A) \in W^{1,2}_{loc}\left(M;\mathbb{R}^l\right)$ be a weak solution of the elliptic system (A.1). This means that, for every $\varphi = (\varphi_A) \in C^{\infty}_c\left(M;\mathbb{R}^l\right)$,

$$-\int_M \sum_{A=1}^{l} \langle \nabla\varphi_A, \nabla u_A\rangle = \int_M \sum_{A=1}^{l} \varphi_A F_A\left(x,u,\nabla u\right).$$

We say that u has the unique continuation property if the validity of (A.4) at some $p \in M$ implies that u is identically zero on M.

We shall prove the following

Theorem A.5. *Let* $(M, \langle,\rangle)$ *be a connected Riemannian manifold and let* $u \in W^{1,2}_{loc}\left(M,\mathbb{R}^l\right)$ *be a weak solution of* (A.1). *Suppose the structural conditions* (A.2) *and* (A.3) *are met. Then* u *has the unique continuation property.*

Our proof of Theorem A.5 is a minor modification of the proof originally given by Kazdan, [87]. To simplify the exposition, we shall limit ourselves to the case of a scalar function $u \in C^2\left(M,\mathbb{R}\right)$. We begin with some elementary lemmas.

Lemma A.6. *Assume that there exist positive constants* R *and* k *such that the function* $K(r) \in C^1\left((0,R]\right)$ *is non-negative and satisfies*

$$\text{(i) } K(R) > 0; \qquad \text{(ii) } K'(r) \le \frac{k}{r}K(r) \ \text{ on } (0,R]. \tag{A.5}$$

Then, there exists a constant $\alpha > 0$ *such that*

$$K(r) \ge \alpha r^k \ \text{ on } (0,R]. \tag{A.6}$$

Proof. Condition (A.5) (ii) implies

$$\frac{d}{dr}\left\{\frac{K(r)}{r^k}\right\} \le 0.$$

Therefore, for every $0 < r \le R$,

$$\frac{K(r)}{r^k} \ge \frac{K(R)}{R^k},$$

proving (A.6) with $\alpha = K(R)/R^k > 0$. □

Thus, if $K(r)$ satisfies the assumptions of Lemma A.6 it cannot have a zero of infinite order at $r = 0$. Given a solution u of (A.1), the strategy of the proof consists in applying the previous lemma to the function defined by

$$K_p(r) = \frac{1}{r^{m-1}}\int_{\partial B_r(p)} u^2, \qquad r_p(x) = \operatorname{dist}(x,p), \tag{A.7}$$

to show that either u is identically zero in a neighborhood of p or u does not vanish of infinite order at p.

The following lemma will be instrumental in proving that the function K_p defined in (A.7) satisfies condition (A.5) (ii) .

Lemma A.7. *Let $0 \le \delta < R$ and assume that $h \in C((\delta, R])$ is non-negative and satisfies*

$$\int_{\delta}^{R} h(s)\,ds \le C$$

for some non-negative constant C. If $z : (\delta, R] \to \mathbb{R}$ is non-negative and satisfies the differential inequality

$$z'(r) \ge -h(r)\,z(r) \quad on\ (\delta, R], \tag{A.8}$$

then

$$z(r) \le e^{C} z(R) \quad on\ (\delta, R].$$

In particular, $z(r)$ is a bounded function.

Proof. From (A.8) we have

$$\frac{d}{dr}\left\{z(r)\,e^{-\int_r^R h(s)ds}\right\} = e^{-\int_r^R h(s)ds}\left\{z'(r) + h(r)\,z(r)\right\} \ge 0.$$

Hence, using $z(r) \ge 0$,

$$e^{-C} z(r) \le z(r)\,e^{-\int_r^R h(s)ds} \le z(R)\,. \qquad \square$$

We will also need to differentiate integrals over geodesic spheres. This is addressed in the next lemma.

Lemma A.8. *Let $w \in C^1(M)$. Then, for $0 < R < \mathrm{inj}(p)$,*

$$\frac{d}{dR}\int_{\partial B_R(p)} w = \int_{\partial B_R(p)} \langle \nabla w, \nabla r\rangle + \int_{\partial B_R(p)} w\Delta r. \tag{A.9}$$

Proof. Let $r(x) = \mathrm{dist}(x, p)$. Applying the divergence theorem to the vector field $Z = w\nabla r$ and using Gauss' lemma we obtain

$$\int_{\partial B_R(p)} w = \int_{B_R(p)} \mathrm{div} Z = \int_{B_R(p)} \langle \nabla w, \nabla r\rangle + \int_{B_R(p)} w\Delta r. \tag{A.10}$$

Note that since $\Delta r = (m-1)r^{-1}\big(1 + o(r)\big)$ as $r \to 0$, the singularity of Δr at p does not affect integrability. By the co-area formula

$$\int_{B_R} u = \int_0^R \int_{\partial B_t} u,$$

for any locally integrable function u. It follows that the RHS, and therefore the LHS, of (A.10) is differentiable. The desired inequality follows by differentiation. $\square$

Remark A.9. Note that, choosing $w = 1$, formula (A.9) reads

$$\frac{d}{dR}\mathrm{vol}\left(\partial B_r\left(p\right)\right) = \int_{\partial B_R(p)} \Delta r.$$

Since $R < \mathrm{inj}\left(p\right)$, $\partial B_R\left(p\right)$ is a smooth, oriented hypersurface with Gauss map ∇r and, for any tangent vector fields $X, Y \in T\partial B_R$, the (scalar) second fundamental form of $\partial B_R\left(p\right)$ is given by $\mathrm{II}\left(X, Y\right) = \mathrm{Hess}\left(r\right)\left(X, Y\right)$. We have already noted that, as a consequence of Gauss' lemma, $\mathrm{Hess}\left(r\right)\left(\nabla r, \nabla r\right) = 0$. Therefore

$$\Delta r = \mathrm{tr}_M \mathrm{Hess}\left(r\right) = \mathrm{tr}_{\partial B_R} \mathrm{Hess}\left(r\right) = \mathrm{tr}_{\partial B_R} \mathrm{II} = \left(m-1\right) H\left(x\right),$$

with $H\left(x\right)$ the mean curvature function of $\partial B_R\left(p\right)$. Thus,

$$\frac{d}{dR}\mathrm{vol}\left(\partial B_r\left(p\right)\right) = \int_{\partial B_R(p)} \left(m-1\right) H.$$

We point out that the 2-dimensional case of this formula is of special interest in that, by the Gauss–Bonnet theorem, it takes the nice form

$$\frac{d}{dR}\ell\left(\partial B_r\left(p\right)\right) = 2\pi - \int_{B_R(p)} {}^M\mathrm{Sect}.$$

We recall the argument for the sake of completeness. Having parametrized ∂B_R by arc-length as a curve $\gamma\left(s\right)$, we see that H is nothing but the geodesic curvature k_g of γ (with respect to the inward-pointing unit normal $-\nabla r$), that is,

$$k_g = \langle D_{\dot\gamma}\dot\gamma, -\nabla r\rangle = H.$$

Indeed, note that $r \circ \gamma\left(s\right) = R$ implies

$$\begin{aligned} 0 &= \frac{d^2}{ds^2} r \circ \gamma \\ &= \mathrm{Hess}\left(r\right)\left(\dot\gamma, \dot\gamma\right) + \langle D_{\dot\gamma}\dot\gamma, \nabla r\rangle \\ &= \mathrm{II}\left(\dot\gamma, \dot\gamma\right) + \langle D_{\dot\gamma}\dot\gamma, \nabla r\rangle \\ &= H + \langle \nabla r, D_{\dot\gamma}\dot\gamma\rangle. \end{aligned}$$

Now, for R small enough, B_R is homeomorphic (diffeomorphic in fact) to a Euclidean disk. Therefore, applying the Gauss–Bonnet theorem we conclude

$$\begin{aligned} \frac{d}{dR}\ell\left(\partial B_r\left(p\right)\right) &= \int_{\partial B_R(p)} H \\ &= \int_0^{\ell(\partial B_r(p))} k_g\left(s\right) ds \\ &= 2\pi\chi\left(B_R\left(p\right)\right) - \int_{B_R(p)} {}^M\mathrm{Sect} \\ &= 2\pi - \int_{B_R(p)} {}^M\mathrm{Sect}, \end{aligned}$$

as claimed.

Now let $\Omega \subset\subset M$ be a relatively compact domain. Since the injectivity radius $\mathrm{inj}\,(x)$ is a positive and continuous function of x, $\mathrm{inj}(\Omega) = \inf_\Omega \mathrm{inj}\,(x) > 0$; see, e.g., [30]. Fix $R_0 < \mathrm{inj}\,(\Omega)$. For any $p \in \Omega$, let $r_p\,(x) = \mathrm{dist}_M\,(x,p)$ and suppose that u is a non-null solution of (A.1) in $B_{R_0}(p)$. Applying Lemma A.8 to the function $K_p\,(r\,(x))$ defined in (A.7) we obtain, for every $0 < R < R_0$,

$$K_p'\,(R) = -\frac{m-1}{R^m}\int_{\partial B_R(p)} u^2 + \frac{2}{R^{m-1}}\int_{\partial B_R(p)} u\,\langle \nabla u, \nabla r_p\rangle + \frac{1}{R^{m-1}}\int_{\partial B_R(p)} u^2 \Delta r_p.$$

On the other hand, by the Laplacian Comparison Theorem,

$$r_p \Delta r_p \leq (m-1)\big(1 + \frac{k_-}{3} r_p^2 + O(k_-^2 r_p^4)\big) \quad \text{in } B_R(p),$$

where k_- is a lower bound for the sectional curvature of M in $B_p(R)$. Therefore there exists a constant C_1, which depends only on the lower bound for the sectional curvature in an R_0-neighborhood of Ω such that, for every p in Ω, and for every $R \leq R_0$,

$$K_p'\,(R) \leq \frac{2}{R}\left\{\frac{1}{R^{m-2}}\int_{\partial B_R(p)} u\,\langle \nabla u, \nabla r_p\rangle\right\} + C_1 R K_p\,(R)\,. \tag{A.11}$$

It is now clear that Lemma A.6 applies and, thus, u does not possess a zero of infinite order at p, provided we are able to prove that, for a sufficiently small $\bar{R} \in (0, R_0)$, we have

$$K_p\,(\bar{R}) > 0 \tag{A.12}$$

and

$$\left\{\frac{1}{R^{m-2}}\int_{\partial B_R(p)} u\,\langle \nabla u, \nabla r_p\rangle\right\} \leq c K_p\,(R) \;\text{ on } (0, \bar{R}] \tag{A.13}$$

for some constant $c > 0$. This is the content of Lemma A.13 and Lemma A.17 below, where it is in fact proved that if p belongs to a relatively compact set Ω, then both the radius $\bar{R}$ and the constant c in (A.13) depend only on the geometry of M in a neighborhood of $\overline{\Omega}$. A crucial ingredient in their proof is, of course, the fact that (A.1) enjoys the structural assumptions of Aronszajn–Cordes.

The proof of (A.12) and (A.13) depends on a number of technical lemmas. We begin with an integral inequality of independent interest, which is sometimes referred to as a version of "the Heisenberg uncertainty principle".

For ease of notation, we will understand that a point p has been fixed, and we will usually drop explicit reference to it in symbols and formulae.

Lemma A.10. *Let $\Omega \subset\subset M$ be a relatively compact domain and let $R_0 < \mathrm{inj}(\overline{\Omega})$. Then, there exist constants $0 < R_1 \leq R_0$ and $\alpha, \beta > 0$ depending only on m and on the geometry of M in an R_0-neighborhood of $\overline{\Omega}$ such that, for every $u \in C^1\big(B_{R_0}(p)\big)$ and $0 < R < R_1$,*

$$\int_{B_R} \frac{u^2}{r^2} \leq \frac{\alpha}{R}\int_{\partial B_R} u^2 + \beta \int_{B_R} |\nabla u|^2\,, \text{ for } m \geq 3, \tag{A.14}$$

while

$$\int_{B_R} \frac{u^2}{\{r \log(2R/r)\}^2} \le \frac{\alpha}{R} \int_{\partial B_R} u^2 + \beta \int_{B_R} |\nabla u|^2, \ \textit{for } m = 2. \tag{A.15}$$

Proof. Assume first that $m = \dim M \ge 3$ and note that, by Gauss' lemma,

$$u^2 \mathrm{div}\left(r^{-1}\nabla r\right) = u^2\left(r^{-1}\Delta r - r^{-2}|\nabla r|^2\right) = r^{-2}u^2\left(r\Delta r - 1\right), \tag{A.16}$$

whence, integrating over $B_R \setminus B_\epsilon$, using Green's identity, Gauss' lemma again, and letting $\epsilon \to 0$, we obtain

$$\begin{aligned}\int_{B_R} \frac{u^2}{r^2}(r\Delta r - 1) &= \int_{B_R} u^2 \mathrm{div} \frac{\nabla r}{r} \\ &= \int_{\partial B_R} \frac{u^2}{r} - 2\int_{B_R} \frac{u}{r}\langle \nabla r, \nabla u\rangle\end{aligned}$$

for every $0 < R \le R_0$. We apply the Schwarz inequality and the elementary estimate

$$2ab \le \lambda a^2 + \frac{b^2}{\lambda}, \ \lambda > 0,$$

with the choice $\lambda = (m-2)/2$, to estimate

$$-\frac{2}{r}u\langle\nabla u, \nabla r\rangle \le \frac{m-2}{2}\frac{u^2}{r^2} + \frac{2}{m-2}|\nabla u|^2,$$

whence, inserting and rearranging, we deduce that, for every $0 < R \le R_0$,

$$\int_{B_R} \frac{u^2}{r^2}\left(r\Delta r - 1 - \frac{m-2}{2}\right) \le \frac{1}{R}\int_{\partial B_R} u^2 + \frac{2}{m-2}\int_{B_R} |\nabla u|^2.$$

Clearly, to conclude the validity of (A.14) we have to prove that, up to choosing $0 < R_1 \le R_0$ sufficiently small, and depending only on m and on the geometry of M in an R_0-neighborhood of $\bar{\Omega}$, we have

$$\inf_{B_{R_1}(p)} \left(r\Delta r - 1 - \frac{m-2}{2}\right) > 0. \tag{A.17}$$

To this end, fix $k_+ > 0$ so as to satisfy

$$k_+ \ge \sup\{\ {}^M\mathrm{Sect}(x)\ :\ \mathrm{dist}(x, \bar{\Omega}) \le R_0\}$$

and choose $T > 1$ large enough that

$$R_1 = \frac{\pi}{2T\sqrt{k_+}} \le R_0 \left(< \mathrm{inj}\left(\bar{\Omega}\right)\right).$$

Then $B_{R_1}(p)$ is a regular geodesic ball of M and, by Hessian comparison, we deduce

$$\begin{aligned} r\Delta r &\geq (m-1)\, r\sqrt{k_+}\cot\left(\sqrt{k_+}r\right) \\ &\geq (m-1)\,\frac{\pi}{2T}\cot\left(\frac{\pi}{2T}\right) \qquad \text{on } B_{R_1}(p), \end{aligned}$$

where the second inequality follows from the fact that the function $t \to t\cot t$ is decreasing. Since $t\cot t \to 1$ as $t \to 0+$, we conclude that

$$r\Delta r - 1 - \frac{m-2}{m} \geq (m-1)\,\frac{\pi}{2T}\cot\left(\frac{\pi}{2T}\right) = \alpha^{-1} > 0 \qquad \text{on } B_{R_1}(p)$$

provided T is sufficiently large (and therefore R_1 is sufficiently small. This proves the validity of (A.17).

In the case where $m = \dim M = 2$ one argues in a similar fashion, replacing (A.16) with

$$\operatorname{div}\frac{\nabla r}{r\log(R/r)} = \frac{\Delta r}{r\log(R/r)} - \frac{1}{r^2\log(R/r)} + \frac{1}{r^2\log^2(R/r)}$$

and choosing $\lambda = 1/2$ in the above elementary inequality. □

The next lemmas have a technical flavor.

Lemma A.11. *Let Ω, R_0 and R_1 be as in Lemma* A.10. *Given $p \in \Omega$, let $u \in C^2(B_{R_0}(p))$, $r(x) = \operatorname{dist}_M(x,p)$, and define, for $0 < R < R_0$,*

$$A(R) = \frac{1}{R^{m-2}}\int_{\partial B_R} u\,\langle\nabla r, \nabla u\rangle\,, \tag{A.18}$$

$$B(R) = \frac{2}{R^{m-2}}\int_{\partial B_R} \langle\nabla r, \nabla u\rangle^2\,, \tag{A.19}$$

$$I_1(R) = \frac{1}{R^{m-2}}\int_{B_R} u\Delta u, \tag{A.20}$$

$$I_2(R) = \frac{2}{R^{m-2}}\int_{B_R} r\,\langle\nabla r, \nabla u\rangle\,\Delta u, \tag{A.21}$$

$$I_3(R) = \frac{1}{R^{m-3}}\int_{\partial B_R} u\Delta u, \tag{A.22}$$

$$D(R) = \frac{1}{R^{m-2}}\int_{B_R} |\nabla u|^2\,. \tag{A.23}$$

Then

$$\begin{aligned} &A'(R) - B(R) + \frac{1}{R}\big[(m-2)I_1(R) + I_2(R) - I_3(R)\big] \\ &\quad = \frac{1}{R^{m-1}}\int_{B_R}\big[\frac{1}{2}\Delta r^2 - \operatorname{Hess} r^2\big(\frac{\nabla u}{|\nabla u|}, \frac{\nabla u}{|\nabla u|}\big) - (m-2)\big]|\nabla u|^2 \end{aligned} \tag{A.24}$$

and there exists a constant C_0 which depends only on m, R_o and on the geometry on M in an R_0-neighborhood of $\bar{\Omega}$ such that, for $0 < R < R_1$ we have

$$\left|A'(R) - B(R) + \frac{1}{R}\left[(m-2) I_1(R) - I_2(R) + I_3(R)\right]\right| \leq C_0 R D(R). \quad \text{(A.25)}$$

Proof. To start off, we note that, by the divergence theorem,

$$R^{m-2} A(R) = \int_{\partial B_R} \langle u\nabla u, \nabla r\rangle = \int_{B_R} div(u\nabla u) = \int_{B_R} (u\Delta u + |\nabla u|^2) \quad \text{(A.26)}$$

and, therefore, by the co-area formula,

$$A'(R) = \frac{1}{R^{m-2}} \int_{\partial B_R} (u\Delta u + |\nabla u|^2) - \frac{m-2}{R^{m-1}} \int_{B_R} (u\Delta u + |\nabla u|^2). \quad \text{(A.27)}$$

We now want to get rid of the integral on the boundary of $|\nabla u|^2$, which is the only term on the right-hand side which does not appear in (A.19)–(A.23). To this end, we apply the divergence theorem to the vector field

$$W = |\nabla u|^2 r\nabla r - 2r\langle \nabla r, \nabla u\rangle \nabla u.$$

Recalling that, given smooth functions f, g, h,

$$\langle \nabla f, \langle \nabla g, \nabla h\rangle\rangle = \nabla f\langle \nabla g, \nabla g\rangle = \operatorname{Hess} g(\nabla f, \nabla h) + \operatorname{Hess} h(\nabla f, \nabla g)$$

we compute

$$\begin{aligned} \operatorname{div} W &= \frac{1}{2}|\nabla u|^2 \Delta r^2 + \frac{1}{2}\nabla r^2 \langle \nabla u, \nabla u\rangle - \langle \nabla r^2, \nabla u\rangle \Delta u - \nabla u \langle \nabla r^2, \nabla u\rangle \\ &= \frac{1}{2}|\nabla u|^2 \Delta r^2 - \operatorname{Hess} r^2(\nabla u, \nabla u) - \langle \nabla r^2, \nabla u\rangle \Delta u \end{aligned}$$

whence, integrating over B_R, we obtain the following identity, first derived, in a Euclidean setting, by Rellich,

$$\begin{aligned} &R\int_{\partial B_R} \left\{|\nabla u|^2 - 2\langle \nabla r, \nabla u\rangle^2\right\} \\ &\quad = \int_{B_R} \left\{\frac{1}{2}|\nabla u|^2 \Delta r^2 - \operatorname{Hess}(r^2)(\nabla u, \nabla u) - 2r\langle \nabla r, \nabla u\rangle \Delta u\right\}. \quad \text{(A.28)} \end{aligned}$$

Solving for $\int_{\partial B_R} |\nabla u|^2$ we obtain

$$\begin{aligned} &\int_{\partial B_R} |\nabla u|^2 = 2\int_{\partial B_R} \langle \nabla r, \nabla u\rangle^2 \\ &\quad + \frac{1}{R}\int_{B_R} \left\{\frac{1}{2}|\nabla u|^2 \Delta r^2 - \operatorname{Hess}(r^2)(\nabla u, \nabla u) - 2r\langle \nabla r, \nabla u\rangle \Delta u\right\}, \quad \text{(A.29)} \end{aligned}$$

whence, substituting (A.29) into the expression (A.27) for $A'(R)$ and rearranging,

$$\begin{aligned} A'(R) &- \frac{2}{R^{m-2}}\int_{\partial B_R}\langle\nabla r,\nabla u\rangle^2 - \frac{1}{R^{m-2}}\int_{\partial B_R} u\Delta u \\ &+ \frac{m-2}{R^{m-1}}\int_{B_R} u\Delta u + \frac{2}{R^{m-1}}\int_{B_R} r\langle\nabla r,\nabla u\rangle\Delta u \\ &= \frac{1}{R^{m-1}}\int_{B_R}\left(\frac{1}{2}|\nabla u|^2\Delta r^2 - \operatorname{Hess} r^2(\nabla u,\nabla u) - (m-2)|\nabla u|^2\right). \end{aligned}$$

Now (A.24) follows recalling the definitions (A.19)–(A.22).

In order to estimate the right-hand side, let $k_+, k_- > 0$ be such that

$$-k_- \leq {}^M\mathrm{Sect} \leq k_+ \text{ on } \{x : \operatorname{dist}(x,\Omega) \leq R_0\},$$

so that, by the Hessian comparison theorem, in B_{R_1} we have

$$\begin{aligned} &\sqrt{k_+}r\cot(\sqrt{k_+}r)\langle\,,\rangle + (1-\sqrt{k_+}r\cot(\sqrt{k_+}r))dr\otimes dr \\ &\leq \frac{1}{2}\operatorname{Hess}(r^2) \leq \sqrt{k_-}r\coth(\sqrt{k_-}r)\langle\,,\rangle + (1-\sqrt{k_-}r\coth(\sqrt{k_-}r))dr\otimes dr \end{aligned} \tag{A.30}$$

and, tracing,

$$(m-1)\sqrt{k_+}r\cot(\sqrt{k_+}r)+1 \leq \frac{1}{2}\Delta r^2 \leq (m-1)\sqrt{k_-}r\coth(\sqrt{k_-}r)+1. \tag{A.31}$$

We now observe that, as $r \to 0$,

$$\begin{aligned} \sqrt{k_-}r\coth(\sqrt{k_-}r)\langle\frac{\nabla u}{|\nabla u|},\frac{\nabla u}{|\nabla u|}\rangle &+ (1-\sqrt{k_-}r\coth(\sqrt{k_-}r))\langle\frac{\nabla u}{|\nabla u|},\nabla r\rangle^2 \\ &= \sqrt{k_-}r\coth(\sqrt{k_-}r) + O\big(1-\sqrt{k_-}r\coth(\sqrt{k_-}r)\big) \\ &= 1 + O\big(k_-r^2\big) \end{aligned}$$

and that an analogous estimate holds replacing k_- with k_+ and coth with cot. The term $\frac{1}{2}\Delta r^2$ can be dealt with in a similar way, thus showing that

$$\left|\frac{1}{2}\Delta r^2 - \operatorname{Hess}\left(\frac{\nabla u}{|\nabla u|},\frac{\nabla u}{|\nabla u|}\right) - (m-2)\right| \leq C_0 r^2 \quad \text{in } B_{R_1} \tag{A.32}$$

where C_1 depends only on m, R_0 and $k_\pm$. Using this estimate in (A.24), and recalling the definition of $D(R)$ yields (A.25). □

We next proceed to estimate the quantities introduced in the previous lemma.

Lemma A.12. *Let p, R_0 and R_1 be as in Lemma* A.10 *and let u be a C^2-solution on $B_{R_0}(p)$ of the differential inequality*

$$|\Delta u| \leq a\frac{f(r)}{r^2}|u| + b\frac{f(r)}{r}|\nabla u| \tag{A.33}$$

where $a, b > 0$ are constants and f is a C^1 function satisfying the structural conditions (A.2). Then, there exist positive constants $c_1, \ldots, c_5$, depending only on m and on the geometry of M in an R_0-neighborhood of $\bar{\Omega}$, such that, for every $0 < R \leq R_1$,

$$|I_j(R)| \leq c_j f(R)\left[K(R) + D(R)\right], \; j = 1, 2, \tag{A.34}$$

$$|I_3(R)| \leq c_4 f(R)\left[K(R) + D(R) + \sqrt{RK(R)B(R)}\right], \tag{A.35}$$

$$|I_3(R)| \leq c_5 f(R)\left[K(R) + D(R) + RB(R)\right], \tag{A.36}$$

$$\frac{1}{R^{m-3}} \int_{\partial B_R} |\nabla u|^2 \leq RB(R) + c_3\left[K(R) + D(R)\right]. \tag{A.37}$$

Proof. We consider only the case $m \geq 3$. We start with the first of (A.34). In what follows we shall make repeated use of (A.14), (A.33) and of conditions (A.2) on f. We have

$$\begin{aligned}
|I_1(R)| &= \frac{1}{R^{m-2}} \left| \int_{B_R} u \Delta u \right| \\
&\leq \frac{1}{R^{m-2}} \int_{B_R} a \frac{f(r)}{r^2} u^2 + \frac{1}{R^{m-2}} \int_{B_R} b \frac{f(r)}{r} |u||\nabla u| \\
&\leq \left(a + \frac{b}{2}\right) \frac{f(R)}{R^{m-2}} \int_{B_R} \frac{u^2}{r^2} + \frac{b}{2} \frac{f(R)}{R^{m-2}} \int_{B_R} |\nabla u| \\
&\leq \frac{\alpha\left(a + \frac{b}{2}\right) f(R)}{R^{m-1}} \int_{\partial B_R} u^2 + \frac{\left(\beta a + \frac{b}{2} + \beta \frac{b}{2}\right) f(R)}{R^{m-2}} \int_{B_R} |\nabla u|^2 .
\end{aligned}$$

This shows the validity of the first of (A.34). The estimate for $|I_2(R)|$ is completely analogous.

We now show (A.37). From (A.29) and (A.32) in Lemma A.11, we have

$$\begin{aligned}
\frac{1}{R^{m-3}} \int_{\partial B_R} |\nabla u|^2 \leq{}& \frac{2}{R^{m-3}} \int_{\partial B_R} \langle \nabla r, \nabla u \rangle^2 \\
&+ \frac{C_0 R + (m-2)}{R^{m-2}} \int_{B_R} |\nabla u|^2 - \frac{2}{R^{m-2}} \int_{B_R} r \langle \nabla r, \nabla u \rangle \Delta u,
\end{aligned} \tag{A.38}$$

and recalling the definition of $I_2(R)$ and the estimate

$$|I_2(R)| \leq c_2 f(R)\left[K(R) + D(R)\right]$$

in (A.34) we conclude that

$$\begin{aligned}
\frac{1}{R^{m-3}} \int_{\partial B_R} |\nabla u|^2 &\leq RB(R) + [C_0 R^2 + (m-2)]D(R) + c_2 f(R)[K(R) + D(R)] \\
&\leq RB(R) + c_3[K(R) + D(R)]
\end{aligned}$$

for some constant $c_3 > 0$, and (A.37) is proved.

We now prove (A.35). From (A.33) and Schwarz's inequality we obtain

$$\begin{aligned}
|I_3(R)| &\le \frac{1}{R^{m-3}} \int_{\partial B_R} |u|\,|\Delta u| \\
&\le \frac{af(R)}{R^{m-1}} \int_{\partial B_R} u^2 + \frac{bf(R)}{R^{m-3}} \int_{\partial B_R} \frac{|u|\,|\nabla u|}{r} \\
&\le \frac{af(R)}{R^{m-1}} \int_{\partial B_R} u^2 + \frac{bf(R)}{R^{\frac{m-3}{2}}} \left\{ \int_{\partial B_R} \frac{u^2}{r^2} \right\}^{\frac{1}{2}} \left\{ \frac{1}{R^{m-3}} \int_{\partial B_R} |\nabla u|^2 \right\}^{\frac{1}{2}} \\
&= af(R)\,K(R) + bf(R)\,K(R)^{\frac{1}{2}} \left\{ \frac{1}{R^{m-3}} \int_{\partial B_R} |\nabla u|^2 \right\}^{\frac{1}{2}}.
\end{aligned}$$

We now use (A.37) to get

$$|I_3(R)| \le af(R)\,K(R) + bf(R)\,K(R)^{\frac{1}{2}} \left\{ RB(R) + c_3 (D(R) + K(R)) \right\}^{\frac{1}{2}}$$

from which (A.35) follows immediately. Finally, (A.36) is a simple consequence of (A.35). □

Lemma A.13. *Let u be as in Lemma* A.12 *and let $c_1 > 0$ be the constant in* (A.34). *Choose $0 < R_2 < R_1$ sufficiently small that*

$$c_1 f(R_2) \le \frac{1}{2}. \tag{A.39}$$

Then, there exists $\delta \in [0, R_2]$ such that the function $K(R)$ defined in (A.7) *satisfies*

$$K(R) = \begin{cases} 0 & \text{for } 0 \le R \le \delta, \\ > 0 & \text{for } \delta < R \le R_2. \end{cases} \tag{A.40}$$

Remark A.14. Notice that, since c_1 depends only on m, R_1 and on the geometry of M in an R_0-neighborhood of $\overline{\Omega}$, so does R_2.

Proof. Again, we consider only the case $m \ge 3$. From (A.26) we have

$$A(R) = D(R) + I_1(R).$$

Hence, applying the first of (A.34) we have, for $0 < R \le R_2$,

$$\begin{aligned}
A(R) &\ge D(R) - |I_1(R)| \\
&\ge D(R) - c_1 f(R) (K(R) + D(R)) \\
&\ge \frac{1}{2} D(R) - \frac{1}{2} K(R),
\end{aligned}$$

that is,

$$D(R) \le 2A(R) + K(R). \tag{A.41}$$

Using (A.41) and the definitions of $A(R)$,$D(R)$ and $K(R)$ into the Heisenberg uncertainty principle (A.14), we obtain

$$\begin{aligned}\int_{B_R} \frac{u^2}{r^2} &\le \frac{\alpha}{R}\int_{\partial B_R} u^2 + \beta R^{m-2} D(R) \\ &\le \frac{\alpha}{R}\int_{\partial B_R} u^2 + 2\beta R^{m-2} A(R) + \beta R^{m-2} K(R) \\ &= 2\beta \int_{\partial B_R} u \langle \nabla r, \nabla u\rangle + (\alpha+\beta) R^{m-2} K(R) .\end{aligned} \tag{A.42}$$

Now, if $K(R) = 0$, then $u = 0$ on ∂B_R and therefore the right-hand side of (A.42) vanishes. This forces $u = 0$ on B_R and proves the validity of (A.40). □

Lemma A.15. *Let u be as in Lemma* A.12 *and let C_0 be the constant in inequality* (A.25) *in the statement of Lemma* A.11. *There exists a constant $c_6 > 0$ (depending only on m and on the geometry of M in an R_0-neigborhood of $\bar{\Omega}$) such that, having fixed $\varepsilon > 0$, if $R_3 \in (0, R_1]$ satisfies*

$$C_0 R_3^2 + c_6 f(R_3) < \varepsilon, \tag{A.43}$$

then, for every $0 < R \le R_3$, we have

$$|A'(R) - B(R)| \le \varepsilon B(R) + \frac{\varepsilon}{R}(K(R) + D(R)) . \tag{A.44}$$

Remark A.16. The radius R_3 depends only on m, ε and on the geometry of M in an R_0-neighborhood of $\bar{\Omega}$.

Proof. According to (A.25),

$$|A'(R) - B(R)| \le C_0 R D(R) + \frac{m-2}{R}|I_1| + \frac{1}{R}|I_2| + \frac{1}{R}|I_3|,$$

whence, applying Lemma A.12, we deduce that for every $0 \le R \le R_1 < R_0$,

$$|A'(R) - B(R)| \le c_6 f(R) B(R) + \frac{C_0 R^2 + c_6 f(R)}{R}[K(R) + D(R)] \tag{A.45}$$

where $c_6 = [(m-2)c_1 + c_2 + c_5]$, and c_j are the constants in Lemma A.12. Since C_0 and c_j depend only on m and the geometry of M in an R_0-neighborhood of $\bar{\Omega}$, so does c_6. The required conclusion follows from (A.43) together with the fact that f is increasing. □

Lemma A.17. *Let u be as in Lemma* A.12. *There exists $R_4 \in (0, R_1]$ and a constant $c_7 > 0$, depending only on m and on the geometry of M in an R_0-neighborhood of $\bar{\Omega}$, such that, if $0 < R \le R_4$, then*

$$A(R) \le c_7 K(R) . \tag{A.46}$$

Proof. Let C_0, c_1 and c_6 be the positive constants of Lemmas A.11, A.12 and A.15, respectively. Choose $R_4 \in (0, R_1]$ in such a way that

$$\text{(i) } c_1 f(R_4) \leq \frac{1}{2}; \qquad \text{(ii) } C_0 R_4^2 + c_6 f(R_4) \leq \frac{1}{2}. \tag{A.47}$$

Next, recall that it follows easily from the definitions of $A(R)$ and $K(R)$ that

$$\text{if } K(R) = 0 \text{ for some } 0 < R \leq R_4, \text{ then } A(R) = 0, \tag{A.48}$$

and in this case inequality (A.46) trivially holds for any choice of $c_7 > 0$. Therefore we only need to verify (A.46) for those values of $R \in (0, R_4]$, for which $K(R) > 0$. According to (A.47) (i) and Lemma A.13, there exists $\delta \in [0, R_4]$ such that

$$\{R \in (0, R_4] : K(R) > 0\} = (\delta, R_4].$$

We let $\varphi : (\delta, R_4] \to \mathbb{R}$ be the C^1-function defined by

$$\varphi(R) = \frac{A(R)}{K(R)} + 1.$$

Then the conclusion of the lemma amounts to proving that φ is bounded on $(\delta, R_4]$. To this end, we shall apply Lemma A.7. First of all, we observe that since $c_1 f(R_4) < \frac{1}{2}$, according to (A.41) in Lemma A.13, $D(R) \leq 2A(R) + K(R)$ on $(0, R_4]$ and therefore

$$\varphi(R) \geq \frac{1}{2}\left[\frac{D(R)}{K(R)} + 1\right] > 0, \quad \text{on } (\delta, R_4]. \tag{A.49}$$

If we show that there exist constants $\mu, \nu \geq 0$ such that

$$\varphi'(R) \geq -(\mu\frac{f(R)}{R} + \nu)\varphi(R) \text{ on } (\delta, R_4], \tag{A.50}$$

then, recalling that $f(r)/r \in L^1(0^+)$, we can apply Lemma A.7 with

$$h(R) = \mu\frac{f(R)}{R} + \nu$$

and conclude that φ is bounded on $(\delta, R_4]$. We have

$$\varphi'(R) = \frac{1}{K(R)^2}\left[A'(R)K(R) - K'(R)A(R)\right]. \tag{A.51}$$

Since $\varphi(R) > 0$ and $f(R) > 0$, (A.50) is surely satisfied for any choice of $\mu, \nu \geq 0$ if R is such that $\varphi'(R) \geq 0$. Thus, we only have to consider values of R for which $\varphi'(R) < 0$, i.e.,

$$A'(R)K(R) < K'(R)A(R). \tag{A.52}$$

Since $C_0 R_4^2 + c_1 f(R_4) \leq \frac{1}{2}$ by (A.47) (ii), using Lemma A.15 with $\varepsilon = 1/2$, we have

$$|A'(R) - B(R)| \leq \frac{1}{2} B(R) + \frac{1}{2R}[K(R) + D(R)]$$

and therefore

$$\begin{aligned} B(R) &= A'(R) + [B(R) - A'(R)] \\ &\leq 2A'(R) + \frac{1}{R}[K(R) + D(R)], \text{ on } (0, R_4]. \end{aligned} \tag{A.53}$$

Now, since $K(R) \geq 0$, if we assume that $A'(R)K(R) < K'(R)A(R)$ we may write

$$K(R) B(R) \leq 2K'(R) A(R) + \frac{K(R)}{R}[K(R) + D(R)]. \tag{A.54}$$

Also, since $c_1 f(R_4) \leq \frac{1}{2}$, by (A.47) (i), according to (A.34) in Lemma A.12 we have

$$|I_1| \leq \frac{1}{2}[K(R) + D(R)],$$

whence

$$\begin{aligned} A(R) &= D(R) + I_1(R) \\ &\leq D(R) + \frac{1}{2}[K(R) + D(R)] \\ &\leq 2[K(R) + D(R)]. \end{aligned} \tag{A.55}$$

On the other hand, recalling the definition (A.18) of $A(R)$ and the estimate (A.11) we have

$$K'(R) \leq \frac{2}{R} A(R) + C_1 R K(R) \text{ on } (0, R_4], \tag{A.56}$$

with $C_1 > 0$ depending only on m and on the geometry of M in an R_0-neighborhood of $\overline{\Omega}$. Using (A.55) and (A.56) into (A.54) we deduce that

$$\begin{aligned} K(R) B(R) &\leq \frac{4A(R)^2}{R} + C_1 R A(R) K(R) + \frac{K(R)}{R}[K(R) + D(R)] \\ &\leq \frac{\alpha [K(R) + D(R)]^2}{R}, \end{aligned} \tag{A.57}$$

for some constant $\alpha > 0$. Inserting this into the estimate (A.35) for I_3, namely,

$$|I_3| \leq c_4 f(R)[K(R) + D(R) + \sqrt{RK(R)B(R)}]$$

we see that, if we assume that $A'(R)K(R) < K'(R)A(R)$ we can improve (A.35) of Lemma A.12 to

$$|I_3| \leq \beta f(R)[K(R) + D(R)],$$

for a suitable constant $\beta > 0$. Using this latter and (A.34) in (A.25) of Lemma A.11 we get the corresponding improvement of inequality (A.44) of Lemma A.15:

$$|A'(R) - B(R)| \leq \frac{\gamma f(R)}{2R}[K(R) + D(R)] + C_1 R D(R), \tag{A.58}$$

for some $\gamma > 0$, and therefore

$$A'(R) \geq B(R) - \frac{\gamma f(R)}{2R}[K(R) + D(R)] - C_1 R D(R).$$

Plugging this back into the expression for $\varphi'(R)$ we conclude that

$$\begin{aligned}\varphi'(R) &= \frac{1}{K(R)^2}[A'(R)K(R) - K'(R)A(R)] \\ &\geq \frac{1}{K^2(R)}\big[K(R)B(R) - K'(R)A(R)\big] \\ &\quad - \frac{1}{K(R)}\big[\gamma\frac{f(R)}{R}(K(R) + D(R)) + C_1 R D(R)\big].\end{aligned} \tag{A.59}$$

To conclude, note that by the definitions of $A(R)$, $B(R)$ and $K(R)$, and Schwarz's inequality,

$$\begin{aligned}\frac{2}{R}A^2(R) &= \frac{2}{R^{2m-3}}\Big(\int_{\partial B_R} u\langle\nabla u, \nabla r\rangle\Big)^2 \\ &\leq \frac{2}{R^{m-2}}\int_{\partial B_R}\langle\nabla u, \nabla r\rangle^2 \frac{1}{R^{m-1}}\int_{\partial B_R} u^2 = B(R)K(R)\end{aligned}$$

so that, using (A.56) and (A.55) we may estimate

$$\begin{aligned}K(R)B(R) - K'(R)A(R) &\geq \frac{2}{R}A^2(R) - A(R)\big[\frac{2}{R}A(R) + C_1 R K(R)\big] \\ &= -C_1 R K(R)A(R) \geq -2C_1 R K(R)\big[K(R) + D(R)\big].\end{aligned}$$

Inserting this into (A.59), yields

$$\varphi'(R) \geq -\Big(\gamma\frac{f(R)}{R} + 3C_1 R\Big)\frac{[K(R) + D(R)]}{K(R)},$$

whence, recalling (A.49), the required inequality (A.50) follows with $\mu = 2\gamma$ and $\nu = 6C_1 R_4$. □

We are now in a position to give the

Proof of Theorem A.5. Suppose that u has a zero of infinite order at $p \in M$. Let $B_{R_0}(p)$ be a geodesic ball with $R_0 < \operatorname{inj}(p)$. Pick $\bar{R} > 0$ to be smaller than the value R_4 in Lemma A.17. Note that, according to (A.47), Lemma A.13 applies.

Furthermore, $\bar{R}$ can be chosen so to depend only on m and on the geometry of a compact neighborhood of p. We show that u is identically null in $B_{\bar{R}}(p)$. By contradiction, suppose that this is not the case. Let us consider the function

$$K(R) = \frac{1}{R^{m-1}} \int_{\partial B_R} u^2, \; 0 < R \leq \bar{R}.$$

Note that, by assumption and according to Lemma A.13, we find $0 < \delta < \bar{R}$ such that $K(R) > 0$ on $(\delta, \bar{R}]$. In particular,

$$K\left(\bar{R}\right) > 0. \tag{A.60}$$

Now, by (A.56) of the previous lemma, we have

$$K'(R) \leq \frac{2}{R} A(R) + \alpha K(R) \text{ on } (0, \bar{R}],$$

for some constant $\alpha = \alpha\left(\bar{R}\right) > 0$. Thus, using (A.46) of Lemma A.17 gives that, for a suitable $k \in \mathbb{N}$ sufficiently large,

$$K'(R) \leq \frac{k}{R} K(R) \text{ on } (0, \bar{R}].$$

Combining this latter with (A.60) and applying Lemma A.6 we conclude that

$$\liminf_{R \to 0^+} \frac{1}{R^{k+m-1}} \int_{\partial B_R} u^2 = \liminf_{R \to 0^+} \frac{K(R)}{R^k} > 0,$$

so that u cannot have a zero of infinite order at p. The contradiction shows that u must vanish on $B_{\bar{R}}(p)$.

It remains to show that u is identically null on M. We use a connectedness argument. Let

$$D = \{x \in M : u(x) = 0 \text{ in a neighborhood of } x\}.$$

Then, obviously, $D \neq \emptyset$ is an open set. We have to show that D is closed. Thus, let $x \in D'$, be a limit point of D, and consider any compact neighborhood Ω of x. The radius $\bar{R} > 0$ introduced in the first part of the proof can be required to depend only on m and the geometry of M in a suitable neighborhood of $\bar{\Omega}$. Let us choose a point $y \in D \cap \Omega$ such that $\operatorname{dist}_M(x, y) < \bar{R}$. The function u vanishes in a neighborhood of y, hence, u has a zero of infinite order at y. Applying the first part of the proof we deduce that u must vanish in $B_{\bar{R}}(y)$. Since this ball contains x, we conclude $x \in D$. □

Appendix B

L^p-cohomology of non-compact manifolds

Throughout this section, we will always assume Riemannian manifolds to be *oriented*. The main purpose of the appendix is to introduce the basic definitions concerning the L^p de Rham cohomology and to collect some classical results about its dependence on the geometry and the topology of the manifold under consideration. The role of L^p harmonic forms will be also briefly discussed. For a more detailed account the reader is referred to Pansu's survey paper [126] for the case $p = 2$, and to Chapter 8 of Gromov's seminal book [75], for the general L^p case. See also the very recent notes by Carron, [29], where possible links between L^2-cohomology and the de Rham cohomology with compact support are established.

B.1 The L^p de Rham cochain complex: reduced and unreduced cohomologies

Suppose we are given a Riemannian manifold $(M, \langle\,,\rangle)$ of dimension $m = \dim M$. Set $\Lambda^0(T^*M) = \mathbb{R}$ and, for any integer $k \geq 1$, let $\Lambda^k(T^*M)$ denote the *k-exterior vector bundle* over M. The sections of $\Lambda^k(T^*M)$, $k \geq 0$, are called *differential k-forms* (or differential forms of degree k). Clearly, differential forms of degree 0 are nothing but real functions. The real vector space of the *measurable* differential k-forms on M is denoted by $\Lambda^k(M)$, while the symbol $C^\infty\Lambda^k(M)$ is reserved to the vector space of the smooth ones. In case of smooth differential forms with compact support, we write $C_c^\infty\Lambda^k(M)$. Thus, having fixed any (non-necessarily orthonormal) local frame field $\{V_i\}$ and denoting by $\{\varpi^i\}$ the dual coframe, a k-form ω is locally represented by

$$\omega = \sum_{1\leq i_1<\cdots<i_k\leq m} \omega_{i_1\cdots i_k}\varpi^{i_1}\wedge\cdots\wedge\varpi^{i_k} \tag{B.1}$$

where the $\binom{m}{k}$ local functions $\omega_{i_1\cdots i_k}$ reflect the regularity of ω (e.g., they are measurable if $\omega \in \Lambda^k(M)$) and $\varpi^{i_1}\wedge\cdots\wedge\varpi^{i_k}$ denotes *the alternating, k-linear product* of $\varpi^{i_1},\ldots,\varpi^{i_k}$ defined as follows: for any tangent vectors $X_1,\ldots,X_k$

$$\left(\varpi^{i_1}\wedge\cdots\wedge\varpi^{i_k}\right)(X_1,\ldots,X_k) = \frac{1}{k!}\sum \operatorname{sgn}(\sigma)\,\varpi^{i_1}\left(X_{\sigma(1)}\right)\cdots\varpi^{i_k}\left(X_{\sigma(k)}\right),$$

the sum being extended to all the permutations σ of k elements.

The vector bundle $\Lambda^k(T^*M)$ inherits from $(M, \langle\,,\rangle)$ a *natural Riemannian structure* defined as follows.

For each $x \in M$ we have to prescribe an inner product $(\,,)_x$ on the fibre $\Lambda^k(T^*_x M)$, smoothly varying with x. To this end, let $\{\theta^i\} \subset C^\infty\Lambda^1(U)$ be an orthonormal co-frame on a small domain $U \subset M$, so that,

$$\langle,\rangle|_U = \sum \theta^i \otimes \theta^i. \tag{B.2}$$

We know that, for every $x \in U$,

$$\left\{\theta^{i_1}(x) \wedge \cdots \wedge \theta^{i_k}(x)\right\}_{1\le i_1<\cdots<i_k\le m} \tag{B.3}$$

realizes a basis of $\Lambda^k(T^*_x M)$. We declare that such a basis is orthonormal in $\left(\Lambda^k(T^*_x M), (,)_x\right)$. Thus, if $\omega, \eta \in \Lambda^k(M)$ and, locally, $\omega = \sum \omega_{i_1\cdots i_k}\theta^{i_1} \wedge\cdots\wedge\theta^{i_k}$, $\eta = \sum \eta_{i_1\cdots i_k}\theta^{i_1} \wedge \cdots \wedge \theta^{i_k}$, then

$$(\omega, \eta) = \sum_{1\le i_1<\cdots<i_k\le m} \omega_{i_1\cdots i_k}\eta_{i_1\cdots i_k}.$$

The Riemannian vector bundle $\Lambda^k(T^*M)$ is then supplied with a *compatible connection*

$$D^k : C^\infty\Lambda^k(M) \to \mathbb{C}^\infty\left(\Lambda^k(T^*M) \otimes T^*M\right), \tag{B.4}$$

with $\Gamma\left(\Lambda^k(T^*M) \otimes T^*M\right)$ the space of smooth sections of the tensor product bundle. It is defined via the Levi–Civita connection D of $(M, \langle,\rangle)$ by the following formula:

Let $\omega \in C^\infty\Lambda^k(M)$. Then, for all vector fields $X_1, \ldots, X_k, Y$ of M,

$$\begin{aligned}\left(D^k_Y\omega\right)(X_1, \ldots, X_k) &:= \left(D^k\omega\right)(X_1, \ldots, X_k; Y) \\ &= D_Y(\omega(X_1, \ldots, X_k)) - \sum_i \omega(X_1, \ldots, D_Y X_i, \ldots, X_k).\end{aligned} \tag{B.5}$$

Recall that *compatibility* means

$$D^k(\omega, \eta) = \left(D^k\omega, \eta\right) + \left(\omega, D^k\eta\right),$$

for every $\omega, \eta \in \Lambda^k(M)$.

Via the connection D^k we can introduce a further first-order, $\mathbb{R}$-linear operator called the exterior differential. The *k-th exterior differential* is the degree $+1$ operator

$$d^k : C^\infty\Lambda^k(M) \to C^\infty\Lambda^{k+1}(M) \tag{B.6}$$

defined as follows: take any local frame $\{V_i\}$ of M and denote by $\{\varpi^i\}$ its dual coframe. Then

$$d^k\omega = \sum \varpi^i \wedge D^k_{V_i}\omega. \tag{B.7}$$

Obviously, the definition does not depend on the chosen frame and, in fact, it depends only on the differentiable structure of M, whereas the Riemannian metric $\langle , \rangle$ plays no role. It can be shown that the exterior differential d is an *anti-derivation* in the sense that, for $\omega_1 \in C^\infty \Lambda^k (M)$ and $\omega_2 \in C^\infty \Lambda^h (M)$,

$$d^{k+h} (\omega_1 \wedge \omega_2) = d^k \omega_1 + (-1)^k d^h \omega_2. \tag{B.8}$$

Furthermore, d enjoys the coboundary property

$$d^{k+1} \circ d^k = 0. \tag{B.9}$$

According to (B.9) the family of real vector spaces and linear maps

$$\left\{ C^\infty \Lambda^k (M) \xrightarrow{d^k} C^\infty \Lambda^{k+1} (M) \right\}$$

gives rise to a (algebraic) cochain complex which is usually called *the (smooth) de Rham complex of M*. The corresponding cohomology

$$H^k_{dR} (M) = \frac{Z^k_{dR} (M)}{B^k_{dR} (M)}$$

with

$$Z^k_{dR} (M) = \ker d^k \text{ and } B^k_{dR} (M) = \operatorname{Im} d^{k-1}$$

is the (smooth) *de Rham cohomology of M*.

It is readily seen that such a cohomology theory is a diffeomorphism invariant of M and it is insensitive to the underlying Riemannian structure. Thus, for instance, the Euclidean space $\mathbb{R}^m$ and the hyperbolic space $\mathbb{H}^m$ are the same space from the de Rham cohomology viewpoint. To prove the above mentioned invariance, one usually notes that every smooth map $f : M \to N$ between differentiable manifolds induces a family of homomorphisms

$$(f^*)^k : C^\infty \Lambda^k (N) \to C^\infty \Lambda^k (M)$$

which are called the *pull-back maps.* They are defined by the formula

$$(f^*)^k (\omega) (v_1, \ldots, v_k) = \omega (f_* (v_1), \ldots, f_* (v_k))$$

where $f_* : TM \to TN$ is the usual differential of a smooth map. For the sake of notation, we simply write $f^* \omega = \omega (f_*)$. In view of the commutation rule

$$d^k_M \circ (f^*)^k = (f^*)^{k+1} \circ d^k_N, \tag{B.10}$$

the family $\{(f^*)^k\}_k$ realizes a homomorphism of cochain complexes which, in turn, gives rise to a cohomology homomorphism denoted by $\{(f^\#)^k\}_k$. Here, the k-th map

$$(f^\#)^k : H^k_{dR} (N) \to H^k_{dR} (M)$$

is simply defined by

$$\left(f^{\#}\right)^k[\omega] = [(f^*)^k\omega].$$

The functioriality property of $^{\#}$ asserts that, for any smooth maps between differentiable manifolds

$$M \xrightarrow{f} N \xrightarrow{g} Z,$$

it holds that

$$(g \circ f)^{\#} = f^{\#} \circ g^{\#}$$

and, furthermore,

$$(\mathrm{Id}_M)^{\#} = \mathrm{Id}_{H_{dR}}.$$

Accordingly, if $f : M \to N$ is a smooth diffeomorphism, then $f^{\#}$ is a cohomology isomorphism with inverse $f^{\#-1} = \left(f^{-1}\right)^{\#}$.

We now leave the smooth realm and enter the L^p world. We fix $1 \leq p < +\infty$, and for every $k = 0, \ldots, m$, we define the Banach space

$$L^p\Lambda^k(M) = \overline{C_c^\infty\Lambda^k(M)},$$

the closure being taken with respect to the obvious norm

$$\|\omega\|_{L^p} = \left(\int_M |\omega|^p\right)^{1/p}.$$

Note that, without any further assumption on $(M, \langle,\rangle)$,

$$L^p\Lambda^k(M) = \left\{\omega \in \Lambda^k(M) : \|\omega\|_{L^p} < +\infty\right\}.$$

Let us now consider the Sobolev space

$$W^{1,p}\Lambda^k(M) = \overline{C_c^\infty\Lambda^k(M)},$$

this time the closure being intended with respect to the norm

$$\|\omega\|_{W^{1,p}} = \|\omega\|_{L^p} + \|d\omega\|_{L^p}.$$

Due to the coboundary property, the exterior differential

$$d^k : C_c^\infty\Lambda^k(M) \to C_c^\infty\Lambda^{k+1}(M) \subset L^p\Lambda^k(M)$$

extends to a bounded operator

$$d^k : W^{1,p}\Lambda^k(M) \to L^p\Lambda^{k+1}(M).$$

Explicitly, for every $\omega \in W^{1,p}\Lambda^k(M)$ take any sequence $\{\omega_n\} \subset C_c^\infty\Lambda^k(M)$ such that $\omega_n \overset{W^{1,p}\Lambda^k(M)}{\to} \omega$ as $n \to +\infty$, and define

$$d^k\omega = \lim_{n\to+\infty} d^k\omega_n.$$

As a matter of fact, there is another natural way to extend the action of the exterior differential to non-smooth forms, i.e., by means of the distributional formula

$$(\widetilde{d}^k \omega, \eta) = \int_M \langle \omega, \delta^k \eta \rangle, \ \forall \eta \in C_c^\infty \Lambda^{k+1}(M).$$

Here, $\delta^k : C^\infty \Lambda^{k+1}(M) \to C^\infty \Lambda^k(M)$ stands for the k-th exterior codifferential of M. It is the formal adjoint of d^k in $L^2\Lambda^k(M)$. The exterior codifferential depends on d^k in a rather explicit way. Let $* : C^\infty \Lambda^k(M) \to C^\infty \Lambda^{m-k}(M)$ be the pointwise linear isometry (*the Hodge-$*$ operator*) defined, for any local, oriented, orthonormal coframe $\{\theta^j\}$ by

$$*\theta^{i_1} \wedge \cdots \wedge \theta^{i_k} = \pm \theta_x^{j_1} \wedge \cdots \wedge \theta_x^{j_{m-k}}$$

where $\theta_x^{i_1} \wedge \cdots \wedge \theta_x^{i_k} \wedge \left(\pm \theta_x^{j_1} \wedge \cdots \wedge \theta_x^{j_{m-k}} \right)$ gives the correct orientation of M. Then

$$\delta^k = (-1)^{m(k+1)+1} * d^k * .$$

Coming back to the distributional differential, we are naturally led to introduce a second Sobolev space of differential forms

$$\widetilde{W^{1,p}}\Lambda^k(M) = \left\{ \omega \in L^p\Lambda^k(M) : \widetilde{d}^k \omega \in L^p\Lambda^{k+1}(M) \right\}.$$

A direct application of the Gaffney cut-off trick shows that, on a *complete manifold*,

$$\widetilde{W^{1,p}}\Lambda^k(M) = W^{1,p}\Lambda^k(M).$$

In particular the weak exterior differential and the densely defined one coincide; see, e.g., [160].

From now on, unless otherwise specified, we will always assume that Riemannian manifolds are geodesically complete.

Obviously, as in the smooth case, the exterior differential enjoys the basic identity

$$d^{k+1} \circ d^k = 0 \tag{B.11}$$

so that, the collection of bounded operators d^k's gives rise to the Banach cochain complex

$$\cdots \to W^{1,p}\Lambda^k(M) \overset{d^k}{\to} W^{1,p}\Lambda^{k+1}(M) \overset{d^{k+1}}{\to} W^{1,p}\Lambda^{k+2}(M) \to \cdots.$$

As usual, the corresponding spaces of *cocycles* and *coboundaries*are defined by

$$Z^{k,p}(M) = \ker d^k \text{ and } B^{k,p}(M) = \operatorname{Im} d^{k-1}.$$

Differential forms in the (Banach) space $Z^{k,p}(M)$ are called *weakly closed* while differential forms in the (possibly incomplete) space $B^{k,p}(M)$ are called *weakly exact.*

Note that, for $\omega \in Z^{k,p}(M)$, $\|\omega\|_{\Lambda^{k,p}} = \|\omega\|_{L^p}$ and we have the important equality

$$\overline{B^{k,p}(M)} = \overline{dC_c^\infty \Lambda^{k-1}(M)}$$

whose validity is an immediate consequence of the continuity of d^{k-1} and the definition of $W^{1,p}\Lambda^k(M)$. Furthermore, by (B.11) and the closeness of $Z^{k,p}(M)$, we have the inclusion of subspaces

$$B^{k,p}(M) \subset \overline{B^{k,p}(M)} \subset Z^{k,p}(M).$$

We can therefore introduce two different cohomology spaces: the first one is simply given by the quotient

$$L^p H^k(M) = Z^{k,p}(M) / B^{k,p}(M)$$

and is called the *k-th de Rham space of unreduced L^p-cohomology.* The second one is obtained by replacing $B^{k,p}(M)$ with its closure

$$\overline{L^p H^k(M)} = Z^{k,p}(M) / \overline{B^{k,p}(M)}$$

and it is named the *k-th de Rham space of reduced L^p-cohomology.* We endow both these spaces with the quotient topology. Note that $\overline{L^p H^k(M)}$ inherits the norm $\|\cdot\|_{L^p}$ of $L^p\Lambda(M)$ and, in this way, becomes itself a Banach space. Note also the obvious inclusion

$$\overline{L^p H^k(M)} \subset L^p H^k(M)$$

and the isomorphism

$$\overline{L^p H^k(M)} \simeq L^p H^k(M) / \left\{ [\omega] \in L^p H^k(M) : \|[\omega]\|_{L^p} = 0 \right\}.$$

In particular, every topological isomorphism $L^p H^k(M) \to L^p H^k(N)$ induces a topological isomorphism at the reduced level.

Example B.1. Following M. Gromov, [75], let us consider the standard m-dimensional hyperbolic space $\left(\mathbb{H}^m_{-1}, \mathrm{can}\right)$ which we identify with its Poincaré model

$$\left(\mathbf{B}_1(0), \frac{4 \sum dx^i \otimes dx^i}{\left(1 - |x|^2\right)^2} \right)$$

where $\mathbf{B}_1(0) \subset \mathbb{R}^m$ denotes the Euclidean unit ball. Then

$$kp = m \Longrightarrow \overline{L^p H^k\left(\mathbb{H}^m_{-1}\right)} \neq 0. \tag{B.12}$$

To see this, we first observe that, on a generic m-dimensional Riemannian manifold $(M, \langle , \rangle)$, condition $kp = m$ implies that the Banach space $L^p\Lambda^k(M)$ is a conformal invariant. Namely, let $\widetilde{\langle,\rangle} = \lambda^2(x)\langle,\rangle$ be a second Riemannian metric, pointwise conformal to $\langle,\rangle$. Then, for every k-form ω,

$$|\omega|_{\widetilde{\langle,\rangle}} = \lambda^{-k}|\omega|_{\langle,\rangle}$$

and, furthermore,

$$\widetilde{d\mathrm{vol}} = \lambda^m d\mathrm{vol}$$

where $d$vol and $\widetilde{d\mathrm{vol}}$ denote the Riemannian measures corresponding to the given metrics. As a consequence

$$\int_M |\omega|^p_{\widetilde{\langle,\rangle}}\, \widetilde{d\mathrm{vol}} = \int_M |\omega|^p_{\langle,\rangle}\, d\mathrm{vol}$$

as claimed.

Now, since the hyperbolic metric is pointwise conformal to the Euclidean one and the unit ball $\mathbf{B}_1(0)$ has finite Euclidean volume, we deduce that the inclusion

$$i : \mathbf{B}_1(0) \hookrightarrow \mathbb{R}^m$$

induces, for every $hq = m$, a bounded pull-back homomorphism

$$i^* : \Lambda^h(\mathbb{R}^m) \to L^q\Lambda^h\left(\mathbb{H}^m_{-1}\right).$$

Let

$$p = \frac{m}{k}, \quad q = \frac{m}{m-k}$$

and consider the closed differential forms

$$\begin{aligned}
\omega_1 &= i^*\left(dx^1 \wedge \cdots \wedge dx^k\right) \in Z^{k,p}\left(\mathbb{H}^m_{-1}\right),\\
\omega_2 &= i^*\left(dx^{k+1} \wedge \cdots \wedge dx^m\right) \in Z^{m-k,q}\left(\mathbb{H}^m_{-1}\right).
\end{aligned}$$

We claim that both ω_1 and ω_2 represent non-trivial de Rham cohomology classes in $\overline{L^pH^k\left(\mathbb{H}^m_{-1}\right)}$ and $\overline{L^qH^{m-k}\left(\mathbb{H}^m_{-1}\right)}$, respectively. Indeed, by contradiction, suppose that this is not the case. Without loss of generality we may assume that

$$\omega_1 \in \overline{B^{k,p}\left(\mathbb{H}^m_{-1}\right)} \tag{B.13}$$

so that $d^{k-1}\tau_n \overset{L^p}{\to} \omega_1$, for some sequence $\{\tau_n\} \subset C^\infty_c\Lambda^{k-1}\left(\mathbb{H}^m_{-1}\right)$. Since

$$\begin{aligned}
\left|\int_{\mathbb{H}^m_{-1}} d\tau_n \wedge \omega_2 - \int_{\mathbb{H}^m_{-1}} \omega_1 \wedge \omega_2\right| &\le \int_{\mathbb{H}^m_{-1}} |(d\tau_n - \omega_1) \wedge \omega_2|\\
&\le \|d\tau_n - \omega_1\|_{L^p} \|\omega_2\|_{L^q}
\end{aligned}$$

using Stokes theorem and the conformal invariance of $L^1\Lambda^m\left(\mathbb{H}^m_{-1}\right)$, we deduce

$$\begin{aligned} 0 &= \lim_{n\to+\infty}\int_{\mathbb{H}^m_{-1}} d\tau_n \wedge \omega_2 = \int_{\mathbb{H}^m_{-1}} \omega_1 \wedge \omega_2 \\ &= \int_{\mathbf{B}_1(0)} dx^1 \wedge \cdots \wedge dx^m = \mathrm{vol}_{\mathbb{R}^m}\left(\mathbf{B}_1\left(0\right)\right), \end{aligned}$$

i.e., a contradiction. Therefore ω_1 represents a non-trivial de Rham class in $L^pH^k\left(\mathbb{H}^m_{-1}\right)$, proving (B.12).

Example B.2. The usual proof of the *Poincaré lemma* for the unit Euclidean ball $\mathbf{B}_1\left(0\right) \subset \mathbb{R}^m$ works even in the L^p setting. Accordingly, for the geodesically incomplete manifold $\left(\mathbf{B}_1\left(0\right), \mathrm{can}_{\mathbb{R}^m}\right)$, the following holds:

$$L^pH^k\left(\mathbf{B}_1\left(0\right)\right) = 0$$

for every $k = 1, \ldots, m$ and for every $p \geq 1$. See, e.g., [127], Lemma 8.

As illustrated by Examples B.1 and B.2 above, diffeomorphic manifolds are far from possessing isomorphic L^p cohomology spaces. The reason is that a smooth map, in general, does not pull back L^p forms to L^p forms and, even in this case, the pull-back operator could be unbounded. However, it is easy to show that if a map induces L^p-bounded pull-backs, then the usual commutation rule (B.10) still holds and we get again bounded cochain and cohomology homomorphisms satisfying the contravariant functoriality property, as explained above. Bi-Lipschitz homeomorphisms are the most natural example of maps enjoying these properties, showing that the L^p de Rham cohomology essentially depends on the Riemannian geometry at infinity of the manifold at hand. More precisely, recall that a continuous map $f : (M, \langle\, ,\rangle) \to (N, (\, ,))$ between Riemannian manifolds is said to be a *Lipschitz map* if the following two conditions are met:

(1) f is differentiable almost everywhere (a.e.), i.e., denoting by $d\mathrm{vol}_M$ the Riemannian measure of $(M, \langle, \rangle_M)$, there exists a set $\Omega \subset M$ with $d\mathrm{vol}_M\left(\Omega\right) = 0$ such that, for every $x \in M \setminus \Omega$, the differential map $f_{*x} : T_xM \to T_{f(x)}N$ exists.

(2) There exists a constant $C > 0$ such that $\left|f_{*x}\right|_N(v) \leq C\left|v\right|_M$ for every $x \in M \setminus \Omega$ and every $v \in T_xM$.

One also says that $(M, \langle\, ,\rangle)$ and $(N, (\, ,))$ are *bi-Lipschitz equivalent* if there exists a homeomorphism $f : (M, \langle\, ,\rangle) \to (N, (\, ,))$ such that both f and f^{-1} are Lipschitz maps.

Remark B.3. In general, given a Lipschitz map $f : M \to N$, the pull-back of a p-integrable form on N is not p-integrable on M. Consider for instance the Riemannian universal covering of the flat torus, say

$$\pi : \left(\mathbb{R}^m, \mathrm{can}_{\mathbb{R}^m}\right) \to \left(\mathbb{T}^m = \mathbb{R}^m/\mathbb{Z}^m, \mathrm{can}_{\mathbb{T}^m}\right).$$

Since π is a local isometry, we have

$$|\pi_*(v)|_{\mathbb{T}^m} = |v|_{\mathbb{R}^m},$$

hence π is Lipschitz. Since $\mathbb{T}^m$ is compact, any continuous form is p-integrable, for every $p > 0$. On the other hand, let $dx^1 \wedge \cdots \wedge dx^k \in C^\infty \Lambda^k(\mathbb{T}^m)$ be induced by the coordinate functions $x^1, \ldots, x^m$ of $\mathbb{R}^m$. Then

$$\pi^*\left(dx^1 \wedge \cdots \wedge dx^k\right) = dx^1 \wedge \cdots \wedge dx^k \notin L^p\Lambda^k(\mathbb{R}^m),$$

showing that the L^p-integrability is not preserved.

Making an essential use of the change of variable formula for Lipschitz transformations yields the invariance property alluded to above.

Proposition B.4. *Bi-Lipschitz equivalent Riemannian manifolds have isomorphic de Rham L^p-cohomologies.*

Example B.5. Suppose that the Riemannian manifolds $(M, \langle\,,\rangle)$ and $(N, (\,,))$ are isometric at infinity. This means that there exists a diffeomorphism $f : M \to N$ such that, for some (possibly empty) compact set $K \subset M$, setting $M' = M \setminus K$ and $N' = N \setminus f(K)$, the restriction $f' = f|_{M'} : (M', \langle,\rangle) \to (N', (\,,))$ is a Riemannian isometry. Then,

$$L^pH^k(M) \simeq L^pH^k(N).$$

In this sense, the de Rham L^p-cohomology depends only on the Riemannian geometry at infinity of the manifold.

The proof of the claimed property is a trivial consequence of Proposition B.4.

Example B.6. Suppose we are given a compact Riemannian manifold $(M, \langle\,,\rangle)$. Its Riemannian universal covering is denoted by $P_M : (\widetilde{M}, \widetilde{\langle\,,\rangle}) \to (M, \langle\,,\rangle)$. Then, the L^p de Rham reduced and unreduced cohomologies of $\widetilde{M}$ are diffeomorphism invariants of the base manifold M. Namely, if $(M, \langle\,,\rangle)$ is diffeomorphic to $(N, (\,,))$, then

$$L^pH^k(\widetilde{M}) \simeq L^pH^k(\widetilde{N})$$

where $P_N : (\widetilde{N}, \widetilde{(\,,)}) \to (N, (\,,))$ is the (Riemannian) universal covering of N. This follows from the fact that any diffeomorphism $f : M \to N$ lifts to a Lipschitz diffeomorphism $\widetilde{f} : \widetilde{M} \to \widetilde{N}$. Indeed, the composition $f \circ P_M : \widetilde{M} \to N$ defines a second universal covering of N. Therefore, by uniqueness, there exists a fiber-preserving diffeomorphism $\widetilde{f} : \widetilde{M} \to \widetilde{N}$ such that $P_N \circ \widetilde{f} = f \circ P_M$. Since both P_M and P_N are local isometries we see that

$$|\widetilde{f}_*|_{\widetilde{N}} = |(P_N)_*|_N\, |\widetilde{f}_*|_{\widetilde{N}} = |(P_N \circ \widetilde{f})_*|_N = |(f \circ P_M)_*|_N = |f_*|_N\, |(P_M)_*| = |f_*|_N\,.$$

On the other hand, by the compactness of both M and N, we have

$$|f_*|_N \le C$$

for some $C > 0$. Therefore $\widetilde{f}$ is a Lipschitz map. Obviously, similar considerations hold for the inverse map $\widetilde{f}^{-1}$, completing the proof.

B.2 Harmonic forms and L^2-cohomology

Recall that a differential form $\omega \in C^\infty \Lambda^k(M)$ is said to be harmonic if

$$\Delta_H \omega = 0$$

where $\Delta_H = d\delta + \delta d$ is a degree 0, second-order, differential operator called the *Hodge Laplacian*. Note that, on functions, Δ_H is nothing but $-\Delta$. The real vector space of (smooth) harmonic k-forms of M is denoted by $\mathcal{H}^k(M)$. For the sake of notation, we also set

$$L^p\mathcal{H}^k(M) = \mathcal{H}^k(M) \cap L^p\Lambda^k(M).$$

Let us now consider the space $L^2\Lambda^k(M)$ with the usual scalar product

$$(\omega, \xi)_{L^2} = \int_M \langle \omega, \xi \rangle_{\Lambda^k}. \tag{B.14}$$

As we pointed out earlier, since $(M, \langle\,,\rangle)$ is geodesically complete, the exterior differential of an L^2-form $\omega \in L^2\Lambda^k(M)$ can be defined via the (distributional) formula

$$(d\omega, \tau) := (\omega, \delta\tau)_{L^2}, \ \forall \tau \in C_c^\infty \Lambda^k(M). \tag{B.15}$$

The L^2-form ω is said to be *weakly closed* if $d\omega = 0$ in the distributional sense. Similarly, we can introduce the distributional co-differential of ω by setting

$$(\delta\omega, \tau) := (\omega, d\tau)_{L^2}, \ \forall \tau \in C_c^\infty \Lambda^{k-1}(M) \tag{B.16}$$

and we say that ω is *weakly co-closed* provided $\delta\omega = 0$. The following description of L^2 harmonic forms is essentially due to A. Andreotti and E. Vesentini; see [160] and [42]. Its proof combines *Gaffney-integration by parts* (which is based on the Gaffney cut-off trick) with classical elliptic regularity theory.

Proposition B.7. *Let $(M, \langle\,,\rangle)$ be geodesically complete. Then $\omega \in L^2\mathcal{H}^k(M)$ if and only if $\omega \in L^2\Lambda^k(M)$ satisfies both $d\omega = 0$ and $\delta\omega = 0$ in the weak sense.*

Remark B.8. One may suspect that similar characterizations hold with different degrees of integrability. In general, even for nice (e.g., with bounded geometry) Riemannian spaces, this is false. In fact, D. Alexandru-Rugina, [4], obtains L^p harmonic k-forms on the hyperbolic space $\mathbb{H}^m_{-1}$, $m \geq 3$, which are neither closed nor co-closed.

The next result is called a decomposition theorem by K. Kodaira; see [42], [89].

Theorem B.9. *Let $(M, \langle\,,\rangle)$ be a geodesically complete manifold. Then*

$$\begin{aligned} L^2\Lambda^k(M) &= \overline{d\left(C_c^\infty \Lambda^{k-1}(M)\right)} \oplus \overline{\delta\left(C_c^\infty \Lambda^{k+1}(M)\right)} \oplus L^2\mathcal{H}^k(M), \\ Z^{k,2}(M) &= \overline{B^{k,2}(M)} \oplus L^2\mathcal{H}^k(M). \end{aligned} \tag{B.17}$$

As a consequence, one gets the following L^2 version of the Hodge representation theorem showing that, as in the compact setting, L^2 harmonic forms completely describe the L^2 reduced cohomology of the manifold.

Corollary B.10. *Let $(M, \langle\, ,\rangle)$ be a complete Riemannian manifold. Then, the map*

$$\mathcal{I} : L^2\mathcal{H}^k(M) \to \overline{L^2H^k(M)}$$

defined by

$$\mathcal{I}(\omega) = [\omega]$$

is a linear isometry.

Example B.11. In view of the L^2 representation theorem, we can use classical analytic tools from harmonic function theory to show that the standard Euclidean space $(\mathbb{R}^m, \mathrm{can})$ satisfies

$$\dim \overline{L^2H^k(\mathbb{R}^m)} = 0, \quad \forall k = 0, \dots, m. \tag{B.18}$$

This amounts to showing that every L^2 harmonic k-form vanishes. In fact, we shall prove the following more general statement

$$\dim L^p\mathcal{H}^k(\mathbb{R}^m) = \begin{cases} 0 & \text{if } 1 \leq p < +\infty, \\ \binom{n}{k} & \text{if } p = +\infty. \end{cases} \tag{B.19}$$

To this end, take $\omega \in L^p\mathcal{H}^k(\mathbb{R}^m)$. Using the definition of the Riemannian structure of $\Lambda^k(\mathbb{R}^m)$ one can show that ω can be globally written as

$$\omega = \sum \omega_{i_1\cdots i_k} dx^{i_1} \wedge \cdots \wedge dx^{i_k}$$

where

$$\omega_{i_1\cdots i_k} \in L^p\mathcal{H}^0(\mathbb{R}^m).$$

The desired conclusion (B.19) therefore reduces to a quite standard *Liouville theorem for harmonic functions on* $\mathbb{R}^m$.

First, we consider the case $1 \leq p < +\infty$. For any fixed $\delta > 0$ let

$$v_\delta = \sqrt{(\omega_{i_1\cdots i_k})^2 + \delta^2}.$$

Obviously, $0 \leq |\omega_{i_1\cdots i_k}| < v_\delta$. Furthermore, since $\omega_{i_1\cdots i_k}$ is harmonic we have

$$\Delta v_\delta = \frac{\delta^2 |\nabla \omega_{i_1\cdots i_k}|^2}{v_\delta^3} \geq 0,$$

proving that v_δ is a positive, subharmonic function on all of $\mathbb{R}^m$. The usual mean value inequality tells us that for every $x_0 \in \mathbb{R}^m$ and for every ball $B_R(x_0)$,

$$0 \leq |\omega_{i_1\cdots i_k}|(x_0) \leq v_\delta(x_0) \leq \frac{1}{\mathrm{vol}(B_R(x_0))} \int_{B_R(x_0)} v_\delta.$$

Whence, using Hölder inequality and letting $\delta \to 0+$ we deduce, for every $p \geq 1$,

$$0 \leq \left|\omega_{i_1 \cdots i_k}\right| (x_0) \leq \frac{1}{\operatorname{vol}\left(B_R\left(x_0\right)\right)^{\frac{1}{p}}} \left\{ \int_{B_R(x_0)} \left|\omega_{i_1 \cdots i_k}\right|^p \right\}^{1/p}.$$

From this latter, since $\left|\omega_{i_1 \cdots i_k}\right| \in L^p\left(\mathbb{R}^m\right)$, letting $R \to +\infty$ we conclude that $\omega_{i_1 \cdots i_k}\left(x_0\right) = 0$. This proves the first part of (B.19).

Now we consider the case $p = +\infty$. Since $\omega_{i_1 \cdots i_k}$ is bounded from below, setting

$$u = \omega_{i_1 \cdots i_k} - \inf_{\mathbb{R}^m} \omega_{i_1 \cdots i_k},$$

we have that $u \geq 0$ is a harmonic function on $\mathbb{R}^m$ satisfying

$$\inf_{\mathbb{R}^m} u = 0.$$

The Harnack inequality then tells us that there exists a universal constant $C > 0$ such that, on every ball $B_R(0)$,

$$0 \leq \sup_{B_R(0)} u \leq C \inf_{B_R(0)} u.$$

Letting $R \to +\infty$ we therefore conclude $u \equiv 0$, i.e., $\omega_{i_1 \cdots i_k} \equiv const$, completing the proof of (B.19).

Remark B.12. As already noted, a theorem by S.T. Yau, [168], shows that, on a generic complete Riemannian manifold $(M, \langle , \rangle)$ of infinite volume, the only non-negative subharmonic function u satisfying $u \in L^p(M)$, for some $1 < p < +\infty$, is the null function $u = 0$.

It should be stressed that the endpoint cases $p = 1$ and $p = +\infty$ give rise to substantial problems of non-technical nature. In general, constancy results are false without some further geometric assumption on the underlying manifold.

B.3 Harmonic forms and $L^{p \neq 2}$-cohomology

A natural question is whether or not L^p harmonic forms may be used to describe the L^p-cohomology even for $p \neq 2$.

In general, the answer is negative. Indeed, as we mentioned in Remark B.8, hyperbolic space $\mathbb{H}^m_{-1}$, $m \geq 3$, supports non-trivial L^p-harmonic k-forms which are not closed, for some $p > 2$ and $1 \leq k \leq (m-1)/2$, see [4]. Clearly, these harmonic forms represent no cohomology class at all. As we shall soon see, the range $p > 2$ is, in a sense, sharp.

In the affirmative direction, only partial results are known. Here is a brief account.

B.3.1 The L^p-Hodge decomposition for Euclidean space

The following result is due to T. Iwaniec and G. Martin, [85]. The original proof relies on classical singular integral methods. In particular, L^p-estimates of the Riesz transform are used.

Theorem B.13. *For every $1 < p < +\infty$, we have the topological splitting*

$$L^p\Lambda^k(\mathbb{R}^m) = dW^{1,p}\Lambda^{k-1}(\mathbb{R}^m) \oplus \delta W^{1,p}\Lambda^{k+1}(\mathbb{R}^m),$$

differential and co-differential being understood in the sense of distributions.

Note that, in the above decomposition, L^p harmonic forms do not appear. This is in accordance with the Liouville properties obtained in Example B.11 above. The next vanishing result for the non-reduced cohomology is an immediate consequence of the theorem.

Corollary B.14. *For every $1 < p < +\infty$ and for every $k = 0, \dots, m$,*

$$L^pH^k(\mathbb{R}^m) = 0.$$

B.3.2 The case of manifolds with bounded geometry

Roughly speaking, a manifold is said to have bounded geometry if its Riemannian invariants (metric tensor, Christoffel symbols, Riemann curvature etc.) are locally uniformly controlled. The following definition employs a locally uniform bound of the metric coefficients in normal coordinates.

A complete Riemannian manifold $(M, \langle , \rangle)$ is said to have *bounded geometry at order* $s \in [0, +\infty]$ if the following hold.

(Inj) The injectivity radius of M satisfies $\operatorname{inj}(M) = i > 0$, i.e., for every $x \in M$, the exponential map $\exp_x : \mathbf{B}_i(0) \subset T_xM = \mathbb{R}^m \to B_i(x)$ is a diffeomorphism.

(G_s) For every $x \in M$, fix an orthonormal frame $\{e_j\}$ of $T_xM = \mathbb{R}^m$, and consider the pull-back metric tensor on $\mathbf{B}_i(0)$,

$$g_{ij} = \langle (\exp_x)_* e_i, (\exp_x)_* e_j \rangle.$$

Then, both the matrix g_{ij} and its inverse g^{ij} are bounded in the C^s-topology of $\mathbf{B}_{i/2}(0)$.

Besides these conditions, consider the following further property which turns out to be very important in applications.

(R_k) The covariant derivatives of the Riemann tensor up to order k are uniformly bounded, i.e., $\left\| D^{(j)}\mathrm{Riem} \right\|_{L^\infty(M)} \leq C$, for every $j = 0, \dots, k$ and for some constant $C > 0$.

The following result shows how conditions (Inj), (G_s) and (R_k) are related to each other. See Theorem 4.7 in [33] and Lemma 5.2 in [166] for the injectivity radius estimates; and [49], [145] for equivalent definitions of bounded geometry.

Theorem B.15. *Let $(M, \langle , \rangle)$ be a complete Riemannian manifold.*

(1) *Assume that M satisfies condition* (R_0) *and, for some $r_0 > 0$,*

$$\inf_{x \in M} \mathrm{vol} B_{r_0}(x) > 0.$$

Then M enjoys property (Inj).

(2) *Suppose M satisfies condition (Inj). Then*

$$(\mathrm{R}_s) \Longrightarrow (\mathrm{G}_s) \Longrightarrow (\mathrm{R}_{s-2}).$$

Remark B.16. From the topological viewpoint, it is a fundamental result by R. Greene, [65], that there is no obstruction for a smooth manifold to support a complete Riemannian metric of bounded geometry.

Remark B.17. An interesting aspect of bounded geometry is that we can mimic the "from-local-to-global" procedure which is typical of the compact realm. In fact, very often, for a compact manifold, a global result is achieved from its local version (in a neighborhood of each point) by using a finite covering argument. For a manifold of bounded geometry, the extension is obtained by using infinite coverings with controlled intersection multiplicity. Clearly, since the covering could be infinite, one is forced to start with "uniformly controlled" local objects, and this is certainly the case if the objects at hand depend continuously on the metric and its derivatives.

Now, we come back to L^p harmonic forms. The next theorem is due to D. Alexandru-Rugina, [5]. The idea behind the proof is a reduction procedure to the L^2 situation via a Sobolev imbedding result and a (pairing) argument similar to that due to Gromov described in Example B.1 . The bounded geometry assumption enters in a crucial way in order to obtain that local Sobolev constants are independent of the chosen point.

Theorem B.18. *Let $(M, \langle , \rangle)$ be a complete manifold of bounded geometry at the order $s = 2$ and let $1 < p \leq 2$. Then, there is a continuous imbedding*

$$L^p\mathcal{H}^k(M) \hookrightarrow \overline{L^p H^k(M)}.$$

An obvious consequence of the theorem is that, if $L^p\mathcal{H}^k(M)$ is infinite-dimensional, then so is $\overline{L^p H^k(M)}$. This is the way Alexandru-Rugina employs her result.

It is an interesting problem to investigate the validity of the opposite implication, namely,

$$\dim L^p\mathcal{H}^k(M) < +\infty \Longrightarrow \dim \overline{L^p H^k(M)} < +\infty.$$

B.4 Some topological aspects of the theory

It is well known that the de Rham cohomology of a compact, smooth manifold M is isomorphic to the simplicial cohomology induced by any chosen smooth triangulation of M. In particular, the de Rham spaces turn out to be homotopy invariants of the manifold. Different proofs of this fact are available, ranging from the very concrete construction by H. Whitney, [164], up to the abstract theory of sheaf cohomology, [162]. Both these approaches have been developed in the setting of the L^p cohomology of non-compact spaces. The former was first used in the seminal paper [43] by J. Dodziuk (see also [46]) concerning the L^2 reduced theory of manifolds with sufficiently bounded geometry, and was subsequently used by V.M. Gol'dshtein, V.I. Kuz'minov and I.A. Shvedov to establish the definitive result in the L^p unreduced setting (see Theorem B.22 below). The latter has been recently reconsidered by P. Pansu in his investigation on the quasi-isometric invariance of the L^p (unreduced) cohomology spaces, see, e.g., [127].

We shall focus our attention on the Whitney-type approach developed by Dodziuk, Gol'dshtein, Kuz'minov, and Shvedov. Our purpose is to emphasize the topological content of the L^p de Rham cohomology. This is accomplished by relating the L^p de Rham spaces with a suitable, global simplicial theory on the underlying triangulated manifold: the ℓ^p simplicial cohomology. So, let $(M, \langle , \rangle)$ be a geodesically complete, non-compact Riemannian manifold of dimension m. Recall that a *smooth triangulation* of $(M, \langle , \rangle)$ is a couple $(\mathbf{K}, \mathfrak{t})$ where

(1) $\mathbf{K}$ is a (locally finite) simplicial complex of dimension $\dim \mathbf{K} = m$ which we may assume to be geometrically realized in some Euclidean space.

(2) $\mathfrak{t} : |\mathbf{K}| \to M$ is a homeomorphism such that, for every k-dimensional simplex $\sigma \in K$, the restriction $\mathfrak{t}_\sigma = \mathfrak{t}|_{|\sigma|} : |\sigma| \to M$ is a smooth imbedding. This means that $\mathfrak{t}_\sigma$ extends smoothly in a neighborhood U of $\sigma \subset \mathbb{R}^k$ and, furthermore, the differential $(\mathfrak{t}_\sigma)_{*_x} : T_x U \to T_{\mathfrak{t}(x)} M$ is injective, for every $x \in |\sigma|$.

The Dodziuk–Gol'dshtein–Kuz'minov–Shvedov theory involves the special class of triangulations described in the next definition.

The smooth *triangulation* $(\mathbf{K}, \mathfrak{t})$ of $(M, \langle , \rangle)$ is said to be *of bounded geometry* (BG for short) if the following further conditions are satisfied:

(3) $\mathbf{K}$ is a BG simplicial complex, i.e., a finite-dimensional, locally finite (geometric) simplicial complex for which there is an (absolute) integer $N > 0$ such that every k-dimensional simplex of $\mathbf{K}$ is the face of at most N simplices of dimension $k+1$. Equivalently, the 1-skeleton of $\mathbf{K}$ is a graph with uniformly bounded vertex degrees.

(4) There exists a universal constant $C > 1$ such that, for every k-dimensional simplex $\sigma \in \mathbf{K}$, and for every $v \in T_x |\sigma|$ (in the sense of (2))

$$C^{-1} \langle v, v \rangle_{\mathbb{R}^k} \le \langle (\mathfrak{t}_\sigma)_{*_x} v, (\mathfrak{t}_\sigma)_{*_x} v \rangle \le C \langle v, v \rangle_{\mathbb{R}^k} .$$

The following existence result is attributed to E. Calabi. For a somewhat complete proof, we refer the reader to the paper [8] by O. Attie.

Theorem B.19. *Let the manifold $(M, \langle,\rangle)$ have bounded geometry at the order $s = 2$. Then, there exists a smooth BG triangulation $(\mathbf{K}, \mathfrak{t})$ of M satisfying the further property*

(5) $D^{-1} \leq \mathrm{vol}\,(\sigma) \leq D$ *and* $D^{-1} \leq \mathrm{diam}\,(\sigma) \leq D$

for some universal constant $D > 1$ and for every $\sigma \in \mathbf{K}$ of maximal dimension $\dim \sigma = m$.

In particular, we have

Corollary B.20. *Every Riemannian covering of a compact Riemannian manifold has a BG smooth triangulation.*

Infinite simplicial complexes of bounded geometry are suitable for a global extension to the non-compact realm of the ordinary simplicial cohomology of a compact polyhedron.

Let $\mathbf{K}$ be a BG complex of dimension $\dim(\mathbf{K}) = m$. Set $S_k(\mathbf{K})$ for the (countable) collection of all the oriented, k-dimensional simplices of $\mathbf{K}$. Let also $\left(C^k(\mathbf{K}), \partial^k\right)$ be the ordinary simplicial cochain complex. We fix $q \geq 1$, and, for any k, we introduce the vector space $\ell^p C^k(\mathbf{K})$ *of ℓ^p simplicial cochains* of dimension k, by setting

$$\ell^p C^k(\mathbf{K}) = \left\{\lambda \in C^k(\mathbf{K}) : \|\lambda\|_{\ell^p C^k(\mathbf{K})} < +\infty\right\}$$

where

$$\|\lambda\|_{\ell^p C^k(\mathbf{K})} = \left(\sum_{\sigma \in S_k(\mathbf{K})} |\lambda(\sigma)|^q\right)^{1/q}.$$

Obviously, $\ell^p C^k(\mathbf{K})$ is a Banach space topologically isomorphic to $\ell^p(\mathbb{Z})$. Furthermore, restricting the action of the ordinary coboundary operator ∂^k to k-dimensional ℓ^p-cochains, yields a bounded operator $\partial^k : \ell^p C^k(\mathbf{K}) \to \ell^p C^{k+1}(\mathbf{K})$ (here the BG assumption enters the game). Because of the (restricted) identity

$$\partial^k \circ \partial^{k-1} = 0$$

the family of Banach spaces and bounded operators $\left\{\ell^p C^k(\mathbf{K}), \partial^k\right\}$ defines a cochain-complex , the *complex of the ℓ^p simplicial cochains.* Set $\ell^p Z^k(\mathbf{K}) = \ker \partial^k$ and $\ell^p B^k(\mathbf{K}) = \mathrm{Im}\,\partial^{k-1}$ for the spaces of the ℓ^p cocycles and the ℓ^p coboundaries, respectively. Since this latter space is not necessarily closed, then, as we did in the Sobolev setting of differential forms, we introduce the *unreduced ℓ^p simplicial cohomology* of $\mathbf{K}$,

$$\ell^p H^k(\mathbf{K}) = \ell^p Z^k(\mathbf{K}) / \ell^p B^k(\mathbf{K})$$

and its (Banach) *reduced* version

$$\overline{\ell^p H^k(\mathbf{K})} = \ell^p Z^k(\mathbf{K}) / \overline{\ell^p B^k(\mathbf{K})}.$$

Obviously

$$\overline{\ell^p H^k(\mathbf{K})} \subset \ell^p H^k(\mathbf{K})$$

and, every topological isomorphism $f : \ell^p H^k(\mathbf{K}_1) \to \ell^p H^k(\mathbf{K}_2)$ induces a topological isomorphism between reduced spaces.

Definition B.21. Suppose the complete manifold $(M, \langle , \rangle)$ has a BG triangulation $(\mathbf{K}, \mathfrak{t})$. The ℓ^p simplicial cohomology of the triangulated manifold $(M, \mathbf{K}, \mathfrak{t})$ is defined as the ℓ^p simplicial cohomology of the BG simplicial complex $\mathbf{K}$. Accordingly, we set

$$\ell^p H^j(M, \mathbf{K}, \mathfrak{t}) = \ell^p H^j(\mathbf{K}) \quad \text{and} \quad \overline{\ell^p H^j(M, \mathbf{K}, \mathfrak{t})} = \overline{\ell^p H^j(\mathbf{K})}.$$

We are now in a position to state the far-reaching extension due to Gol'dshtein–Kuz'minov–Shvedov of the classical de Rham theorem. The philosophy underlying the clever and quite complicated proof resemble the original one by Whitney. Indeed, with some oversimplification, their aim is to show that integration of (non-smooth) forms over simplexes establishes the desired isomorphism.

Theorem B.22. *Let $(M, \langle , \rangle)$ be a complete m-dimensional manifold admitting a smooth BG triangulation $(\mathbf{K}, \mathfrak{t})$. Then, for $p \geq 1$ and for any $k \in \mathbb{N}$, there is a topological isomorphism*

$$L^p H^k(M) \simeq \ell^p H^k(M, \mathbf{K}, \mathfrak{t})$$

between unreduced cohomologies.

It is worth noting that, unlike the compact case, the ℓ^p simplicial cohomology is not a continuous homotopy invariant of BG polyhedra. For instance, one can appeal to Theorem B.22 and recall Corollary B.14 and Example B.1. Accordingly, the Riemannian manifolds $\mathbb{R}^m$ and $\mathbb{H}^m_{-1}$ have different ℓ^p simplicial cohomologies.

A natural question is whether there are situations where the topology of the BG polyhedron is reflected in the ℓ^p cohomology of a triangulation. In this respect, we first recall the following result due to Dodziuk, [43].

Theorem B.23. *Let $\tilde{\mathbf{K}}$ be a connected simplicial complex of dimension m and let Γ be a discrete group of simplicial homeomorphisms acting freely on $\tilde{\mathbf{K}}$. Suppose also that the orbit space $\mathbf{K} = \tilde{\mathbf{K}}/\mathbf{\Gamma}$ is an m-dimensional, finite simplicial complex so that the quotient projection is a simplicial map. Then $\tilde{\mathbf{K}}$ is a BG simplicial complex and the ℓ^p simplicial cohomology of $\tilde{\mathbf{K}}$ is a continuous-homotopy invariant of the couple $(\mathbf{K}, \mathbf{\Gamma})$.*

Combining this result with Theorem B.22, and using also the equivariant triangulation theory by S. Illman, [84], yields the following fundamental

Theorem B.24. *Let $(M, \langle\,,\rangle)$ and $(N, (\,,))$ be compact Riemannian manifolds. Denote by $(\widetilde{M}, \widetilde{\langle\,,\rangle})$ and $(\widetilde{N}, \widetilde{(\,,)})$ the respective Riemannian universal coverings. If M and N are (continuously) homotopic, then, for every $k \in \mathbb{N}$ and for every $p > 1$, there is a topological isomorphism*

$$L^p H^k(\widetilde{M}) \simeq L^p H^k(\widetilde{N}).$$

Bibliography

[1] U. Abresch, *Lower curvature bounds, Topogonov's theorem, and bounded topology.* Ann. sci. École Norm. Sup. **218** n. 4 (1985), 651-670.

[2] L.V. Ahlfors, *An extension of Schwarz's lemma.* Trans. Amer. Math. Soc. **43** (1938), 359–364.

[3] L. Ahlfors, L. Sario, *Riemann Surfaces.* Princeton University Press, Princeton, N.J. 1960.

[4] D. Alexandru-Rugina, *L^p-integrabilité des formes harmoniques K-finies sur les espaces hyperboliques réels et complexes.* Rend. Sem. Mat. Univ. Politec. Torino **54** (1996), 75–87.

[5] D. Alexandru-Rugina, *Harmonic forms and L^p-cohomology.* Tensors (N.S.) **57** (1996), 176–191.

[6] L. Ambrosio, X. Cabré, *Entire solutions of semilinear elliptic equations in $\mathbb{R}^3$ and a conjecture of de Giorgi*, J. Amer. Math. Soc. **13** (2000), 725–739.

[7] M. Anderson, *The compactification of a minimal submanifold in Euclidean space by the Gauss map.* Preprint.

[8] O. Attie, *Quasi-isometry classifications of some manifolds of bounded geometry.* Math. Z. **216** (1994), 501–527.

[9] T. Aubin, *Some Nonlinear Problems in Riemannian Geometry*, Springer 1998.

[10] R. Bishop, R. Crittenden, *Geometry of manifolds.* Academic Press, New York, 1964.

[11] P. Bérard, *A note on Bochner type theorems for complete manifolds.* Manuscripta Math. **69** (1990), no. 3, 261–266.

[12] P. Berard, *Remarques sur l'equations de J. Simons*, Differential Geometry (A symposium in honor of M. do Carmo on his 60th birthday), Pitman Monographs Surveys Pure Appl. Math, **52**. Longman Sci. Tech. Harlow, 1991, 47–57.

[13] P. Bérard, M. do Carmo, W. Santos, *The index of constant mean curvature surfaces in hyperbolic* 3*-space*, Math. Z. **234** (1997), 313–326.

[14] H. Berestycki, L. Caffarelli, L. Nirenberg, *Further qualitative properties for elliptic equations in unbounded domains*, Ann. Scuola Norm. Sup. Cl. Sc. **25** (1997), 69–94.

[15] L. Bers, F. John, M. Schechter, *Partial Differential Equations.* Lectures in Applied Mathematics III. Interscience Publishers, New York 1964.

[16] B. Bianchini and M. Rigoli, *Non existence and uniqueness of positive solutions of Yamabe type equations on non positively curved manifolds,* Trans. Amer. Math. Soc. **349** (1997), 4753–4774.

[17] J.-P. Bourguignon, *Les variétés de dimension 4 à signature non nulle dont la courbure est harmonique sont d'Einstein.* Invent. Math. **63** (1981), no. 2, 263–286.

[18] J.-P. Bourguignon,*The magic of Weitzenböck formulas*, Variational Methods Paris (1998), 251–271. Progress in Nonlinear Differential Equations and Applications IV, Birkhäuser 1990.

[19] L. Brandolini, M. Rigoli, A.G. Setti, *Positive solutions of Yamabe-type equations on the Heisenberg group*, Duke Math. J. **91** (1998), 241–296.

[20] L. Brandolini, M. Rigoli and A.G. Setti, *Positive solutions of Yamabe type equations on complete manifolds and applications,* Jour. Funct. Anal. **160** (1998), 176–222.

[21] T. Branson, *Kato constants in Riemannian geometry.* Math. Res. Lett. **7** (2000), 245–261.

[22] L. Caffarelli, B. Gidas, J. Spruck, *Asymptotic symmetry and local behavior of semilinear elliptic equations with critical Sobolev growth.* Comm. Pure Appl. Math. **42** (1989), 271–297.

[23] M. Cai, G.J. Galloway, *Boundaries of zero scalar curvature in the ADS/CFT correspondence.* Adv. Theor. Math. Phys. **3** (1999), 1769–1783.

[24] D.M.J. Calderbank, P. Gauduchon, M. Herzlich, *Refined Kato inequalities and conformal weights in Riemannian geometry.* J. Funct. Anal. **173** (2000), 214–255.

[25] H.-D. Cao, Y. Shen, S. Zhu, *The structure of stable minimal hypersurfaces in* $\mathbb{R}^{m+1}$. Math. Res. Lett. **4** (1997), no. 5, 637–644.

[26] G. Carron, *Une suite exacte en* L^2*-cohomologie*, Duke Math. J. **95** (1988), 343–373.

[27] G. Carron, *Inégalités isopérimétriques de Faber-Krahn et conséquences.* Actes de la Table Ronde de Géométrie Différentielle (Luminy, 1992), 205–232, Sémin. Congr., 1, Soc. Math. France, Paris, 1996.

[28] G. Carron, L^2*-Cohomolgie et inégalités de Sobolev.* Math. Ann. **314** (1999), 613–639.

[29] G. Carron, L^2*-harmonic forms on non compact manifolds*, arXiv: 0704.3194.

[30] I. Chavel, *Riemannian Geometry – A Modern Introduction.* Cambridge University Press, Cambridge, 1993.

[31] S.Y. Cheng, S.T. Yau, *Differential equations on Riemannian manifolds and their geometric applications*, Comm. Pure Appl. Math. **XXVIII** (1975), 333–354.

[32] J. Cheeger, D. Gromoll, *The structure on manifolds with non-negative Ricci curvature.* J. Diff. Geom. **6** (1971), 119-128.

[33] J. Cheeger, M. Gromov, M. Taylor, *Finite propagation speed, kernel estimates for functions of the Laplace operator, and the geometry of complete Riemannian manifolds.* J. Diff. Geom. **17** (1982), no. 1, 15–53.

[34] S.Y. Cheng and S.T. Yau, *Differential equations on Riemanninan manifolds and their geometric applications.* Comm. Pure Appl. Math. **28** (1975), 333–354.

[35] S.S. Chern, *Complex manifolds without potential theory.* Universitext. Springer-Verlag, New York-Heidelberg, 1979.

[36] S.S. Chern, *On the holomorphic mappings of Hermitian manifolds of the same dimension.* Proc. Symp. Pure Math. **11**, Entire Functions and related parts of Analysis (1968), 157–170.

[37] P. Chernoff, *Essential self-adjointness of powers of generators of hyperbolic equations.* J. Funct. Anal. **12** (1973), 401-414.

[38] T. Colding and W.P Minicozzi, *An excursion into geometric analysis.* Surv. Diff. Geom. **IX** (2004), 83–146.

[39] M. Cwikel, *Weak type estimates for singular values and the number of bound states of Schrodinger operators*, Ann. Math. **106** (1977), 93–100.

[40] M. Dajczer, L. Rodriguez, *Rigidity of real Kaehler submanifolds.* Duke Math. J. **53** (1986), no. 1, 211–220.

[41] E.B. Davies, *Spectral Theory and Differential Operators.* Cambridge University Press, Cambridge, 1995.

[42] G. de Rham, *Differentiable manifolds.* Springer Verlag, 1984.

[43] J. Dodziuk, *de Rham-Hodge theory for L^2-cohomology of infinite coverings.* Topology **16** (1977), no. 2, 157–165.

[44] J. Dodziuk, *L^2-harmonic forms on rotationally symmetric Riemannian manifolds.* Proc. AMS **77** (1979) 395–400.

[45] J. Dodziuk, L^2-harmonic forms on complete manifolds, Semin. Differential Geometry, Ann. Math. Stud. **102** (1982), 291–302.

[46] J. Dodziuk, *Sobolev Spaces of differential forms and the de Rham-Hodge isomorphism.* J. Diff. Geom. **16** (1981), 63–73.

[47] J. Eells, L. Lemaire, *Selected topics in harmonic maps.* CBMS Regional Conference Series in Mathematics, **50** (1983).

[48] J. Eells, L. Lemaire, *A report on harmonic maps.* Bull. London Math. Soc. **10** (1978), 1–68.

[49] J. Eichhorn, *The boundedness of connection coefficients and their derivatives.* Math. N. **152** (1991), 145–158.

[50] J. Eschenburg and E. Heintze, *An elementary proof of the Cheeger-Gromoll splitting theorem*, Ann. Glob. Analysis and Geometry **2** (1984), 141–151.

[51] H. Federer, *Geometric Measure Theory.* Springer Verlag, Berlin, Heidelberg, New York, 1969,

[52] M.J. Ferreira, M. Rigoli, R. Tribuzy, *Isometric immersions of Kähler manifolds.* Rend. Sem. Mat. Univ. Padova **90** (1993), 25–38.

[53] D. Fischer–Colbrie, *On complete minimal surfaces with finite Morse index in three manifolds*, Invent. Math. **82** (1985), 121–132.

[54] D. Fischer–Colbrie, R. Schoen, *The structure of complete stable minimal surfaces in 3-manifolds of non-negative scalar curvature*, Comm. Pure Appl. Math. **XXXIII** (1980), 199–211.

[55] A. Friedman, *Partial Differential Equations*, Holt, Rinehart and Winston, Inc, New York 1970.

[56] P.M. Gaffney, *A special Stokes' theorem for complete Riemannian manifolds*, Ann. Math. **60** (1954), 140–145.

[57] S. Gallot, *Isoperimetric inequalities bases on integral norms of Ricci curvature*, Asterisque **157-158** (1988), 191–216.

[58] S. Gallot, D. Meyer, *Opérateur de courbure et Laplacien des formes différentielles d'une variété Riemannienne.* J. Math. pures et appl. **54** (1975), 259–284.

[59] M. Giaquinta, *Multiple Integrals in the Calculus of Variations and Nonlinear Elliptic Systems*, Annals of Math. Studies 105, Princeton University Press, Princeton, N.J. 1983.

[60] D. Gilbarg, N. Trudinger, *Elliptic Partial Differential Equations of Second Order*, second edition, Springer-Verlag, Berlin 1983.

[61] S.I. Goldberg, *An application of Yau's maximum principle to conformally flat spaces.* Proc. Amer. Math. Soc. **79** (1980), 268–270.

[62] V. Gol'dshtein, M. Troyanov, *The Kelvin-Nevanlinna-Royden criterion for p-parabolicity.* Math Z. **232** (1999), 607–619.

[63] V.M. Gol'dshtein, V.I. Kuz'minov, I.A. Shvedov, *The de Rham isomorphism of the L^p-cohomology of noncompact Riemannian manifolds.* (Russian) Sibirsk. Mat. Zh. **29** (1988),34–44; translation in Siberian Math. J. **29** (1988), 190–197.

[64] V.M. Gol'dshtein, V.I. Kuz'minov, I.A. Shvedov, *The integration of differential forms of classes $W^{p,q}$.* Sibirsk. Mat. Zh. **23** (1982), 63–79.

[65] R. Greene, *Complete metrics of bounded curvature on noncompact manifolds.* Arch. Math. **31** (1978), 89–95.

[66] R. Greene and H.H. Wu, *Function Theory on Manifolds Which Possess a Pole*, Lecture Notes in Math. n. 699, Springer Verlag, Berlin, 1979

[67] R. Greene, H.H. Wu, *Embeddings of open Riemannian manifolds by harmonic functions.* Ann. Inst. Fourier, Grenoble **25** (1975), 215–235.

[68] R. Greene, H.H. Wu, *Whitney's imbedding theorem by solutions of elliptic equations and geometric consequences.* Proc. Symp. Pure Math. **27** (1975), 287–296.

[69] R. Greene, H.H. Wu, *C convex functions and manifolds of positive curvature.* Acta Math. **137** (1976), 209–245.

[70] Ph. Griffiths, *Entire holomorphic mappings in one and several complex variables.* Annals of Mathematics Studies, No. 85. Princeton University Press, Princeton, N. J., 1976.

[71] A. Grigor'yan, *Dimension of spaces of harmonic functions*, Math. Zametki **48** (1990), 55–61, Engl. transl. Math. Notes **48** (1990), 1114–1118.

[72] A. Grigor'yan, *Analytic and geometric background of recurrence and non-explosion of the Brownian motion on Riemannian manifolds*, Bull. Amer. Math. Soc. **36** (1999), 135–249.

[73] A. Grigor'yan, *Stochastically complete manifolds and summable harmonic functions*, (in Russian) Izv. An SSSR, ser. matem., bf 52 no. 5 (1988), 1102–1108. Engl. transl. Math. USSR Izvestiya **33** (1989), 425–423.

[74] D. Gromoll, W. Meyer, *On complete open manifolds of positive curvature.* Ann. Math. **90** (1969), 75–90.

[75] M. Gromov, *Asymptotic invariants of infinite groups.* Geometric group theory, Vol. 2 (Sussex, 1991), 1–295, London Math. Soc. Lecture Note Ser., **182**, Cambridge Univ. Press, Cambridge, 1993.

[76] R. Gulliver, *Index and total curvature of complete minimal surfaces.* Geometric measure theory and the calculus of variations (Arcata, Calif., 1984), 207–211, Proc. Sympos. Pure Math., **44**, Amer. Math. Soc., Providence, RI, 1986.

[77] R. Gulliver, *Minimal surfaces of finite index in manifolds of positive scalar curvature.* Calculus of variations and partial differential equations (Trento 1986), Lect. Notes Math. 1340, Springer, Berlin (1988), 115–122.

[78] P. Hartman, *On homotopic harmonic maps.* Can. J. Math. **19** (1967), 673–687.

[79] T. Hasanis, *Conformally flat spaces and a pinching problem for the Ricci tensor.* Proc. Amer. Math. Soc. **86** (1982), 312–315.

[80] E. Hebey, *Sobolev Spaces on Riemannian Manifolds*, Lecture Notes in Mathematics, **1635**. Springer-Verlag, Berlin, 1996.

[81] L.I. Hedberg, *Two approximations problems in functions spaces.* Ark. Math. **16** (1978), 51–81.

[82] L.I. Hedberg, *Spectral synthesis in Sobolev spaces, and uniqueness of solutions of the Dirichlet problem.* Acta Math. **147** (1981) 237–264.

[83] D. Hoffman, J. Spruck, *Sobolev and isoperimetric inequalities for Riemannian submanifolds.* Comm. Pure Appl. Math. **27** (1974), 715–727.

[84] S. Illman, *Existence and uniqueness of equivariant triangulations of smooth proper G-manifolds with some applications to equivariant Whitehead torsion.* J. reine ang. Math. **524** (2000), 129–183.

[85] T. Iwaniec, G. Martin, *Quasiregular mappings in even dimensions.* Acta Math. **170** (1993), 29–81.

[86] W. Jäger and H. Kaul, *Uniqueness and stability of harmonic maps and their Jacobi fields*, Manuscripta Math. **28** (1979), 269–291.

[87] J.L. Kazdan, *Unique Continuation in Geometry.* Comm. Pure Appl. Math. **XLI** (1988), 667–681.

[88] S. Kobayashi, K. Nomizu, *Foundations of differential geometry.* Vol. II. Interscience Tracts in Pure and Applied Mathematics, No. 15 Vol. II Interscience Publishers John Wiley & Sons, Inc., New York-London-Sydney 1969

[89] K. Kodaira, *Harmonic fields in Riemannian manifolds.* Ann. Math. **50** (1949), 587–665.

[90] N. Kuiper, *On conformally flat spaces in the large*, Annals of Math. **50** (1949), 916-924.

[91] N. N. Lebedev, *Special Functions and Their Applications*, Dover N.Y. 1972

[92] J.M. Lee, *Introduction to smooth manifolds.* Graduate Texts in Mathematics 218, 2003. Springer-Verlag, New York.

[93] L. Lemaire, *Harmononic mappings of uniformly bounded dilations.* Topology **16** (1977), 199–201.

[94] P. Li, *On the Sobolev constant and the p-spectrum of a compact Riemannian manifold.* Ann. Scient. Ec. Norm. Sup. **13** (1980), 451–469.

[95] P. Li, R. Schoen, L^p *and mean value properties of subharmonic functions on Riemannian manifolds.* Acta Math. **153** (1984), no. 3-4, 279–301.

[96] P. Li, *On the structure of complete Kähler manifolds with nonnegative curvature near infinity.* Invent. Math. **99** (1990), 579–600.

[97] P. Li, *Lecture Notes on Geometric Analysis.* Lecture Notes Series No. 6, Research Institute of Mathematics, Global Analysis Research Center, Seoul National University, Korea (1993)

[98] P. Li, M. Ramachandran, *Kähler manifolds with almost nonnegative Ricci curvature.* Amer. J. Math. **118** (1996), no. 2, 341–353.

[99] P. Li, L.-F Tam, *Positive harmonic functions on complete manifolds with non-negative curvature outside a compact set*, Ann. Math. **125** (1987), 171–207.

[100] P. Li, L.-F Tam, *Symmetric Green's functions on complete manifolds*, Amer. J. Math. **109** (1987), 1129–1154.

[101] P. Li, L.-F. Tam, *The heat equation and harmonic maps of complete manifolds*, Invent. Math. **105** (1991), 1-46.

[102] P. Li, L.-F Tam, *Harmonic functions and the structure of complete manifolds*, J. Differential Geom. **35** (1992), 359–383.

[103] P. Li and J. Wang, *Complete manifolds with positive spectrum.* J. Diff. Geom.**58** (2001), 501-534.

[104] P. Li, J. Wang, *Minimal hypersurfaces of finite index.* Math. Res. Let. **9** (2002), 95–103.

[105] P. Li, J. Wang, *Stable minimal hypersurfaces in a nonnegatively curved manifold.* J. reine angew. Math. **566** (2004), 215–230.

[106] P. Li, S.T. Yau, *On the Schrödinger equation and the eigenvalue problem.* Comm. Math. Phys. **88** (1983), 309–318.

[107] P. Li, S.T. Yau, *Curvature and holomorphic mappings of complete Kähler manifolds* Compositio Math. **73** (1990), 125–144.

[108] A. Lichnerowicz, *Applications harmoniques et variétés kähleriennes.* 1968/1969 Symposia Mathematica, Vol. III (INDAM, Rome, 1968/69), 341–402.

[109] E. Lieb, *The number of bound states of one-body Schroedinger operators and the Weyl problem.* Geometry of the Laplace operator . Proc. Sympos. Pure Math., **XXXVI** (1979), pp. 241–252.

[110] T. Lyons, D. Sullivan, *Function theory, random paths and covering spaces.* J. Diff. Geom. **18** (1984), 229–323.

[111] Y.-C. Lu, *Holomorphic mappings of complex manifolds.* J. Diff. Geom. **2** (1968), 299-312.

[112] N. Mok, *Bounds on the dimension of L^p holomorphic sections of vector bundles over complete Kähler manifolds of finite volume.* Math Z. **191** (1986), 303–317.

[113] J. A. Morrow and K. Kodaira, *Complex manifolds.* Holt, Rinehart and Winston, New York, 1971.

[114] W.F. Moss, J. Pieprbrink, *Positive solutions of elliptic equations,* Pacific J. Math **75** (1978), 219–226.

[115] J. Munkres, *Elementary differential topology.* Annals of Mathematics Studies, No. 54. Princeton University Press 1966.

[116] M. Nakai, *On Evans potential.* Proc. Japan Acad. **38** (1962), 624–629.

[117] T. Napier, M. Ramachandran, *Structure theorems for complete Kähler manifolds and applications to Lefschetz type theorems.* Geom. Funct. Anal. **5** (1995), no. 5, 809–851.

[118] A. Newlander, L. Nirenberg, *Complex analytic coordinates in almost complex manifolds.* Ann. of Math. (2) **65** (1957), 391–404.

[119] L. Ni, *Vanishing theorems on complete Kähler manifolds and their applications*, J. Differential Geom. **50** (1998), 89–122.

[120] L. Ni, *Gap theorems for minimal submanifolds in* $\mathbb{R}^{n+1}$. Comm. Anal. Geom. **9** (2001), no. 3, 641–656.

[121] L. Ni, Y. Shi and L.F. Tam, *Poisson equation, Poincarè-Lelong equation and curvature decay on complete Kähler manifolds.* J. Diff. Geom. **57** (2001), 339–388.

[122] M. Okumura, *Hypersurfaces and a pinching problem on the second fundamental tensor.* Amer. J. Math. **96** (1974), 207–213.

[123] M. Okumura, *Submanifolds and a pinching problem on the second fundamental tensors.* Trans. Amer. Math. Soc. **178** (1973), 285–291.

[124] H. Omori, *Isometric immersions of Riemannian manifolds.* J. Math. Soc. Japan **19** (1967) 205–214.

[125] Y. Otzu, *Topology of complete open manifolds with nonnegative Ricci curvature.* Geometry of Manifolds (Matzumoto 1988) 205–302. Perspect. Math. **8**, Academic Press, Boston MA, 1989.

[126] P. Pansu, *Introduction to L^2 Betti numbers.* Riemannian geometry (Waterloo, ON, 1993), 53–86, Fields Inst. Monogr., **4**, Amer. Math. Soc., Providence, RI, 1996.

[127] P. Pansu, *Cohomologie L^p: invariance sous quasiisomètries.* Preprint 2004.

[128] P. Petersen, *Riemannian Geometry.* Springer Verlag, 1997.

[129] P. Petersen, G. Wei, *Relative volume comparison with integral curvature bounds.* GAFA **7** (1997), 1031–1045.

[130] S. Pigola, M. Rigoli and A.G. Setti, *Volume growth, "a-priori" estimates and geometric applications.* GAFA **13** (2003), 1302–1328.

[131] S. Pigola, M. Rigoli, A.G. Setti, *Maximum principle on Riemannian manifolds and applications*, Memoirs of the AMS, **174** no. 822.

[132] S. Pigola, M. Rigoli, A.G. Setti, *Vanishing theorems on Riemannian manifolds and geometric applications*, J. Funct. Anal. **229** (2005), 424–461.

[133] M.H. Protter, H.F. Weinberger, *Maximum principles in differential equations.* Springer-Verlag, New York, 1984.

[134] A. Ratto, M. Rigoli, and A. G. Setti *A uniqueness result in PDE's and parallel mean curvature immersions in Euclidean spaces*, Complex Variables **36** (1996), 221–233.

[135] A. Ratto, M. Rigoli and L. Veron, *Scalar curvature and conformal deformations of noncompact Riemannian manifolds.* Math. Z. **225** (1997), 395-426.

[136] J. Rawnsley, *f-structures, f-twistor spaces and harmonic maps.* Lecture Notes in Math. **1164**, Springer, Berlin, 1985.

[137] M. Reed, B. Simon, *Methods of Modern Mathematical Physics. I Functional Analysis.* Academic Press, San Diego, 1980.

[138] D. Reed, B. Simon, *Methods of Modern Mathematical Physics. II Fourier Analysis, Self Adjointness.* Academic Press, San Diego, 1972.

[139] M. Rigoli, A.G. Setti, *Energy estimates and Liouville theorems for harmonic maps.* Intern. Jour. Math. **11** (2000), 413–448.

[140] M. Rigoli, A.G. Setti, *Liouville type theorems for $\varphi-$subharmonic functions.* Rev. Mat. Iberoam. **17** (2001), 471–520.

[141] G.V. Rozenbljum, *Distribution of the discrete spectrum of singular differential operators.* (Russian) Dokl. Akad. Nauk SSSR **202** (1972), 1012–1015.

[142] T. Sakai, *Riemannian Geometry.* Transl. Math. Monographs **149**. American Mathematical Society, 1997.

[143] J.H. Sampson, *Applications of harmonic maps to Kähler geometry.* Complex differential geometry and nonlinear differential equations (Brunswick, Maine, 1984), 125–134, Contemp. Math., **49**, Amer. Math. Soc., Providence, RI, 1986.

[144] L. Sario, M. Nakai, *Classification theory of Riemann surfaces.* Die Grundlehren der mathematischen Wissenschaften, Band 164 Springer-Verlag, New York-Berlin 1970.

[145] T. Schick, *Manifolds with boundary and of bounded geometry.* Math. Nachr. **223** (2001), 103–120.

[146] R. Schoen, S.T. Yau, *Harmonic maps and the topology of stable hypersurfaces and manifolds with non-negative Ricci curvature.* Comment. Math. Helv. **51** (1976), no. 3, 333–341.

[147] R. Schoen, S.T. Yau, *Lectures on differential geometry.* Conference Proceedings and Lecture Notes in Geometry and Topology, Volume I, 1994. International Press.

[148] R. Schoen, S.T. Yau, *Lectures on Harmonic Maps.* Lectures Notes in Geometry and Topology, Volume II, 1997. International Press.

[149] R. Schoen, S.T. Yau,*Compact group actions and the topology of manifolds with non-positive curvature.* Topology **18** (1979), 361–380.

[150] C. Schott, *L^p theory of differential forms on manifolds.* Trans. AMS **347** (1995), 2075–2096.

[151] Y.-B. Shen, X.-H. Zhu, *On stable complete minimal hypersurfaces in* $\mathbb{R}^{n+1}$. Amer. J. Math. **120** (1998), no. 1, 103–116.

[152] Y.T. Siu, *The complex-analyticity of harmonic maps and the strong rigidity of compact Kähler manifolds.* Ann. of Math. **112** (1980), 73–111.

[153] J. Spruck *Remarks on the stability of minimal submanifolds of* $\mathbb{R}^n$, Math. Z. **144** (1975), 169–174.

[154] N. Steenrod, *The topology of fibre bundles.* Princeton University Press, 1951.

[155] R.S. Strichartz, *Analysis of the Laplacian on the complete Riemannian manifold*, J. Funct. Anal. **52** (1983), 48–79.

[156] C.-J. Sung, L.-F. Tam, J. Wang, *Spaces of harmonic functions*, J. London Math. Soc. **61** (2000), 789–806.

[157] M. Tani, *On a conformally flat Riemannian space with positive Ricci curvature.* Tôhoku Math. J. (2) **19** 1967 227–231

[158] J. Tysk, *Finiteness of index and total scalar curvature for minimal hypersurfaces.* Proc. Amer. Math. Soc. **105**(1989), no. 2, 429–435.

[159] P. Tolksdorf, *Regularity of a more general class of quasilinear elliptic equations*, J. Differential Equations **51** (1984), 126–150.

[160] E. Vesentini, *Lectures on Levi convexity of complex manifolds and cohomology vanishing theorems.* Tata Institute of Fundamental Research, Lectures on Mathematics, No. 39 (1967).

[161] X. Wang, *On conformally compact Einstein manifolds.* Math. Res. Lett. **8** (2001), 671-688.

[162] F. Warner, *Foundations of differentiable manifolds and Lie groups.* Graduate Texts in Mathematics 94, 1983. Springer Verlag.

[163] G.W. Whitehead, *Elements of homotopy theory.* Graduate Texts in Mathematics 61, 1978. Springer-Verlag.

[164] H. Whitney, *Geometric integration theory.* Princeton University Press.

[165] H.H. Wu, *The Bochner technique in differential geometry.* Math. Rep. **3** (1988), no. 2, i–xii and 289–538.

[166] D. Yang, *Convergence of Riemannian manifolds with integral bounds on curvature. I.* Ann. scient. Éc. Norm. Sup. **25** (1992), 77–105.

[167] S.T. Yau, *Harmonic functions on complete Riemannian manifolds*, Comm. Pure Appl. Math. **28** (1975), 201–228.

[168] S.T. Yau, *Some function theoretic properties of complete Riemannian manifolds and their applications to geometry*, Indiana Univ. Math. J. **25** (1976), 659–670.

[169] S.T. Yau, *A general Schwarz lemma for Kähler manifolds.* Amer. J. Math. **100** (1978), no. 1, 197–203.

[170] S. Zhu, *The classification of complete conformally flat manifolds of nonnegative Ricci curvature,* Pacific J. Math. **163** (1994), 189-199.

[171] S.-H. Zhu, *A volume comparison theorem for manifolds with asymptotically nonnegative curvature and its applications.* Amer. J. Math. **116** (1994), 669–682.

Index